Arming the Periphery

Arming the Periphery

The Arms Trade in the Indian Ocean during the Age of Global Empire

Emrys Chew

palgrave
macmillan

First published 2012 by
PALGRAVE MACMILLAN

Palgrave Macmillan in the UK is an imprint of Macmillan Publishers Limited, registered in England, company number 785998, of Houndmills, Basingstoke, Hampshire RG21 6XS.

Palgrave Macmillan in the US is a division of St Martin's Press LLC, 175 Fifth Avenue, New York, NY 10010.

Palgrave Macmillan is the global academic imprint of the above companies and has companies and representatives throughout the world.

Palgrave® and Macmillan® are registered trademarks in the United States, the United Kingdom, Europe and other countries

ISBN 978-0-230-35485-2

A catalogue record for this book is available from the British Library.

Library of Congress Cataloging-in-Publication Data

Chew, Emrys.
    Arming the periphery : the arms trade in the Indian Ocean during the
    age of global empire / by Emrys Chew, Nanyang Technological
    University, Singapore.
        pages cm.
    Includes bibliographical references.
    ISBN 978–0–230–35485–2
    1. Weapons industry–Indian Ocean region–History. 2. Arms
    transfers–Indian Ocean region–History. I. Title.

HD9743.A4I435 2012
382'.456234091824–dc23   2012009567

10   9   8   7   6   5   4   3   2   1
21   20   19   18   17   16   15   14   13   12

Transferred to Digital Printing in 2013

Contents

Maps

Note on the cover illustration

Armed with traditional *jezails* and British Enfield rifles, the Khan of Lalpura and his band of tribal warriors are seen here with Anglo-Indian Captain (later Colonel Sir) Robert Warburton. Set against the mountainous backdrop of the Khyber Pass on India's Northwest Frontier, the scene is captured vividly by Victorian 'photo journalist' John Burke, at the height of the 'Great Game' and start of the Second Anglo-Afghan War in 1878. It illustrates a pivotal, yet delicately poised, collaborative relationship between the leading Mohmand chief and (at his right shoulder, seated behind the rock) the British political officer in the Khyber. Warburton's own story recalled the romance of imperialism typified by Rudyard Kipling's *Kim*, being himself the offspring of cross-cultural relations between Shah Jahan Begum, purportedly a niece of former Amir Dost Mohammed, and the British Army Lieutenant-Colonel Robert Warburton, a prisoner of the First Afghan War (1839–42). Forty years on, the line of communication to India via the Khyber remained just as vital to Major-General Sir Frederick Roberts' field force in Kabul. The Mohmand tribesmen at the frontier were initially guarded and only intermittently hostile. But within a year, Roberts had arrested Yakub Khan's wife—the Khan of Lalpura's daughter—on charges of espionage. Alienated if not antagonized, the Mohmands commenced their systematic harassment of the British line of supply near Jalalabad.

Preface

Global arms transfers have been surveyed comprehensively and analysed rigorously from the contemporary angles of international relations and security studies, but are seldom the subject of major historical research. This study arose from the need to redress lacunae both at the inter-disciplinary level and in the existing historiography, thus establishing broader connections with economic, military, imperial, and global historical writing.

A suitably eclectic approach was adopted across the historical discipline. For the commercial and industrial aspects of the study, the use of statistical evidence from government files, business archives, and trade journals was vital. Apart from customs records available at the Public Record Office, valuable data was extracted from papers and periodicals in the archives of the Birmingham Proof House.[1] The 'Barbara Smith Archive' of the Birmingham Reference Library was also useful, incorporating a vast annotated bibliography of reference sources on the industrial history of nineteenth-century Birmingham. *The Ironmonger* magazine—found in its entire series alongside contemporary newspaper articles at the Newspaper Library in Colindale—yielded an especially rich vein of statistical and anecdotal material, offering numerous insights into the gun trade and associated industries (such as ivory). *Arms and Explosives*, the trade journal of the small arms industry, formed another high-quality source in a similar area.[2]

For the imperial and military dimensions, the bulk of archival evidence was drawn from libraries in London and Cambridge. Of primary importance were the records of the East India Company and India Office, located in the Oriental and India Office Collections of the British Library, along with the records of the Admiralty, the Board of Ordnance, and the Foreign, Colonial, and War Offices, located in the Public Record Office, London. These files contained sources encompassing East India Company and British Government munitions orders and arms transfers to Africa, India, and Southeast Asia; the papers of prominent colonial officials; and correspondence on the slave-and-arms traffic in Africa and the Gulf region. The University Library at Cambridge furnished equally wide-ranging material: from nineteenth-century parliamentary papers, travelogues, military despatches, and service magazines; to contemporary reports on global armament, disarmament, and security arrangements. The Royal

Commonwealth Society Collection, located in the Cambridge University Library, distilled precious material on the gun-running exploits of Charles Stokes and the international crisis in the Congo that followed his summary execution by the Belgian colonial authorities.[3] These metropolitan and colonial sources were essential for generating a portrait of the multi-layered, many-segmented phenomenon that was the first truly global arms trade.

Equally indispensable, however, was the primary source material accessed during research trips to the 'periphery' itself, most notably Singapore. This included printed Malay and Burmese documents, Straits Settlements records (both published and unpublished), detailed contemporary travelogues (like those edited by J. R. Logan) and associated sources, supplying altogether vital information about opium trafficking, gun-running, and marauding activities in nineteenth-century Southeast Asia. Admittedly, whilst every attempt was made to combine sources from metropolitan, peripheral, and indigenous standpoints, the sheer geographical scope of the survey across the span of several centuries, coupled with the secretive and controversial aspects of the subject, have produced inevitable unevenness in the depth of coverage in some areas.[4] Nonetheless, a serious effort was undertaken to forge a new synthesis out of a considerable variety of source material and, particularly where material from the indigenous perspective was lacking, to draw upon recent works of historical anthropology concerning Indian Ocean societies. My hope is that this approach will demonstrate how crucial the Indian Ocean context was for the expansion of the arms transfer system from Europe, and how critical it proved in shaping the course of cross-cultural and colonial encounters, whether from the commercial, political, or military perspective.

Through the modern facilities and high level of cooperation afforded by the various archives, there was also wide access to an extensive secondary literature. I am indebted to scholarly works on business history and wider economic and industrial development; comparative studies on imperialism, colonial warfare, and other overseas history; and surveys of weapons technologies and production (both old and new), in addition to the military and cultural use of firearms, and the arms transfer system of the present day. Again I am confident that this volume has attempted, as other studies have not, to bring together all these disparate elements in a new historical synthesis based upon interdisciplinary, core-periphery, and cross-cultural approaches.

I want to express sincere appreciation to the people who inspired and facilitated this grand endeavour. I am especially grateful to the late Clive Trebilcock, who agreed to supervise this study at the University of Cambridge and offered valuable advice in the course of preparing the original doctoral thesis, particularly with regard to the use of economic, industrial, and military historical data. I am indebted to the insights of two other Cambridge dons, the imperial and global historians Christopher A. Bayly and Anthony G. Hopkins, who first navigated me to material on British arms trade connections with India and West Africa during the eighteenth century, which then became a starting-point for my own forays into the field.

Special thanks are due to the following: Roger Hancock, formerly Proof Master of the Birmingham Proof House, for privileged access to its archives; William Garside, from the Department of Economic History at Birmingham University, for pointing me to the 'Barbara Smith Archive' and other sources in the concrete jungle of Birmingham's industrial frontier; Terry Barringer, formerly librarian of Royal Commonwealth Society Collection, who first introduced me to the colourful career of 'Charlie' Stokes, the lay missionary who became an ivory-and-arms trader; William Storey, from Millsaps College in Mississippi, who showed interest in my work from his own research in the history of the gun trade and gun control in nineteenth-century South Africa and New Zealand; Michael Charney, from the University of Michigan, who supplied useful leads on the arms trade in nineteenth-century Burma; Eric Tagliacozzo, from Cornell University, whose outstanding work on the history of smuggling across Southeast Asia's 'porous borders' overlapped in places with my own research; and Randolf Cooper, whose study of gun production and deployment in the context of the Anglo-Maratha Wars was especially illuminating. I remain grateful for funding via the British Government's Overseas Research Students (ORS) Award Scheme, a Cambridge Commonwealth Trust Fees Scholarship, and travel grants from the University of Cambridge, all of which eased the financial pressure by subsidizing expenses arising from my research trips.

At the 'periphery', too, I obtained valuable assistance from numerous individuals. They include Jill Quah, formerly head librarian of the National University of Singapore (NUS) Central Library, and Lily Tan, formerly Senior Director of the National Archives of Singapore, who provided key access to the use of microfilms and reprographics. I am grateful to Lee Li Kheng, cartographer at the NUS Department of Geography, for her help in rendering the maps used in this book, and to Susan Ang, lecturer in the NUS Department of English Language and Literature, for her constructive comments on earlier drafts of the manuscript. I also acknowledge the generosity of colleagues at the S. Rajaratnam School of International Studies (RSIS), Nanyang Technological University, Singapore: in particular, the

Dean of the School, Ambassador Barry Desker, for his support; Bernard Loo and Alan Chong, my colleagues in the Military Studies Programme, and Lieutenant-Colonel Joshua Ho, Sam Bateman, and Jane Chan, previously my colleagues in the Maritime Security Programme, for their good humour; the librarian, Chong Yee Ming, for her efficiency; and my fellow historians Kwa Chong Guan, Ang Cheng Guan, Kumar Ramakrishna, Joey Long, Ong Wei Chong, and Ho Shu Huang, for their encouragement.

Special mention should be made, too, of the professional and technical expertise rendered by the publishing team at Palgrave Macmillan: specifically, in Basingstoke, the editors Jenny McCall and Clare Mence, whose support and skills proved essential through critical stages of the production process; and, in Petaling Jaya, the copy-editing unit managed by Shirley Tan, who patiently and painstakingly undertook the various rounds of revision and refinement that yielded the final typescript.

Finally, it is impossible to measure the contribution of close friends and family. From my loving parents, Ernest and Aileen, I gained an early appreciation of things historical and literary, which made accessible many frontiers of knowledge. It is to them, in particular, that I wish to dedicate this book. Having crossed myriad frontiers to write this book, I am keenly aware of my limitations. Despite the scholarly and kindly attentions of many people, any mistakes that may remain are entirely my own.

Singapore
May 2012

Introduction

The arms trade is today so widespread that it creates anxiety as much as it compels attention. It impacts private individuals staring down the barrel of a gun, as well as public officials duty-bound to regulate if not eradicate it. As strategic analysts and decision-makers rise to the challenge of comprehending security in post-Cold War and post-9/11 scenarios, they find themselves drawn to matters of military industrial policy: how states decide what weapons to develop or discontinue; and how best to produce, procure, or prevent proliferation in line with changing threat perceptions.[1]

Notwithstanding the vastly diminished threat of global conflict through superpower confrontation, it was apparent by the early 1990s that the world's fastest developing region—the Indo-Pacific—had been experiencing a sustained build-up of modern conventional weapons for a decade. The overall trade in large weapons systems may have declined significantly after the mid-1980s, but the trade in small arms and light weapons continues to flourish. The disposal of surplus weapons by arms-producing states has inundated both formal and informal channels; this trend, reinforced by the mushrooming of black markets across regions, would suggest that the true value of the international arms trade far exceeds the official figures. The weapons have aggravated regional disputes and ethno-religious conflicts, arming 'non-state actors' that include local warlords, regional resistance groups, and global terrorist networks. Unsurprisingly, Hollywood films such as *Lord of War* (2005) and *Blood Diamond* (2006) have dramatized this trafficking of weapons, alongside the illicit drugs and gemstones that finance their purchase—such contraband fuels the pursuit of profit and power that would perpetuate cycles of conflict amid a climate of insecurity.

Yet there is a sense in which this keen competition to buy or sell weaponry is hardly novel, whether in the context of 'gun cultures' or 'small wars', driven or shaped by crises at the local, national, regional, or global level. These contemporary problems related to small arms proliferation bring to mind the international arms trade of an earlier age, when transfers of light weaponry gained far wider currency than the larger, more costly heavy armaments. From the diffusion of muzzle-loading muskets through breech-loading rifles to light weapons of repeating and rapid-firing capability in the later phases, its historical manifestation had repercussions transcending the boundaries of regions, continents, and oceans. It proliferated well beyond the 'metropolis' of arms-producing states in the West, interpenetrating indigenous societies across the 'periphery' of the wider world. Comparisons and connections between that past and our present arms transfer system are intriguing; it was precisely during the nineteenth century that the modern arms trade was transformed, by stages, into a truly worldwide phenomenon.

This is a book primarily about the arms trade in the 'long' nineteenth century, the period 1780–1914 that many scholars associate with the birth of the modern world. European historians have applied this term to the era of tumultuous change between the 'double revolution' of the late eighteenth century (the Industrial Revolution in Britain, the French Revolution of 1789) and the Great War.[2] Imperial and international historians have accepted this periodization as well, but mainly to define an epoch of European expansion and crisis, from the end of the first British Empire and 'swing to the East' in the 1780s to the 'high' imperialism of the 1880s–90s, which climaxed in the outbreak of the First World War.[3] But whatever the reasons for taking the late eighteenth century as their starting-point, few would likely disagree that the period should end in 1914. Although there were continuities across the great divide of 1914–18, there are good reasons for treating those years as a watershed. Total war on a global scale magnified every stress and strain throughout Europe and the non-European world—social, political, economic, cultural, and military—that led to national revolution in the metropolitan 'core' and nationalist resistance at the imperial periphery. Even at the start of the twenty-first century, the First World War presents itself as a pivotal event in modern history, the first great catastrophe of the twentieth century that ruptured the international order of states and empires.

Equally, it makes good sense to view the sweeping developments of the long nineteenth century as integral to the drama of the arms trade. In addition to evolving mechanisms, there are now established norms for tackling the arms trafficking and associated crime exacerbated by the negative effects of this cross-border interconnectedness the contemporary world has labelled 'globalization'. But it was through the engines and factories of Europe's industrialization that arms production was first modernized; through the tentacles of imperial expansion and indigenous crisis that arms procurement was first 'globalized'; and through the ensuing crescendo of violence augmented by arms proliferation that arms control was first internationalized.

The organization of the book

Chapter 1 begins by framing the debate on international arms transfers from an historical perspective. It suggests comparisons between the contemporary arms transfer system and its precursor by tracing patterns of continuity and change in the ongoing cycles of military innovation, production, and diffusion. It then proceeds to chronicle the emergence of an increasingly interdependent small arms trade, linking the European core of the industrial world economy to its clientele in the wider periphery of Africa and Asia. In so doing, it attempts to connect the mercantile and maritime arenas of the Mediterranean-Atlantic and Indian Ocean in a trailblazing epoch of industrial enterprise and imperial expansion. Such a comparative approach combines the horizontal linkage of 'lateral history' with the long-term view of 'vertical history', which the eminent French historian Fernand Braudel called the *longue durée*.[4] Without this critical plane of analysis that provides essential context across both space and time, our understanding of the workings of the international arms transfer system would be greatly impoverished, if not dangerously incomplete.

Chapter 2 situates the metropolitan origins of the arms trade in the struggle for mastery of pre-industrial Europe, before tracing the course and contours of its development through an age of industry. Here, the British example stands out, even as the island nation was transformed progressively into the leading industrial and imperial power of the Western metropole. Britain's Industrial Revolution in the late eighteenth century, ahead of all continental rivals, translated into a commanding lead in industrial production. Still, it was the country's proto-industrial base and its capacity for industrious enterprise—most notably evident in Birmingham—that gave the local gun trade its decisive advantage over potential competitors. Ultimately, when tested by the vicissitudes of war and peace in the nineteenth century, the strangely ambivalent relationship between the British state and its private sector both stunted the productive capacity and blunted the innovative thrust of the country's small arms manufacturers. Britain's initial competitive advantage in the production of such weapons was also eroded by the subsequent industrial transformation of the United States and continental Europe, accompanied by the fluid transfer of military technology between states. Chapter 2 charts the long-term trajectory and fluctuating fortunes of the British small arms trade: from the craft origins and eventual relocation of private arms production from London to Birmingham during Britain's Industrial Age, through the peculiarities of the British system of state and private arms production, to domestic responses to the twin challenges of the new technology and foreign competition.

Chapters 3 and 4 shift the focus of enquiry to the ramifications of arms trade networks, spreading from manufacturers in the Euro-Atlantic core to markets beyond Europe. Whereas industrialization generated much of the infrastructure for the modernization of weapons production, the arms transfer system was internationalized through the globalizing dynamics of Western imperialism and indigenous crisis. Indeed, markets provided by indigenous societies in the non-European world proved indispensable to the survival (if not success) of small arms manufacturers in Europe, especially during periods of relative peace and slack demand. These chapters offer a means of examining connections between pre- and post-industrial worlds, along with the cross-border movements of people, commodities, knowledge, and techniques that characterized the early history of globalization. Around the Indian Ocean, the mercantile markets of towns and cities far inland were complemented by bustling bazaars that operated in port-cities along the ocean littoral. Uniquely sustained over many centuries by the annual monsoon cycle, and animated by a lively subculture of bargaining and 'value added' services, these cosmopolitan centres became a vital 'arms trade nexus' between local communities and the outside world.

Chapter 3 explores the interplay between the arms trade and indigenous patterns of commerce, production, and state formation in the western Indian Ocean zone, extending from East Africa into the Middle East and South Asia. Of key interest would be the arms traffic proliferating across the Swahili coast, Zanzibar and Oman, the Northwest Frontier and Afghanistan, which was facilitated by port-cities like Muscat. Chapter 4 examines the arms traffic interpenetrating the trade and politics of societies in the eastern Indian Ocean zone, extending from South Asia into Southeast Asia. A particular focus would be arms transfers around the Eastern Archipelago, circulated via transhipment centres including Singapore and Jolo.

Chapter 5 concludes by showing how the ongoing evolution of indigenous small arms technologies and manufacturing, galvanized by the revolutionary impact of military diffusion from the Euro-Atlantic core, meant that Western military forces did not always enjoy techno-logical superiority or tactical advantage. Homemade or imported weaponry in the hands of native opponents could prove just as effective operation-ally (if not more so) in the context of local geography, military culture, strategy and tactics. Over the second half of the twentieth century into the dawn of the twenty-first, the West was still leaning—via armed con-flict with Indian Ocean societies—that the technological and organiza-tional superiority of Western arms did not guarantee victory in warfare.

Locating the arms trade in space and time

Inasmuch as the arms trade straddles a complex series of structures and patterns with local, national, regional, and global interconnections, it lends itself to careful study in the intermediate analytical space between global histories (including the early history of globalization) and the histories of regions and their communities (sometimes included under 'area studies'). Indeed, from the broad perspective of imperial and inter-national histories, the arms trade must be located within that wider cul-tural sphere of frontier encounters and exchanges, which encompassed elements of conflict as much as collaboration. Kaleidoscopic interactions between the European explorers and empire-builders, merchants and mis-sionaries, soldiers and settlers, on the one hand, and diverse native popu-lations outside Europe, on the other, altogether facilitated significant transfers of military hardware. Firearms were bartered for various conces-sions or offered for mutual advantage, as when indigenous collaborators were equipped to fight common enemies. These Europeans also encoun-tered forms of indigenous resistance and warfare, sometimes assisted by rival Europeans. Amid the swirling maelstrom, the arming of the native would shape the entire course of cross-cultural collision.

A web of activity and intrigue, involving European metropolitan gov-ernments, colonial regimes, chartered companies, and gun-runners, was spun out across the commercial networks and political systems of the Indian Ocean arena. For centuries, this transoceanic milieu constituted a vast interregional arena that many still regard as precursor of the modern Indo-Pacific regional system. The volatile trade and politics of East Africa, the Middle East, South Asia, and Southeast Asia remain the context for much of today's illicit trafficking, but it is instructive to explore the ways in which the European interlopers, laden with military merchandise and clandestine cargoes, were first drawn into—then came to dominate—regional markets around the Indian Ocean littoral.[5] By combining metro-politan and indigenous components, the military-fiscal sinews of the arms trade would grow more sophisticated, adapting to the requirements of the emerging industrial world economy. A well-articulated arms transfer system would develop over the long nineteenth century: at one end, the cunning dealers who devised multifarious ways of supplying 'warlike stores' made in the workshops of Britain and the wider Western hemisphere; at the other, a vibrant Indian Ocean complex, in which an international arms bazaar underpinned by time-honoured traditions of bargaining and bartering now serviced the escalating demand for weaponry.

International arms transfers would enable new patterns of indigenous state formation around the Indian Ocean, even as they empowered the global projection of increasingly strident forms of European imperialism. The finely balanced, layered concept of sovereignty shared by pre-colonial Indian Ocean polities was displaced progressively by Western notions of indivisible law and monolithic sovereignty imported under colonial conditions from Europe. Yet the delineation of separate colonial spheres—circumscribing territorial sovereignty and maritime jurisdiction—remained a matter of ongoing accommodation and readjustment at the periphery. The chronological and spatial transition from 'frontier' to 'border' and 'boundary' did not always proceed in smoothly linear fashion; in reality, frontier zones remained turbulent and poorly defined long after Europeans had mapped frontiers and demarcated boundaries.[6] Likewise, there would be a blurring of distinctions between categories of 'traders' and 'traffickers', since state-sponsored arms transfers and non-state smuggling occurred simultaneously against the 'see-saw' progress of arms control legislation and sheer patchiness of law enforcement. Arms consignments often continued their surreptitious movement along clandestine routes, across porous borders, over both land and sea.

The Indian Ocean arms trade would prove instrumental in both the limitation and expansion of European power and influence, against the changing temper of indigenous state formation, warfare, and resistance. Insofar as they were instruments of colonial authority, these European agencies and arms transfers would project ideologies and images of empire, forcefully, even violently. Conversely, an arming of the western and eastern Indian Ocean zones would alter the complexion of myriad societies outside Europe, affect the pace and pattern of indigenous state formation, and achieve a paradigmatic shift in native ways of war and resistance. Arms proliferation and armed conflict—including 'asymmetric' colonial warfare—reached such epidemic proportions by the late nineteenth century that Europe's great powers would be jolted into formulating unprecedented cooperative measures to curtail the arms trade in

Africa and Asia, where it had intersected the traffic in equally controversial commodities like slaves, ivory, and opium. In the Brussels Conference Act of 1890 and the Jolo Protocol of 1897, we see the first major attempts at international arms control and multilateral security arrangements; in effect, the origins of what would become a worldwide securitization of commodities, from arms transfers to the trafficking of human cargoes, drugs, and other contraband.

Foreshadowing the future?

There are powerful continuities from the historical narrative of the arms trade that still impact our world, interwoven with such persistent themes as nationalism, globalism, religious fundamentalism, and terrorism. Arms trafficking financed by drug money and other contraband still presents a formidable challenge to international security. The arms trade still engenders creeping militancy, contributing to the fragility and erosion of states incapable of handling developments within their societies. It still empowers aggressive nationalism, genocide, civil wars, and insurgencies, along with acts of terror and piracy. In many instances, there remains a proven correlation between the availability of small arms and explosives, on the one hand, and instability or even state collapse, on the other.

The stakes have been raised on the global stage featuring state actors with volatile nuclear capabilities, supported by a cast of non-state actors exploiting rapid Internet communication to perpetrate organized crime or terror. Across a levelled post-9/11 landscape, and amid long-running indigenous tensions, we witness the capacity of local warlords, resistance fighters, or jihadists to 'recycle' violence from the Indian Ocean heartland to the urban frontiers of the Western hemisphere. The bomb blasts of July 2005 and July 2006 (in London and Bombay, respectively) would only add a further explosive dimension to the repercussions of arms proliferation and asymmetric warfare in the global village. The terror attacks of November 2008 were unleashed upon Bombay by armed militants from across the Arabian Sea. These contemporary episodes echo in terrible new forms the business of arms transfers and armed conflict that first took the world by storm during the nineteenth century. It becomes clearer yet that the 'post-colonial' arms trade of the present builds upon the foundations of its colonial predecessor, while presenting fresh challenges to international stability and security.

Ultimately, to discern the past patterns and structures is to understand better the international arms trade of future generations. While appreciating the cross-cultural constraints under which armed forces and security agencies must operate, it is imperative to understand what drives the arms trader and the armed terrorist, learn how they derive socio-economic and financial support, track their modes of operation, and trace their lines of communication. Such insights, borne out over time, still hold the key to the successful dismantling of global terrorist networks linked to arms transfers and their pervasive culture of violence.

1

The Arms Trade and Global Empire in the Indian Ocean

I urge you to beware the temptation of pride...
to ignore the facts of history and the aggressive impulses of an evil empire,
to simply call the arms race a giant misunderstanding and thereby remove yourself from the struggle between right and wrong and good and evil.

Ronald Reagan, 8 March 1983

International arms transfers in historical perspective

The 'evil' Soviet Empire may have fallen, but a reconfiguration of the international order after 1991 has again opened Eurasia to potential arms races that revolve around Russia, the Middle East, India, and China. A resurgence of regional crises and ethno-religious conflicts in the aftermath of the Cold War, and the prominence of terrorist attacks and various forms of unconventional warfare in more recent times, have further renewed concerns about the international arms trade. Statistics indicate that international arms transfers in the late 1980s, just prior to the collapse of the Soviet Union, involved as many as 50 states as suppliers and 120 as recipients. World military expenditure, in general decline through the 1990s, has been climbing since the start of the new millennium. Crucially, international arms production was not retarded over the past two decades; the financial value of global arms exports was approximately $48 billion in 1989 and still above $50 billion in 2007.[1] Transfers of large 'high-tech' weapons systems have undergone a relative decline, but the trade in small arms and light weapons shows little sign of receding, fuelled by regional tensions and ethno-religious strife frequently involving non-state participants. The disposal of surplus conventional weapons by arms-producing states has indeed augmented arms trans-fers—via 'white', 'grey', and 'black' markets—suggesting that the true value of the international arms trade far surpasses given estimates.[2]

A clear pattern of production and proliferation has emerged in support of these contemporary trends. The Soviet Empire's disintegration created opportunities for illicit arms trading in the grey and black markets, whe-ther by state-controlled arms companies or criminal groups with access to weapons stockpiles located in the former Soviet Union and Eastern Europe. Belarusian, Russian, and Ukrainian officials were implicated in selling arms to states like Iraq through the 1990s until 2000, thereby con-travening United Nations sanctions. The final destination of those weapons was concealed through counterfeit end-user certificates, and facilitated by various front companies opened in Jordan, Syria, and Malaysia that acted as the official buyers, with illicit consignments funnelled through inter-national trading entrepôts like Singapore.[3] Seemingly legitimate domestic arms exports have thus served as a cover for illegal transfers. In 2002, a criminal group targeted the largest arsenal of the Ukrainian Army, manag-ing the theft of 190 firearms, over 40 missile launchers, plus ammunition and explosives.[4] By combining the operations of the traditional underworld with legitimate commercial structures, government bureaucracy, and repre-sentatives from the military and security services, such criminal groups have regularly supplied militants in Africa and Asia.[5]

Meanwhile, Europe as a whole has consolidated its extensive share of the world's legitimate arms export industry, being a major stakeholder in the white market of the international arms bazaar. Between 1994 and 2001, the European Union's 15 member states alone exported conven-tional weapons and military equipment worth nearly $10 billion (or 32 per cent of the world total). Strong government support for several large deals has enabled many EU countries to compete in a global market dominated by the United States.[6] Again, at the receiving end were devel-oping countries in Africa and Asia; and, in particular, states along the Indo-Pacific rim such as China, South Korea, and Malaysia. Since 1995, the annual reports of the Stockholm International Peace Research Insti-tute (SIPRI) and International Institute for Strategic Studies (IISS) confirm that Asian governments have been the world's top arms buyers—recipients of over 40 per cent of global arms imports.[7]

Whereas the trade in large weapons systems may be decelerating, the distribution of light weaponry has become increasingly problematic. Regardless of whether they were obtained through the white, grey, or black markets of the international arms bazaar, or recycled from pre-vious conflicts, at least one independent analyst has commented that the suffering inflicted by small arms is 'almost entirely limited to the periphery of Europe and to the developing world, particularly Africa and South Asia'.[8] In those conflict-prone regions, the direct correlation between the availability of small arms and explosives, on the one hand, and political instability or even state collapse, on the other, has been underscored by the United Nations.

For a start, the term 'small arm' (or 'light weapon') is really a misnomer. These basic pieces of military equipment have been responsible for the vast majority of deaths in armed conflicts since the UN was set up in 1945. In 1999 alone, there were 27 major armed conflicts in 25 countries around the world. Over 1,000 people died in 14 of those conflicts; the high incidence of intensive conflict occurred in only two other years of that decade; and every conflict was empowered overwhelmingly by the use of light weaponry.[9] Such compact, portable weapons still possess tremendous capacity for inflicting chaos and death. The Kalashnikov AK-47 assault rifle can fire up to 30 rounds in less than three seconds, with each bullet lethal up to a range of 500 metres. The M-16 assault rifle features additional semi-automatic and fully automatic capabilities, firing up to 950 rounds per minute. At close quarters, the Heckler and Koch AM 180 sub-machine gun is even deadlier, with a rate of fire of 1,500 rounds per minute. The Russian-made RPG-7 (rocket-propelled grenade) can penetrate 330-millimetre thick armour up to 500 metres away, while the American-made Claymore landmine is able to propel up to 700 steel balls in a 60-degree arc to devastating effect within a 50-metre radius. Recognizing the need for a global effort to tackle small arms proliferation linked to various forms of armed conflict, the UN convened a major conference at New York in 2001 to address all aspects of the illicit small arms trade.[10]

The burden of responsibility for trading in such death-dealing merchandise would lead EU member states to adopt an ethical Code of Conduct for Arms in 1998. This was an attempt to formulate common arms export regulations that could help stem the supply of weapons from the Western hemisphere. Unfortunately, some member states would circumvent or ignore national export criteria and the EU Code, exploiting loopholes, omissions, and weaknesses to let arms slip through the net. EU enlargement in 2004 would even expand the scope and number of companies manufacturing and exporting munitions, with more than 400 companies in 23 out of the 25 member states still producing small arms and light weapons, in total only slightly fewer than in the United States. This has presented further opportunities and dangers for European arms controls.[11]

At the other end of the arms transfer system in Africa and Asia, there is every indication that the problem of small arms proliferation has deep perennial roots. The impact of surplus weapons recycled from past generations is still felt amid 'killing fields' and 'gun cultures' shaped by organized crime, political movements, insurgencies, and sectarian violence. In a wide arc of crisis ranging from societies around the western Indian Ocean (parts of the Congo, Rwanda, Burundi, Uganda, Sudan, Somalia, Iraq, Afghanistan, and Pakistan) to those flanking the eastern Indian Ocean (parts of India, Bangladesh, Nepal, Sri Lanka, Burma, Cambodia, Indonesia, and the Philippines), government security forces have long encountered weaponry as potent as what they themselves may access, including assault rifles and rocket-propelled grenades. Small arms have riddled or ripped apart the social fabric wherever entrenched gun cultures have developed, fuelling violent crime, banditry or piracy, and 'turf wars' revolving around syndicates, sectarian feuds, or secessionist movements. For transcontinental arms control blueprints such as the Declaration of the Bamako Conference in 2000, or regional gun control associations such as the South Asia Small Arms Network, the task is to disarm the periphery systematically: promoting collaborative efforts between state and society aimed at reducing the demand for weapons, if not stopping their proliferation altogether.[12]

Admittedly, it is a task made more daunting by the ongoing activities of arms trade intermediaries. Adept at dealing, facilitating, and trafficking through the operation of both grey and black markets, such intermediaries continue to have access to an intricate network of subcontractors, suppliers, and shipping agencies across the Western 'metropolis', in addition to links with corrupt officialdom and criminal organizations at the non-Western 'periphery'. Dealers buy and sell directly large quantities of arms; brokers facilitate deals without actually taking possession of arms shipments; and smugglers (or gun-runners) engage personally in trafficking. All are aided and abetted by transportation agents and cargo handlers capable of making vessels or aircraft virtually disappear while delivering arms shipments; corruption payments to complicit authorities providing access to weapons stockpiles; documentation that is fabricated by corrupt officials, forged, or falsified, to circumvent export licensing and customs regulations; and financial institutions or arrangements that enable arms acquisition through direct purchases in cash or credit, or through commodity exchanges involving items of high value such as drugs, diamonds, and other precious or intoxicating substances.[13]

All in all, the contemporary world sees an interlocking pattern: arms production in the Western hemisphere; arms proliferation in Africa and Asia; links between profit and violence through the recycling of arms and armed conflict; and attempts at arms control against cloak-and-dagger activities in the shadowy underworld of grey and black markets. For policy-makers and private individuals alike, this jigsaw puzzle of an arms transfer system does raise fundamental questions about military industrial policy and collective security. How do states decide which weapons to produce or procure, and which ones to discontinue, dismantle, or dispose of? When, and how far, should privatization, innovation, civil-military integration, or internationalization be pursued to keep strategic military industries viable in the long run? When should 'dual-use' technology be introduced to allow for the conversion of equipment from wartime (military) to peacetime (civilian) applications? What can be done to curb proliferation along clandestine routes, or improve security across porous borders?

The contemporary emphasis is perhaps warranted, given the urgency of these pressing issues, but it has also skewed the debate in favour of developments within living memory, usually from the perspective of international relations or current affairs.[14] Such imbalance is compounded by the immediacy of high-profile terrorist attacks and other forms of 'asymmetric' warfare in a new information age. These are mostly associated with events flowing from the Al-Qaeda attack of 11 September 2001 on the World Trade Center, New York, and the subsequent international war against terrorism: a pattern of cross-cultural collision enmeshing America and Afghanistan, before engulfing other metropolitan centres (Madrid and London) and parts of the non-Western periphery (the Middle East, South Asia, and Southeast Asia; from Istanbul to Bali and Bombay). Notwithstanding the weapons of improvised character and indigenous manufacture, there is little doubt that these attacks were empowered, at least in part, by weapons and explosives derived through international arms transfers.[15]

But the real danger lies in distortion: viewing the 'modern' phenomenon of arms transfers in deracinated form—severed from deeper roots in the historical past—and thereby failing to discern long-term patterns (or, at best, paying only cursory attention thereto). One example is Anthony Sampson's survey of the modern arms industry. Excellent though it is in covering recent developments, including some key components and controversies associated with contemporary arms sales, its attempt at historical analysis only goes as far back as the Crystal Palace Exhibition of 1851:

It was in the mid-nineteenth century, in the wake of the first industrial revolution, that the modern armaments industry began to take

shape, inspired and pressed forward by a handful of inventive entrepreneurs who developed the science of explosives and guns.[16]

Yet the concept of the 'arms bazaar' is not, strictly speaking, a 'modern'
one. Whether in ancient Greece, the seventeenth-century Dutch Republic,
or eighteenth-century India, arms producers and dealers have for over
two millennia supplied the demands of local, regional, and international
markets.[17] The Indian Ocean, with its cosmopolitan bazaar culture, has
commanded historical significance as a geo-strategic arena for international arms transfers at least since Vasco da Gama's arrival in India in
1498. But the 'post-colonial' interest generated by the arms trade since
the Indian Ocean ceased to be a British lake 'east of Suez' after 1968 has
again obscured the earlier connections.[18] As new 'global' histories are composed across the threshold of a new millennium, offering vital long-term
perspectives on world affairs, it proves just as timely to bring to the contemporary understanding of international arms transfers a more balanced,
nuanced historical approach.

Even here, the construction of an historical approach that combines
the horizontal linkage of 'lateral history' with the long-term view of
'vertical history' is not without difficulty. The history of weapons technology is fraught with ambiguity. Mary Kaldor has described modern
armaments, arms technology, and arsenals as 'baroque': 'the offspring
of a marriage between private enterprise and the state, between the
capitalist dynamic of the arms manufacturers and the conservatism
that tends to characterize armed forces and defence departments in
peacetime'.[19] But at what stage, and under what circumstances, does
weapons technology become 'baroque' within the overall arms production system? Moreover, Kaldor's characterization may be misleading,
given the historical examples where it was the state—not the private
trader—that proved the more enterprising and less conservative in
pioneering military industrial development. Specific instances in the
history of small arms production illustrate how governments did not merely
gild the lily of arms technology through the elaboration of existing military hardware, but spearheaded significant technological innovation and
industrial renovation even when the nation was not at war. Whether in
late eighteenth-century Europe or early nineteenth-century America, the
patterns of continuity and change that characterized military technological development reflect such ambivalence. Anxieties attended the rise of
large-scale arms manufacturing within a scenario both subtle and complex; isolated rural societies found themselves 'floundering between
the two worlds of agrarian pastoralism and industrial progress', the

responses of their gunmakers to state-led technological change proving no less 'hesitant and equivocal'.[20]

The history of arms transfers is equally problematic. It may be conceded that today's international arms trade is unsurpassed in terms of scope and scale. But when, and how, did this new global pattern emerge? Against what historical precedents can it be meaningfully compared? Early accounts of inter-state arms transfers remind us that interregional arms trading could scarcely be considered 'new', having been employed at least since the Peloponnesian War (431–404 BCE) to achieve the political, military, and economic objectives of states and sovereigns.[21] Furthermore, the trading of weaponry by 'state actors' discounts the long-standing role played by 'non-state actors'—religious figures and resistance fighters, bandits and pirates, revolutionaries and terrorists—along the expansive margins of the arms transfer system, betwixt the interstices of state power, where the proliferation of light weaponry could constitute a particular threat. Arms transfers of every description have been motivated across many centuries, and not mere decades, by the almost timeless pursuit of wealth, power, and victory in warfare.[22]

A typology can assist in making historical sense of the structures and the shifting patterns of military innovation, production, and diffusion. First, the hierarchy of state systems may be stratified by its grasp of technology according to four types of military production and consumption:

- *innovators* (top-level producers; at the forefront of technology);
- *adaptors* (upper-level producers; close to the innovation frontier; capable of reproducing and perhaps enhancing newer technology);
- *replicators* (lower-level producers; capable of duplicating older technology); and
- *operators* (bottom-level producers; basic consumers).

Next, the dynamics of the arms transfer system may be characterized by delineating the main phases of transformation—economic, political, and military—accompanying each 'wave' of weapons innovation, production, and diffusion:

- (I) Disequilibrium and disintegration. 'Revolutionary' technological innovation occurs, resulting in the emergence of centres of advanced arms production that monopolize new techniques based on their underlying factor endowments and pursuit of profit and power. Advantages that weapons based on the new technologies could bring then motivate states and various elites to acquire them, thus causing the new weaponry (and the ability to produce it) to diffuse throughout the system. Such arms procurement, with or without recourse to armed conflict, brings about corresponding changes in the military and political balances. Traditional elites or long-established powers with insufficient factor endowments and political motivation lose influence and authority, even as new elites arise and the external world order alters.

- (II) Competition and accommodation. States that are not initially within the top level soon attempt to eliminate the near-monopoly in the means of producing advanced weapons that top-level states possess, by acquiring the new technologies themselves. Notwithstanding the acceleration of an arms race in many instances, the uneven distribution of socio-economic and political endowments results in the imperfect diffusion of new technologies (and the ability to reproduce them) throughout the system, giving rise to the intermediate upper and lower levels of producers. With varying degrees of success, regional elites interpose themselves as intermediaries between the old sources of internal authority and the new sources of global authority. Such 'successor states' use new weaponry to supplant or contain the internal old order, and to set limits to the advance of the external new order.

- (III) Stratification. The diffusion of military technology is neither smooth nor linear, and the system remains highly stratified. Although the rate of incremental innovation in top-level states slows in the later stages of the process (leading to 'baroque' arms production), the barriers to movement between levels remain relatively high as the intense competition between a slowly-growing number of suppliers increases the importance of underlying factor endowments as the prime determinant of a state's location in the global military hierarchy.

- (IV) Reintegration and new equilibrium. At the top of the hierarchy, the integration of the great (or global) power into the region is progressive because the new technological dynamic usually leaves that power with superior weaponry. Also, the arms trade creates possibilities for other kinds of trade, acting as a trade multiplier that is the motor for the accumulative process. The most significant modifier to this process of integration usually proves to be the constraints placed on the new technology: where the new military equipment of the global power is not always deployable (or deployed effectively) in an alien geopolitical context; and where local geopolitical and cultural factors reinforce forms of indigenous resistance among 'non-state actors'. At the bottom of the hierarchy, weak states attempt to acquire the means of producing military technology—either to fortify their positions in relation to strong states, or to exercise their own form of local dominance—but the success of such attempts is nearly always limited. Once (IV) is in place, the evolutionary process awaits the next 'revolution'.

Global history suggests that the continuous evolutionary dynamic of the international arms transfer system has been punctuated and propelled by 'revolutionary' bursts of military technological change. Arms procurement and proliferation have occurred whenever weapons of a 'usable' technology could be supplied from one centre to meet some specified demand elsewhere, usually in the context of crisis. States have projected their military and political power according to their command of technological and economic resources, but the emergence of new technologies and new centres of production has tended eventually to devolve power from the established sources of authority toward new elites.[23]

To attempt a preliminary sketch outlining the dynamics of the international arms transfer system over space and time, we must first trace the evolution of firearms technology to its origins in the East, beginning with the use of incendiaries in Taoist ritual during the period of the Song dynasty in China (960–1279). Firearms technology progressed from incendiaries through the stages of flame-throwing weapons (fire-lances); explosives (bombs launched from trebuchets); rockets; and, finally, barrel guns and cannon. The 'Gunpowder Revolution' of the thirteenth century set off the first significant wave of the arms transfer system: the discovery of a fast-burning nitrate-rich gunpowder mixture, of sufficient propellant force to discharge a projectile from a narrow metal bore, prepared the way for the manufacture of cannon and small arms. Through the Mongols, knowledge of this new technology would spread from China to India, the Islamic world (Arabs, Turks, Persians), and Europe. But such technology was not yet universally or even generally applied in the military cultures of China, Europe, or the Islamic states; it had yet to demonstrate its superiority over traditional arms, or integrate into established military structures and tactics in each region.[24]

Early modern Europe, seething with inter-state competition, appeared more willing than either the Chinese or Indo-Islamic world to adopt firearms and attempt innovation using the new technology. In the West, the introduction of the yew longbow or steel crossbow in the hands of a skilled archer had already exposed the vulnerability of the armed knight on horseback.[25] Whatever the limitations of early firearms, Europe's reception and application of firearms technology soon enabled the ascent of new elites. Over time, the loci of firearms production, innovation, and diffusion would shift from Italy to the Low Countries, England, and Sweden, in some cases pushing the technological frontier from mere adaptation to cutting-edge innovation.[26] The Europeans concentrated on light and mobile artillery, though it was in handguns that they excelled: the arquebus, operating on a simple, cheap and fairly fool-proof matchlock mechanism, remained the principal small arm until its replacement by the flintlock musket in the late seventeenth century.

Across the East, the relatively more stable, closed empires of Turkey, Persia, India, and China maintained technological and trading dom-inance. But where traditional modes and means of warfare prevailed, it was in heavy siege artillery rather than small arms that they invested huge military-fiscal resources. Only in the warring states of mainland Southeast Asia—Burma, Siam, and Vietnam—did the new technology in firearms find widespread application in an array of handguns and light weapons.[27] With the new patterns of state formation came the first ripples of European incursion in maritime parts of Asia, reinforced by military production and diffusion. Such developments accompanied the disintegration of the Malacca Sultanate from the sixteenth century, when Portuguese and Dutch imperial ventures interacted with indi-genous societies in a state of flux, paving the way for prolonged crisis in the Malay world and galvanizing the lower levels of military produc-tion and consumption.[28] Unlike the evolution of a relatively standard-ized 'Western way of war' in early modern Europe, with its growing reliance on firearms, there was no monolithic 'Asian way of war' since firearms were incorporated into warfare in a manner consistent with the spectrum of pre-existing military cultures. Firearms were not neces-sarily drivers but rather indicators of change, where the policies of a victorious elite rather than their procurement of guns determined the specific pattern and pace of military, political, and social transforma-tion. Such was the case with Mughal India's calculated programme of 'military-fiscalism' from the sixteenth century, or Tokugawa Japan's deliberate reversal from the gun to the sword from the seventeenth century, intended to maintain a grip on political authority and social hierarchy.[29]

The arms transfer system was transformed by a second wave of major innovation in the eighteenth and nineteenth centuries: the series of far-reaching industrial and technological developments that included the 'Breech-loader Revolution'. In the West, Britain, France, Germany, the Low Countries, and the United States retained or gained prominence in arms production, innovation, and diffusion. In the East, however, the frontiers of military production and technology kept pace with the West only up to the 1850s; thereafter, the asymmetry widened as mass-produced, rifled, breech-loading weapons were perfected in Europe and America, while military technology and production stagnated across Asia. The great Islamic empires of the Ottomans, the Safavids, and the Mughals came under increasing pressure as provincial elites and regional powers gained the ascendancy. Among the latter were Oman, Mysore, the Marathas, the Sikhs, and the Afghans, along with the Bugis, the Tausug, and other seafaring peoples of Southeast Asia. As power was redistributed, these groups interposed themselves as power-broking intermediaries between the old sources of internal authority and the new sources of global authority. Local societies were quick to appreciate the importance of firearms for survival and geopolitical manoeuvring in these regions, for the learning curve on new forms of weaponry and their procurement seems to have been shorter than that on any other commodity.[30] But while the new elites used this new weaponry to contain or displace the internal old order, or to check the advance of the external new one, none of them ultimately proved capable of breaking into the upper echelons of arms production.

The period of intense international competition and conflict that witnessed two World Wars (1914–45) engendered a third wave of major innovation, the 'Scientific Revolution'. It combined the internal combustion engine with the possibilities of modern chemical and electrical engineering technologies, and subsequently developed technologies to support strategic nuclear weapons. Although Britain, France, and Germany were able to remain at the forefront of technological development for a time, the advanced science involved in this innovation cluster meant that military research and development and weapons production would be conducted on a scale that excluded all but the very largest states from being innovators. With the exhaustion and eclipse of the old European empires from the early twentieth century, the United States and the Soviet Union were soon poised to achieve pre-eminence in military technology and production.

This third revolution in military technology represented a quantum change in the capabilities that opposing sides in the post-war world deployed. Through the dynamics of the arms transfer system and grey-market trading, it also enabled the superpowers and top arms-producing states to arm insurgent groups involved in conflicts throughout satellite states. After 1945, countries like Israel, Brazil, and China would eventually progress into the ranks of the world's major arms producers, while new states in the Third World and regional elites around the Indo-Pacific emerged to become the top arms consumers.[31] But highlighting the extent of arms transfers through the grey market is the example of clandestine support provided by the US Central Intelligence Agency to Afghanistan, via Pakistan, during the Soviet occupation of 1979–89. This conduit facilitated the transfer of an estimated $2 billion in small arms, light weapons, and other military supplies to the *mujahidin*, with consequences far beyond the original scope of the arms-supplying state.[32] With the jihad in Afghanistan assuming a life of its own and spawning a second generation of *mujahidin* who called themselves the Taliban, it was only a matter of time before their support of Osama bin Laden and Al-Qaeda would lead to the attacks on America and a whole new cycle of asymmetric warfare.

From the end of the Cold War to the onset of a global 'War on Terror', the arms transfer system would again evolve to accommodate the changing world order. Metropolitan Europe would strengthen its position in the hierarchy of arms-producing states in the Western hemisphere, with countries such as Britain, France, Germany, Italy, and Sweden resuming their historical role as suppliers of small arms. These arms transfers from Europe, along with weaponry recycled through the Cold War or acquired from arsenals in the former Soviet Empire, are still consigned via intermediaries to markets across the frontiers of Africa and Asia. In terms of weapons technology, it is debatable whether 'smart' technologies belong to a 'baroque' phase of the 'Scientific Revolution', or form the vanguard of a new 'Digital Revolution' that would drive the next wave of the arms transfer system. What is not in doubt is the fact acknowledged by the recent concept of a 'Revolution in Military Affairs': small arms, light weapons, and explosives—imported, recycled, or home-made—continue to prove lethal in the context of small wars, terrorist attacks, and other forms of asymmetric conflict today.[33]

Global arms trading in the age of industry and empire

In surveying the patterns of military innovation, production, and diffusion that have systematized international arms transfers, it becomes apparent that something decidedly transformative occurred over the period 1780–1914. The interlocking events and movements of that 'long' nineteenth century were unprecedented in both qualitative and quantitative terms, making their impact at the level of the locality, nation, region, and globe. Well before the supposed commencement of the current phase of globalization from 1945, the growing interconnectedness of social, economic, and political changes that characterized the birth of the modern world also began to define the character of the arms trade as we know it, and especially the trade in light weaponry. The industrial transformation and imperial expansion that made the great powers greater yet, and the emergence of a capitalist world system through cycles of war and peace, generated the preconditions and framework of a technologically sophisticated small arms trade which could, for the first time, be described as global.

Until the nineteenth century, the small arms trade of Europe had been piecemeal and proto-industrial in character. The workshop industries of the early modern European states, typified by centres such as Liège or Birmingham, produced various types of muzzle-loaders that were intended primarily for domestic consumption rather than large-scale export. The type of smoothbore flintlock musket popularly known as 'Brown Bess'— first adopted by the British Army in 1706, and appearing in later 'baroque' incarnations as the 'India Pattern' of 1794, or the 'Light Infantry' or 'New Pattern' of 1814—was still widely used in military circles, though increasingly criticized by contemporaries for being 'comparatively deficient' in technological terms.[34]

It was the potent combination of 'industrious revolutions' and industrialization, altering basic patterns of consumption and production across the long nineteenth century, which first enabled the modernization of military production. The vital socio-economic developments that attended the Industrial Revolution, occurring amid a wider struggle for mastery in Europe, galvanized the entire supply side of the arms transfer system.[35] From clusters of workshops to chains of factories, state-led initiatives spearheaded military innovation and industry across the Western hemisphere, pushing arms trade and technology by stages into higher gear. New technologies in metallurgy and steam power were applied to weapons and warfare, with devastating consequences. The progressive commercialization of war and military service, along with arms manufacture and supply, would climax in the construction of a military-industrial complex, capable of producing anything from small arms and ammunition to heavy armaments, explosives, pontoon-bridges, and warships.[36]

For small arms production, the changes would be far-reaching even if the transition from flintlock to percussion ignition occurred only on paper in 1836, and the changeover from smoothbore to rifle was delayed for nearly two more decades. Following a period of general peace in Europe (1815–53), the outbreak of the Crimean War spurred the rapid introduction and deployment of the rifled-musket, with the consequent rapid decline in the use of the 'Brown Bess'-type musket among regular troops.[37] Among the key innovations in the first phase of 'precision' small arms were the Dreyse (1848), the Minié (1851), the new Enfield (1853), the Snider and the Chassepot (1866), the Martini-Henry and the Mauser

(1870–71), and the Gras (1874).[38] By 1878, all the European powers had rearmed their infantry with the Chassepot, Snider, and Martini-Henry breech-loaders, all using metallic cartridge cases and steel barrels instead of the iron used in the old muskets. This changeover immediately liberated huge quantities of obsolete weapons for export, mostly percussion guns using caps instead of the flintlock mechanism. Then came the second phase: the single-shot breech-loader had barely been perfected when the repeating or magazine rifle was introduced, again with minimal lags in the procurement of the new weaponry and technology. Ottoman Turkey deployed the Winchester, an American invention, in the Russo-Turkish War of 1877–78. France followed with the Lebel (1885), Austria with the Mannlicher (1886), Russia with the Mosin-Nagant (1891), and the British War Office with the Lee-Metford (1888) and Lee-Enfield (1895). The Maxim gun appeared in 1891; and, by this stage, so too had several smokeless powders, the development of cordite (1889) being perhaps the most significant. The use of smokeless powder was generally adopted in all these new small-bore magazine rifles or machine-guns, which thus threw on the market a huge supply of black powder from the arsenals of Europe.[39]

Unparalleled both in terms of the quantity produced and the quality achieved, the repeating firearms belonged to a new industrial age and a new generation of mass-manufactured weapons having the potential to wreak mass destruction. Rifling, the sliding bolt, and the magazine spring are mass-produced elements of high sophistication; and the new magazine rifles achieved even greater accuracy, penetration, and range of fire. But the new mitrailleuses of the late 1870s, made at the Witten manufactory in the Prussian Rhineland, possessed sheer rapidity of firepower—at the rate of 800 to 1,400 bullets per minute—with each projectile capable of penetrating three iron plates three-eighths of an inch thick and afterwards burying itself an inch deep in an iron plate placed behind the former.[40] By 1880, *The Ironmonger* was reporting the arrival of yet another lethal light weapon:

> All the most terrible and deadly weapons of war hitherto introduced appear to have been thrown into the shade by an improved Gatling gun ... capable of firing 1,000 shots per minute, and killing a man or a horse at a mile range. The gun has a compact appearance, can be taken to pieces and easily carried about, can be applied to military or naval use, and the mechanism of that is simplicity itself. ... By the use of this implement three men can do the work of 300 riflemen.[41]

The Maxim gun of the 1890s then proved to be an 'epoch-making' weapon, as Basil Collier has noted, spawning a whole generation of automatic weapons which rendered prepared positions 'well-nigh proof against orthodox infantry assaults unless they were first reduced by exceptionally severe or prolonged artillery bombardments'. Such trends were reinforced by 'improvements in the design of small arms and standards of musketry, and by a growing awareness of the value of entrenchments and barbed-wire entanglements in the light of experience in the South African, Russo-Japanese and Balkan wars'.[42] Finally, cordite was a breakthrough: in addition to providing propulsion of high power for the new weaponry, it was smokeless and stable.[43]

The 'Breech-loader Revolution' of the 1850s–80s would have profound repercussions on the arms transfer system and warfare. In terms of military innovation and production, it paved the way for a progressively modern arms industry across the production centres of the Western hemisphere; the cluster of technologically intensive armament innovations between 1890 and 1914 would stimulate further industrial 'spin-off'.[44] In terms of military diffusion, it led to the forging of new patterns of international trade and politics; the world would experience the consequences of the adoption of the new weaponry by European armies and the 'dumping' of outmoded firearms in Africa and Asia, in particular the critical role played by all these weapons in establishing colonial authority and empowering indigenous resistance.

In that respect, the burgeoning arms trade emerged not only from the matrix of an industrial Prometheus unbound, but also from the wings of an imperial Juggernaut. The long nineteenth century witnessed Western imperial expansion on a global scale; and inasmuch as arms transfers were integral to that experience of global empire, it was empire that brought about the globalization of arms industry and technology. As Daniel Headrick has observed,

> in the European colonial empires we find many of the elements common to all cases of technology transfer: the distinction between geographic relocation and cultural diffusion, the role of importers and exporters of technology, the cultural and economic matrix of the importing society, and the politicizing of technology.[45]

In the early phases of Western expansion, European soldiers entered the services of Asian kingdoms and helped spread the knowledge of firearms. This was partly responsible for aggravating the intense political rivalry between states. But the transfer of weapons and military technology between the European seaborne empires and the states of Africa and Asia was relatively sparing. Small arms were supplied primarily for the protection of European people and property. As a military-strategic commodity, they were only on a secondary basis to be sold outright, exchanged for other commodities, or presented as gifts to indigenous rulers. Under the constraints of the prevailing mercantilist ethic, the Portuguese in the Zambezi hinterland and the East Indies, the Dutch at the Cape and the Indonesian Archipelago, and the French and the English in coastal India, were major importers of muskets yet reluctant arms dealers.[46]

What altered this set of cross-cultural, cross-border relationships was the ebb and flow of great power rivalry, which gathered pace during the second half of the eighteenth century and fed into the mainstream of late nineteenth-century 'new imperialism'. Imperial activity tended to destabilize frontiers: great power competition distorted judgements and encouraged pre-emptive strikes; the men on the spot, fired by personal ambition, fomented their own convenient crises. By augmenting and accelerating arms transfers from metropolitan centres in the Western hemisphere, the new imperial dynamic would eventually arm the periphery with the most modern, rather than the most obsolete, of weapons. The imperial encounter would turn the small arms trade into an ever-ramifying international experience.

The metropolitan arms trade evolved into a multi-layered, many-segmented 'imperial game' about profit and power-broking. Its products were sought as standard, if illicit, stock-in-trade inasmuch as they were means of establishing political and social hegemony. Its players were numerous and international. Profits accrued to politicians and private traders, potentates, petty chiefs, and pirates. Power—and firepower of an unprecedented sort—was redistributed both at home and abroad, to the extent of exporting and importing armed conflict around the world. Many forms of conflict occur arguably even without arms transfers. Moreover, the causes of conflict in the imperial and colonial context are to be located in labyrinthine political and socio-economic processes. Yet it is clear that transfers of weaponry and technology played a mutually reinforcing role in the causation and conduct of the imperial encounter:

> A complex process like imperialism results from both appropriate motives and adequate means ... the technological changes that made imperialism happen, both as they enabled motives to produce events, and as they enhanced the motives themselves.[47]

Over the course of the nineteenth century, firearms demonstrated a widening comparative technological advantage wielded by the Western powers (as top- and upper-level producers) over non-European peoples (as lower- and bottom-level producers), even as they triggered off cycles of indigenous war and resistance. Arms transfers and technology 'enabled' and 'enhanced' cross-cultural conflict by sustaining the part-myth, part-reality of Western superiority, while also supporting and shaping the internal economic and political changes of non-European societies. Michael Adas has argued that the ideology of civilizing mission gave the Europeans— and pre-eminently the British—'the sense of righteousness, self-assurance, and higher purpose which as much as the firepower of their Maxim guns made it possible for them to dominate most of humankind through much of the nineteenth century'.[48] Equally, among the 'tools' of empire, and the 'mechanisms' of scientific, technological, and ideological development, it might be argued that those instruments of war and the military-industrial complex would be a crucial catalyst—if not sufficient cause—of Western expansion into the wider world.[49]

It was out of such collision and accommodation between an expanding Europe and widely differing societies in the non-European world that much of the structure of the modern world was generated. In particular, witness the transformation of the long-distance mercantile voyages, caravan routes, and circuits of commerce that constituted the pre-Columbian world economy from the Mediterranean-Atlantic to the Indian Ocean.[50] The instruments and impulses of both industrialization and imperialism imposed uniform patterns of rule, 'law', and 'progress' on societies distant from one another, while integrating vast areas of the globe previously bound into interregional unities into a growing industrial world economy.[51] In the manifestation of both colonial authority and indigenous ambitions in Asia and Africa, and in the movement of world trade, we find the makings of a small arms trade authentically global in character.

Much has been written about transfers of European trade-guns to the periphery, though either in connection with empire-building in North America and the Atlantic slave trade of the seventeenth and eighteenth centuries, or the 'Scramble for Africa' in the late nineteenth century.[52] From the historiography, we learn that Birmingham 'trade-guns' were already making an impact from the early seventeenth century, spreading inland from 'factories' along the West African coast and on the Senegal.[53] Chartered in 1672, the Royal African Company was to become a bonanza for the gun-slave trade cycle, particularly after the Treaty of Utrecht concluding the War of the Spanish Succession in 1713 awarded Britain a vast slave trade monopoly. The Spanish *Asiento* was essentially a 30-year contract to supply West African slaves to

Spain's West Indian colonies, providing a minimum annual quota of 144,000 slaves, with a potential for further smuggling activities in traditionally closed Spanish markets in America.[54] The records of the Royal African Company underscore the magnitude of English participation. Between 1673 and 1704, when the Company shipped some 66,000 firearms (and over 9,000 barrels of gunpowder) to the Gold and Slave Coasts, Senegambia, Sierra Leone, the Ivory Coast, and Angola, in exchange for luxuries like gold, slaves, and ivory. By 1730, some 180,000 guns per year were being imported into the Gold and Slave Coasts.[55] Government legislation relating to the export of firearms and ammunition from England, in addition to the extant private records of gunmakers and merchants, confirm that guns and gunpowder had become major commodities in the eighteenth-century Atlantic trade. From such early commercial inroads, and at least until the mid-nineteenth century, the trade-gun (or 'slave-gun') supplied to the 'armies' of the African rulers who gathered slaves would become the most numerous and important product of the Birmingham gun trade.[56]

From this imperial perspective, the establishment of the Birmingham Proof House could be viewed as the culmination of conditions in Birmingham before 1813, when 'immense numbers of guns were made, with the knowledge and certainty, that if they were ever fired out of, they were certain to burst in the discharge'.[57] Turned out rapidly and bound for coastal markets from western to southern Africa, these scandalously dangerous, cheap trade-guns would attract a great deal of negative publicity. Proofing legislation would be introduced deliberately to establish some measure of safety and quality control. Within a decade or so, one contemporary observer could claim that 'the guns now made, and which our merchants now exchange for the gums and ivories of Africa, are as sound and secure as the musket used by the English soldier, or as the fowling piece used by the English gentleman in the sports of the field'.[58] The native customers did not seem to disagree, for Britain exported some 52,540 guns and pistols to the Gold and Slave Coasts and the Bight of Benin in 1829, while Cape Coast received 2,781 cases of muskets in 1840. In 1844, some 80,530 muskets, 596 pistols, and six shotguns reached Africa's shores.[59]

The arms trade was to become an important precondition of European colonization in Africa. Commander John Glover, acting Governor of Lagos, was already treating the gun-slave trade in Porto Novo from 1864 to 1872 as a pretext for British forward policy and annexation.[60] Massive arms transfers in the 1870s, arising from the exploitation of Southern African diamond fields, were a key incentive to Lord Kimberley's and Lord Carnarvon's confederative schemes from the metropolis.[61] Indirectly, this flowed from a decision taken by most European powers to rearm with metallic-cartridge breech-loaders from the late 1860s, and thereafter the willingness of certain states to dispose of munitions no longer used by their armies.[62] But the demands of the diamond fields for African labour in the 1870s—demands apparently satisfied by permitting the labourers to purchase guns for self-defence against hostile native tribes or white settlers like the Boers—also greatly increased the availability of firearms, especially in southern Africa.

It was just a matter of time before such arms transfers also empowered the cross-cultural and inter-tribal conflicts of the European partition of Africa. The Irish historian and politician W. E. H. Lecky was critical of these developments:

> Experience has already shown how easily these vague and ill-defined boundaries may become a new cause of European quarrels, and how often, in remote African jungles or forests, negroes armed with European guns may inflict defeats on European soldiers which will become the cause of costly and difficult wars.
>
> The seizure of the Transvaal dragged us into war with the Zulus, with Secocoeni, with Bechuanas, Korannas, Griquas—naked barbarians, but barbarians with guns with which we had ourselves provided them.[63]

The initial, relative absence of colonial authority or centralized indigenous rule simply encouraged European adventurers and concession-hunters to pursue rivalries in the 'backyards' of other peoples. The proliferation of firearms and violation of the traditional order created power vacuums that would elicit official intervention. Europeans in Africa became increasingly anxious about Africans acquiring the newer rifles to fight their local wars and soon acted to prevent this.

Various arms control arrangements ensued, all of which showed limitations. The Sand River Convention of 1852 typified the many unsuccessful local initiatives falling short of concerted international action. In southern Central Africa, missionaries and explorers like David Livingstone continued to supply Tswana chiefs with guns for self-defence against encroaching whites, while many white traders and farmers—Boers and English—openly flouted the regulations. Boer government officials, too, traded firearms with Tswana or Ndebele neighbours.[64] Subsequently, even concerted international action scarcely seemed adequate. The Brussels Conference Act of 1890 signified the first major attempt at international arms control and multi-lateral security arrangements. It stipulated a broad spectrum of

measures to diminish inland wars between Africans via European media-
tion, banning firearms imports throughout tropical Africa, as far as the
Limpopo. Despite some positive results, the gun-running and concession-
hunting persisted in the long run, with intensifying conflicts among rival
groups of Europeans and their African allies or opponents.

These distinctive emphases and episodes in the historiography still
have power to illuminate, but should not blind us to the fact that the first
transfers of Western arms moved further eastward. Through the expeditionary forces along the Zambezi and the activities of the *Estado da India*
in the Eastern Archipelago, the Portuguese had unwittingly pioneered
Europe's small arms trade in scattered territories from East Africa to East
Asia as early as the sixteenth century.[65] The Dutch East India Company
(*Vereenigde Oostindische Compagnie*, or VOC) established a major gunpowder factory in Pulicat, on India's Coromandel Coast, which supplied
VOC settlements in Batavia, Malacca, and Ceylon, and supported the
Company's vast network of trade and dominion throughout much of the
seventeenth century. For various commercial and strategic reasons, the
French traded firearms to local rulers from Mozambique to coastal India
during the second half of the eighteenth century.[66]

Western great power rivalry led to armed confrontation at the Asian
periphery as far back as the mid-eighteenth century. The Anglo-French
duel for global empire was conducted with blazing intensity in southern India during the War of the Austrian Succession (1740–48), erupt-
ing again during the Seven Years' War (1756–63), and persisting as a
security threat until the French Revolutionary and Napoleonic Wars
(1792–1815). In the early stages, neither European power possessed
sufficient manpower or military resources to achieve its objectives on
its own. Instead, just as state-sponsored arms exports and cash subsidies comprised a major component of wartime diplomacy in Europe,
massive arms shipments were supplied by the European trading companies to indigenous allies in South Asia: 'hardly a ship came', in the
1760s, 'that did not sell them cannon and small arms'.[67] The East India
Companies of both France and Britain bartered weapons for local commodities and concessions, or sold military services to Indian princes to
further their military-strategic ends.

But where the French behemoth had exported weapons and mobil-
ized troops on a grand scale, the British leviathan deployed economic
resources for military-strategic objectives on an even greater scale. By
the dawn of the long nineteenth century, the British arms trade had
become the largest in the world, inundating not only the Americas but
also many parts of Europe and Asia with sophisticated muskets.[68] Thus,

although Mysore was a client of the French and a recipient of French flintlocks, it is ironic that as many as two-thirds of the weapons used by the great warlord Haidar Ali and his son Tipu Sultan in the wars of the 1780s against the English East India Company were actually of English manufacture. At the onset of the French Revolutionary Wars, the British Army overseas used muskets made by the East India Company factories.[69] Domestic small arms production expanded rapidly, especially after 1804, when the Board of Ordnance established its own office at Birmingham for proofing, while embarking simultaneously on a new phase of production at the Tower of London. British and Indian factories would thus supply massive quantities of munitions to southern European and Asian regimes during the early nineteenth century. The export of this British arms surplus proved critical in arming the forces of the state at an international level, thereby ensuring its victory over a host of 'non-state actors': internal dissidents, peasant rebels, and armed plunderers.[70]

Official figures suggest a pattern of production and proliferation indicative of the growing value of this eastbound trade. By 1815, at the close of the Napoleonic Wars, the British Board of Customs reported an annual export of 6,704 firearms to (mainly West) African shores, assigned a real value of £5,577. But the volume and value of firearms exported to Asia (notably India, the Eastern Archipelago, and China) was recorded as being much higher: a total of 151,572 weapons, worth £103,463.[71] By 1897, a sizable quantity of firearms—amounting to 58,060 muskets, rifles, and shotguns—was still being exported to West Africa (worth £32,897), yet it was clear that Indian Ocean markets were absorbing weapons of markedly higher quality: 10,576 (worth £36,500) exported to Portuguese and French East Africa; 14,501 (worth £30,615) to Arabia (Muscat); 11,399 (worth £28,821) to Persia; 574 (worth £2,354) to Zanzibar and British East Africa; and 5,564 (worth £28,394) to the Indian subcontinent and British East Indies.[72]

At first glance, these statistics point to changes within the arms industry. The 'Breech-loader Revolution' of the 1850s–80s achieved greater product differentiation within the arms trade: the ongoing mass production of cheap trade-guns, alongside the disposal of older generations of military firearms and surplus munitions, even as the earlier breech-loaders were replaced in stages by more sophisticated repeaters. More fundamentally, however, these figures suggest that the British small arms industry was still heavily reliant on non-European markets and sales of African trade-guns into the closing decades of the long nineteenth century—a necessary survival strategy whenever the demand for military firearms slackened. According to one contemporary observer, it was 'due to the stimulus that has recently been given in the development of that part of the world in the way of colonization' that Birmingham's trade-gun sales had improved, which must undoubtedly help compensate for the attrition of all branches of the industry.[73] Birmingham may have manufactured by 1907 some 20 million guns for the African market alone, a statistic corroborated by the annual proof returns of the Birmingham Gun-barrel Proof House.[74]

Annual Proof Returns from Birmingham, 1892–1910

Year	African Barrels	Military Barrels (Breech-loaders)	Total Proofs
1892	81,820	11,779	379,086
1893	115,146	7,058	335,271
1894	75,313	9,219	299,273
1895	80,819	24,188	328,791
1896	51,725	32,025	324,898
1897	66,643	40,171	402,115
1898	119,834	18,613	392,939
1899	139,642	11,443	375,513
1900	124,155	5,706	390,268
1901	77,220	8,094	262,361
1902	83,640	4,635	282,749
1903	109,158	4,157	330,800
1904	69,661	8,177	304,969
1905	110,615	7,094	337,457
1906	62,632	16,964	370,528
1907	57,431	29,538	371,435
1908	37,244	25,037	326,697
1909	18,058	13,615	340,176
1910	65,693	8,564	420,239

Source: *Arms and Explosives*, June 1893 to June 1911.

These statistics also appear to confirm a changing pattern of distribution, which implied an increasing potential for damage and destruction beyond the metropolis. While cheap trade-guns still predominated among the firearms penetrating West Africa, the bulk of more expensive, lethal military precision-arms and sporting rifles were proliferating across Indian Ocean societies. Elsewhere, too, Customs records enumerate the export of substantial quantities of light weaponry (including machine-guns), heavy armaments, ammunition, and explosives to markets in that periphery. The numbers tallied probably underestimate the true scope and significance of arms transfers from the metropolis, discounting the lucrative illicit trade in the Indian Ocean that eluded official detection and control.

As European private entrepreneurs superseded the old, monopolistic trading companies in a freewheeling capitalist world, mercantilist commerce was cannibalized by an even more assertive industrial capitalism arising from Europe. The new dynamic would provide much of the economic drive and entrepreneurial activity behind the creation of European colonial states and trading settlements, which in turn facilitated the supply of European-made tools of war. Still, the impulses of industry and empire originating from metropolitan Europe go only halfway toward explaining the historical evolution of the small arms trade. Just as important is the need to consider the growing demand linked to widespread usage by Indian Ocean societies at various stages of crisis and transformation. To appreciate more fully how that Indian Ocean 'periphery' contributed to the emergence of an international arms bazaar, it is first necessary to delineate some of its key historical features and frontiers.

Global arms trading in the Indian Ocean arena

The geographical configuration of the Indian Ocean has always resembled a gigantic water basin. It is framed by the continental landmasses of Africa (to the west), Asia (to the north), Australia (to the east), and Antarctica (to the south). It is also connected to the world's other oceanic divisions by several strategic waterways: on the western side, the Cape of Good Hope and the Red Sea; and, on the eastern side, the funnel-like Straits of Malacca, leading to the Indonesian and Philippine Archipelagos, and then the South China Sea beyond. Centrally positioned, the Indian subcontinent has broadly dominated the ocean's northern waters, further dividing them into the Arabian Sea and the Bay of Bengal. The final geographical shape and size of the Indian Ocean have largely depended on the degree to which one might incorporate a Southern Ocean, stretching towards the Antarctic.[75]

While precise geographical boundaries remain a matter of debate, there is barely any dispute over historical evidence pointing to the Indian Ocean as the world's first 'cosmopolitan' maritime arena. The earliest networks of seaborne commerce and cross-border interaction were directly enabled by the compact and closed character of the Indian Ocean, with its narrow entrances and exits, vast and varied hinterlands, and climatic conditions conducive to both coastal and high-seas navigation. Unlike the Mediterranean (always controlled by people of its coasts), the Atlantic ('created' by people from one part of its coasts), or the Pacific (as a concept constructed only fairly recently by people from very distant lands), the Indian Ocean since ancient times was an important geo-strategic arena of interregional unities held together informally by trade winds and diplomatic relations. Collectively, Asians, Africans, and Europeans participated in and were, until the late eighteenth century, controlled by a sophisticated system of commerce and politics underpinned by the annual monsoon cycle.[76] The combined 'infrastructure' and internal dynamics of the Indian Ocean arena would be exploited when it came to transferring military hardware and technology across porous border areas.

It was the Europeans—the Portuguese, and after them the Dutch, the English, and the French—who gave the ocean its name, after the land they were all striving to explore for its fabulous riches.[77] But for much of the early modern period, the dominant players in that transoceanic milieu were really the great imperial states of monsoon Asia, whose long-term expansionary power as 'gunpowder empires' often rested as much on military technology and resources as political organization and wealth. As K. N. Chaudhuri put it,

> it is impossible to dismiss or overlook the close relationship between the means of warfare and the emergence of centralized imperial states in Asia ... Technological innovations juxtaposed with the basic social urges created almost by chance, as it were, the ground-work of larger historical movements: examples of the random interacting with the long term.[78]

In the wake of China's 'Gunpowder Revolution' from the thirteenth century, other Asian metallurgists quickly mastered the technique of casting large cannon from bronze or iron once the use of artillery had been established in siege warfare. The Ottomans perfected the large siege cannon at the start of the early modern period, and the Mughals were not far behind. In small arms production, Ottoman Turkey was by the sixteenth century manufacturing muskets considered the equal of the Spanish musket, then the finest in Europe; its leadership of the arms industry in Asia would be acknowledged until the seventeenth century. From the 1520s, the Ottoman Empire established a state monopoly of firearms manufacture and imports, superintended and stored in special depots and state arsenals. Internationally, the Ottomans played a pivotal role in the transfer of firearms and firearms technology to various Asian and North African states. Turkish guns and gunners were supplied directly to allies in Turkestan and the Crimea, as well as Gujarat, Aceh, and Ethiopia; indirectly, the Turks drove their rivals—the Ak Koyunlu and the Safavids in Persia, and the Mamluks in Egypt—to procure them from the Europeans.[79] On the Indian subcontinent, skilled artisans commissioned by the great Mughal Emperor Akbar (r.1556–1605) were fashioning matchlocks of superb quality, with gun barrels forged out of tempered Damascus steel. Mughal heavy armaments and small arms, produced and kept in state foundries and arsenals from the seventeenth century, served as instruments through which opponents could be overawed and brought within the orbit of Mughal 'military-fiscalism'.

As the new technology was transferred from China and the great Islamic empires, and from Europe on a more limited basis at this time, competent upper- and lower-level producers of firearms were soon emerging elsewhere in the Indian Ocean arena. Among the feuding states of mainland Southeast Asia, notably Burma and Siam, the persistence of mutual threat acted as a catalyst for both indigenous weapons production and the incorporation of the new weaponry into military organization and tactics. Burma had its own foundries and gunpowder mills, but preferred to replace its obsolete guns and obtain European gunpowder through periodic arms purchases from English and French traders at the coast.[80] The Thais, for their part, became so renowned for the production of artillery pieces, handguns, and gunpowder that the Japanese shogun Tokugawa Ieyasu procured muskets and gunpowder from them in the years 1606–8. A pair of seventeenth-century cannon, which the Thai king Narai presented to Louis XIV of France, remained sufficiently potent to be used for the storming of the Bastille in 1789. When the Thai capital Ayuthia fell to Burmese invaders in 1767, its armoury was found to contain over 10,000 serviceable muskets. Vietnam, preoccupied with many internal wars, initially acquired firearms from European gunsmiths and shipwrecks but later came to manufacture them locally for use on a relatively large scale. The Vietnamese were considered among the fastest of any nation in the loading and firing of their muskets, requiring only four motions to load and fire compared with the 20 motions required by an Englishman of the same period.[81]

Among the maritime states of the Eastern Archipelago, Aceh developed a reputation for the production and deployment of large cannon, swivelguns, and arquebuses through its association with Turkey. The Javanese kingdom of Mataram had developed the capacity by the mid-seventeenth century to manufacture some 800 muskets every three months. The Bugis and Makassar people of Sulawesi, the Minangkabau from the interior of Sumatra, and the Malays from around the Archipelago, also adapted quickly to the introduction of firearms owing to the frequency of wars between them. Bugis and Minangkabau artisans were making muskets of such straight bore and fine inlay work by the eighteenth century that they drew praise from European observers, while Malay gunfounders were compared favourably with German gunmakers, then considered the acknowledged leaders in the field.[82]

Still, arms industries in Asia evolved more gradually than those of their Western counterparts, given the generally stylized forms of warfare and reliance on traditional weaponry in established war strategies. In China, a Confucian prejudice against things military meant that the mandarins considered it degrading to devote their time and energy to the development of firearms. In the Muslim land empires, light cavalry armed with recurved bows remained highly effective in their nomadic environment. Before the eighteenth century, firearms occupied a curiously low position in the weapons hierarchy of Mughal military culture. The Mughals relegated the use of small arms to the infantry, while the Mughal artillery branch held little prestige or hope for advancement, being largely staffed by Portuguese and other foreigners under contract. For Southeast Asians, even outdated guns were sufficient for their war requirements, and there was little incentive to enter into a technological arms race with Europeans who had progressed from the matchlock to the more reliable wheel-lock and flintlock mechanisms. These Asian cultural perceptions of firearms would remain essentially unchanged until the political and economic upheavals of the eighteenth century.[83]

It was then that a sequence of interlocking crises triggered new patterns of military production and diffusion across monsoon Asia. The resplendent 'gunpowder empires' eventually became victims of their own success, falling prey to imperial overstretch. Old imperial centres, hollowed-out to a mere shell of their former glory through the ascendancy of provincial elites, soon faced even greater challenges: tribal breakouts in the resurgence of great warrior coalitions from Arabia, Central Asia, and Southeast Asia; the building-up of new successor states across Indian Ocean regions; and the breaking-in of sustained European capital flows and commerce.[84] These changes opened up frontiers of opportunity for concession-hunting and gun-running: small arms were bartered for other commodities or concessions; presented as gifts to cultivate relations with indigenous rulers; and supplied to local allies for political and military-strategic objectives.[85] Various authorities would introduce restrictions in various places, at various times throughout this period. Yet the weapons manufactured by state factories and private firms continued to multiply, delivered by a combination of quasi-state agencies and free agents who operated across increasingly clandestine commercial networks, linking metropolitan Europe to markets in the Indian Ocean. It was a new, mutually reinforcing pattern: general political crisis, uneven economic development, and sporadic religious revivalism across these Indian Ocean societies would engender further opportunities and demand for munitions. Consequently, the new levels of profit and violence flowing from such arms transfers would play their part in reducing the Indian Ocean complex (and progressively more of its hinterland) to a component of a much larger industrial world economy.[86]

Far from being undifferentiated and monolithic, these patterns of military production and diffusion were at every point determined by the particular features of indigenous societies, including their level of advancement and sophistication. There were variable geopolitical factors and underlying factor endowments, along with variegated responses to the new weapons and their supporting technologies from Europe. Hence, the successor states and provincial elites of Mughal India would yield examples of successful arms procurement and production, where in some cases the products were qualitatively comparable to imported Western firearms until the mid-nineteenth century. But an arguably superior manufacturing tradition was gradually reduced to a more dependent one, as several of these successor states with 'hybrid arsenals' were initially able to present serious resistance before eventually succumbing to the 'hybrid armies' of a British Raj. In contrast, societies in East Africa, the Middle East, and the Eastern Archipelago showed greater reliance on arms imports in both qualitative and quantitative terms, thus paralleling one another in their economic dependence on exchanges of local produce, luxury products, human cargoes, and weapons consignments.

The waning of Mughal hegemony by the mid-eighteenth century opened up autonomous spaces that were readily exploited by the lower ranks of India's 'hierarchy of kings' and other forms of indigenous capitalism. The revenue and military entrepreneurs, the big bankers, and the warrior peasant lords of the villages all derived wealth from commodity trade and speculated in money profit; most needed cash to buy muskets, cannon, elephants, and other symbols of power and status. The advanced nature of indigenous military technology and production did not preclude the hiring of European military advisers or the purchase of Western military equipment. Internal arms races on the subcontinent drove native rulers to seek whatever perceived organizational or technological advantages they might derive from external sources of power in order to acquire an edge, however marginal, over local rivals. The English East India Company was drawn into—and benefited greatly from—this turbulent scenario of war and opportunity: playing off one native state against another, selling its own services and supplies cross-culturally in the 'all-India military bazaar'.[87]

From the 1740s onward, an upsurge in military activity on the subcontinent by armies of the Indian rulers and the European East India Companies meant an ever-expanding market for munitions that the country captains eagerly supplied—to both enemies and allies of the English East India Company. For instance, against the volatility of indigenous state formation and resistance in western India, the security of the spice trade and further expansion of British commercial interests influenced the Bombay presidency's decision to provide the Mysorean warlord Haidar Ali with military aid and equipment in 1765, at a time when Mysore was in danger of being overrun by the Marathas. Then, when Mysorean power was in the ascendancy, the Bombay presidency contemplated bribing Haidar's son Tipu Sultan into acquiescence with military supplies, although Mysore would become a client of the French and a recipient of French arms.[88]

From the 1780s, arms procurement and production on the subcontinent would rise to fever pitch. The end of the American Revolutionary War (1782–83) and the passing of Pitt's Commutation Act (1784) consolidated British control of the wider country trade and encouraged the activities of powerful, capital-rich agency houses and agents adept at both the commodity trade—in spices, tea, opium, indigo, calicoes, cotton piece goods, raw silk, ceramics, and saltpetre—as well as the arms trade.[89] Arms and ammunition had already figured prominently among the stock-in-trade bartered to petty rajas for spices, but the stakes were raised when great magnates like Mahadji Scindia and the Nizam of Hyderabad were drawn into this volatile military market. In 1784, the submission of the Mughal emperor to the 'protection' of Mahadji Scindia, the greatest of Maratha warlords, following a brief rearguard action in defence of the heartland of Delhi, signified a watershed in the devolution of political power to the Mughal successor states.

An internal arms race ensued, coupled with the drive for modernization. Scindia proceeded to establish his own 'military-industrial complex' near Agra. The Maratha ordnance factories emphasized adaptation rather than innovation, but incorporated relatively sophisticated indigenous technology and involved local manufacturers. These developments so alarmed the English East India Company that it forbade Britons to serve as gunners with the Marathas and sought to curtail the trade in muskets. Scindia, assisted by French and Portuguese military advisers, went on to create one of the finest armies in India—including the 27,000-strong brigade known as the 'Deccan Invincibles'—supplied from the arsenal at Agra. By combining these new weapons with new battlefield tactics, the Marathas were able to mount a formidable challenge to the British between 1775 and 1818, over the course of three Anglo-Maratha Wars.[90]

A broadly similar pattern of crisis and proliferation emerges among the other societies of the western Indian Ocean region. While the struggle for control of the South Asian military economy stimulated arms production to an almost unique level of sophistication among the Mughal successor states, the weakening of Mughal, Ottoman, and Safavid authority had also subjected these empires to attacks by powerful but unstable warrior coalitions of Persian, Afghan, or Central Asian origin. As early as 1720, tribal revolts by the Ghilzai Afghans of Kandahar and the Abdalis of Herat had eradicated Safavid influence from much of southeastern Afghanistan. The sacking of key cities in Persia and India further disrupted Eurasia's major caravan trade routes. The Afghans sacked the Safavid capital Isfahan in 1722. The armies of the Persian, Nadir Shah, plundered Mughal Delhi in 1739. A new Afghan kingdom emerged to consolidate its hold over western Persia, Afghanistan, Sindh, and the Punjab, while Northern India, invaded four times by the Afghans in 1747 and 1759–61, was menaced again in 1797. India's Northwest Frontier remained in perpetual flux as it was populated steadily by warlike Pathan and Baluch tribes.[91] In the ensuing struggle among successor states and tribal confederacies, gun-running would become a profitable adjunct to the traditional caravan trade. Whereas Afghan warriors could rely on their traditional long rifled-musket (*jezail*) well into the nineteenth century, most would amass for use a combination of locally-made and imported firearms.

The crisis of the land-locked interior was tied to the fortunes of the Indian Ocean littoral. With Portuguese power in decline by the seventeenth century, and Safavid Persia engulfed by crisis in the eighteenth century, smaller states along the Arabian coast started competing for the profits of the mercantilist system and primacy of the Gulf's waters. The tribal confederacy and continental theocracy of Oman emerged to extend political control over the coast and assume a larger role in maritime trade. This participation in maritime trade was to fuel mounting social tension and conflict that transformed the very basis of Oman's tribal society. A fiercely contested civil war, compounded by incursions from neighbouring Persia, led to the collapse of Yarubi rule and its replacement by the more secular Busaidi dynasty in 1749. The new ruling house would draw strength from buoyant seaborne trade rather than territorial or spiritual overlordship, thus enabling Oman's transformation from a raiding naval power into an expansionist commercial empire. Its twin centres at Muscat (retaken from the Portuguese in 1650) and Zanzibar (subjugated by 1744) would prosper from the arms traded through Omani bazaars.[92]

Through the agency of Omani expansionism and Indian merchant capitalism, the gun, slave, and ivory trade of the Swahili coast and East Central Africa would be integrated progressively from the 1770s into the Indian Ocean trading complex and the world capitalist system. Under the aegis of the French, the East African slave trade became an economic bonanza. Apart from Mozambique, the French had sought to develop Kilwa as a slaving centre. By 1785, the prosperity of Kilwa had convinced Oman's Busaidi dynasty to tighten their grip on the East African trade, so that a new and more determined Omani governor was despatched to manage local affairs. Coastal commerce in small arms, slaves, and ivory was to be concentrated at Zanzibar, while Indian merchant capitalists (especially *banians*, a specific class of Hindu merchant) were employed to farm the customs there. Small arms provided the means of securing both the ivory and the slaves, which were then channelled into supporting the wider regional economy. Capitalizing on the disruptions caused to the mercantile marine by shockwaves from the Napoleonic Wars, the Omani further expanded their shipping by taking over much of the carrying trade in the western Indian Ocean previously dependent on British and French vessels.[93] In embryonic form, too, were the makings of the 'Arms Emporium of the Middle East': galvanized by late nineteenth-century Islamic revivalism, Muscat's future role as a gateway for arms traffic to tribal militants in Afghanistan and India's Northwest Frontier was all but assured, becoming a persistent source of insecurity to the Government of British India.[94]

A similar set of developments was reshaping the trade and politics of societies further eastward. In mainland Southeast Asia between the sixteenth and nineteenth centuries, arms transfers contributed to the political centralization and economic development of states such as Burma, Thailand, Laos, Cambodia, and Vietnam.[95] Even more striking, though, was the transformation of maritime Southeast Asia. After the Malacca Sultanate collapsed in 1511, the Eastern Archipelago witnessed a dramatic pattern of dislocation and relocation of political authority and maritime trade. The Riau-Johor kingdom and other successor states of the disintegrating Malay imperium suffered their own interlocking crises: from 1699, the trauma of regicide and the tensions of a Malay-Bugis power struggle; and then, with the fall of Riau in 1784, a final collapse of centralized authority followed by the formation of new Malay states amid the turbulence of Islamic fundamentalist protest and anti-colonial resistance.[96]

The fall of Riau, in particular, opened a new chapter of disorganization and warfare in the Malay world, revolving around the Riau-Johor Empire 'below the wind'. But there was also a new configuration of entrepôt trade, accompanied by raiding and slavery, in the 'Sulu zone' connecting the Indonesian and Philippine Archipelagos. European commercial expansion into China further stimulated the growth of a slave economy in Southeast Asia. Slave labour was needed to procure the exotic local products much sought after in Chinese markets, which could then be exchanged for the tea that the Europeans wanted. Much of the slave traffic crystallized around organized markets and depots in the Sulu region, with Jolo Island at the core of this redistributive network becoming the most important slaving centre by 1800. Collectively, these changes fuelled a massive demand for the firearms used in the capture of slaves. Consequently, arms transfers would be aided and abetted by the opportunism of local mercenaries, marauders, and merchants: warrior-traders, including the Bugis and the Tausug; 'sea gypsies' (*orang laut*) and the Iranun and Balangingi 'pirate wind'; and entrepreneurial groups like the Straits Chinese.[97]

Whatever the regional variations across the interior or along the coast, on land or at sea, the political economy of the Indian Ocean coped with these 'globalizing' pressures through reorientation and accommodation. Pre-existing trade networks would readjust to meet the changing patterns of production and consumption across indigenous societies, in addition to the commercial ambitions of various interlopers. The hollowing-out effect of the regional crises and the coalescence of new commercial and political interests would engender, by the early nineteenth century, a different circuit of port-cities. Where Kilwa, Mocha, Aden, Basra, Hormuz, Diu, Cambay, Surat, Goa, Calicut, Colombo, Madras, Masulipatnam, Malacca, and Aceh had once functioned as the great Indian Ocean emporia, new entrepôts arose to join or supplant them, notably those that would become established gateways for small arms traffic: Mombasa, Zanzibar, Muscat, Singapore, and Jolo, among others.[98] Although the Indian Ocean was being turned into a primarily 'British Lake', Sugata Bose has observed how the 'huge asymmetry in economic power relations on a world scale led Indian and Chinese capitalists to build their own lake in the stretch of ocean from Zanzibar to Singapore'. These Indian and Chinese networks of trade and finance combined to form a new and distinctive 'bazaar nexus', linking European capital to diverse local communities.[99] Notwithstanding the advent of steamships, traditional sailing vessels such as dhows and prahus would continue to ply the routes between port-cities, complementing the caravan trade that operated further inland. Together they facilitated, by time-honoured yet innovative methods, the movement of arms consignments from the fluid frontiers of coastal areas through the porous borders of the hinterland.

A nineteenth-century triangular trade in guns, slaves, and ivory would redefine the political economy of the western Indian Ocean. The collapse of imperial rule in Southwest Asia, the ambitions of local chiefs, and Britain's growing dominance in the emerging industrial world economy opened a period of profound upheaval for the Persian Gulf region. The Gulf trade witnessed a mercantile renaissance, in which pearls, dates, cotton, wool, and opium were exchanged for imported foodstuffs, dyes, manufactured goods, and firearms. But labour shortages also occurred, generating fresh demand for imported manpower in the Gulf ports, coastal villages, and local militia. Small arms supplied the means of securing the ivory so highly prized in Western markets, in addition to the slave labour that supported the regional economy. The East African slave trade, reinforced by the trade in small arms and ivory, provided the temporary workforce in the western Indian Ocean hinterland until the First World War.[100]

The political economy of the eastern Indian Ocean underwent a similar kind of transformation, with one significant variation: it was primarily a trade in guns, slaves, and drugs. Again, the disintegration of regional imperial rule in South Asia and Southeast Asia, the emergence of successor states, and British ascendancy in the world economy paved the way for a burgeoning trade in Chinese tea, Indian opium and textiles, and assorted marine, forest, and plantation products that fluctuated in importance until the industrial take-off of tin and rubber. The consumption of tea was addictive, but even more so the use and abuse of opium. Drug money from Indian opium would feature critically in the commercial equation, helping to finance expanding gun-slave trade networks across the entire Indian Ocean. Gun-running and slave-raiding persisted in the Eastern Archipelago as part of a wider route of advancement; the weapons secured the slaves, and the slaves supplied the manpower for the wider acquisition of wealth, status, and power. Reliance on slave labour would decrease only gradually as foreign indentured labour and entrepreneurial talent—primarily Chinese coolies and businessmen—flooded the region to work the plantations, mines, and ports, thus generating alternative sources of revenue.[101]

The gun-slave trade cycles running through both halves of the Indian Ocean does suggest certain parallels with the infamous triangular trade of the Atlantic. Between the sixteenth and nineteenth centuries, the human cargoes and arms consignments of that well-documented Atlantic trade were part and parcel of a vast commercial web linking Britain, West Africa, the West Indies, and North America. The main commodity bartered for slaves was firearms, an exchange neatly encapsulated by one African ruler: 'You have three things we want: powder, musket, and shot; and we have three things you want: men, women, and children.'[102] Whereas distilled spirits, textiles, and other metal goods were also traded, only imported firearms so directly enabled slave-raiding and other traditional modes of power-broking. What has been covered far less extensively in the historiography is the increasingly clandestine Indian Ocean triangular trade of the long nineteenth century. How were exchanges of various commodities, later classified as 'contraband'—alcohol, drugs, gemstones, ivory, human cargoes, and arms consignments—more precisely bound up with Western expansion, on the one hand, and the interlocking crises of Africa and Asia, on the other?

Before the nineteenth century, the main market for the ivory of East Africa had been the Indian subcontinent. Alcohol, cloth, beads, and iron had been the main stock-in-trade for this ivory; small arms were transferred more sparingly. Apart from military-strategic objectives and monopolistic tendencies within the mercantilist system, economic rationale had stressed the damage to the ivory trade that could be caused by hostile Africans armed with muskets. During the nineteenth century, however, the European market not only consumed half of East Africa's annual ivory exports directly from Zanzibar, but also received about half of the annual export to India, which was re-exported to Europe. To balance the trade, arms were increasingly sought and supplied from Europe. In Mozambique, slaves and ivory were still traded for beads, iron, and alcohol, but increasingly exchanged for cloth, gunpowder, and muskets. Skill in hunting and ivory extraction was a well-trodden path to wider political authority, and the business of hunting elephants in the interior of Mozambique was dominated by firearms technology.[103]

From the 1770s, slaves and ivory originating from East Central Africa began to pour into the region via Zanzibar and Muscat. Where pearling, fishing, and the carrying trade had persisted as the mainstays of economic life on the Pirate (later Trucial) Coast, the profits were reinvested in matchlocks and gunpowder from Persia, as well as slaves from Zanzibar. The trade of Bushire gave wider prominence to the local arms trade. Musket-barrels made at Shiraz were sold throughout the Gulf. Above all, Zanzibar, Oman, and Muscat—politically and administratively unified until 1861—were being transformed into the western Indian Ocean's main conduits for small arms transfers. At least until the 1870s, the rulers of Zanzibar and Oman presided over a thriving commercial network reaching far into East Africa; its caravans of slaves and ivory, exchanged for cargoes of muskets and rifles, would not only sustain the great western Indian Ocean trade nexus centred on Zanzibar, but also supply the growing demands of an Afro-Arab world in turmoil.[104]

Over the span of a century, between 1777 and 1876, an estimated 1.2 million slaves were exported from Indian Ocean and Red Sea ports in East Africa; millions more were moved around the coast or interior of Africa. Slightly over 300,000 slaves were shipped to Middle Eastern and Indian destinations from the Swahili coast, and a little under 500,000 across the Red Sea and Gulf of Aden. The Persian Gulf, Yemen, and Hijaz were the main destinations for these human cargoes, with the possibility of further re-export to Egypt, Syria, Anatolia, and Northwest India. This external slave trade continued to flourish for as long as it remained lucrative.[105] Indeed, the institution of slavery was strengthened as slave labour was redirected within the interior to produce new, 'legitimate' exports. A profitable pearl-fishing industry and a prosperous trading community would finance the acquisition of British-made firearms and generate further demand for slaves across the Arabian Peninsula and among the tribes of lower Iraq. Arab merchants in Zanzibar and Oman were giving way to the continuous interpenetration of Indian capital, particularly from Gujarat, but Arab traders from the ports of Trucial Oman would actively reinvest the proceeds of piracy, pearling, and the date trade into the gun-slave traffic.[106]

A similar trading configuration would emerge in the Eastern Archipelago. In maritime Southeast Asia, arms transfers were driven by a 'pirate wind' issuing from the general crisis of the Malay world. Marauding was by the late eighteenth century a common sideline of Malay chiefs, who equipped their fleets with brass guns, muskets, gunpowder, and shot. They sought to enslave weaker parties and to plunder—arms and ammunition, as well as articles of luxury—with payment for human cargoes frequently made in terms of munitions. In pre-colonial Singapore, the Temenggong of Johor would supply marauding bands with arms and ammunition, along with cash advances, in return for a share of the plunder.[107] If anything, the establishment of Singapore as a British free port would further facilitate the transhipment of munitions: British muskets acquired by Bugis traders, for instance, were resold elsewhere in the Archipelago.

Borne by the easterly 'pirate wind' of the southwest monsoon, the slave-raiding marauders from the Mindanao and Sulu Archipelagos—

the Iranun and the Balangingi—would become by the 1830s the most dreaded in the region.[108] The disruption of the traditional economy by European trade monopolies had driven these seafaring communities to slave-raiding on a large scale, often in conjunction with the raiding and trading activities of regional states. Such 'piracy' took the form of an annual cycle whose very regularity illustrates the extent to which these maritime pursuits were bound up with the socio-economic structures of the region. Marauding functioned as an integral, legitimate, and time-honoured means of enlarging and consolidating the power bases of rival chiefs during periods of turbulence.[109] In the face of nineteenth-century European expansion, the full range of marauding activities would merge with traditional entrepôt trade, slave-raiding, and arms smuggling on a rising tide of profit and violence. The institution of slavery was also reinforced: between 1770 and 1870, some 200,000–300,000 slaves were conveyed in Iranun and Balangingi vessels to the Sulu Sultanate alone, where they were needed for the extraction of exotic local produce highly valued in southern Chinese markets.[110] Acknowledging the sheer scale of gun-running and slave-raiding along that frontier of the eastern Indian Ocean, the Jolo Protocol of 1897 would eventually attempt to do for the waters around the Philippine Archipelago—between the South China Sea and the Pacific Ocean—what the Brussels Conference Act did for tropical Africa: to introduce and to enforce, on a multilateral platform involving European as well as native powers, some measure of prohibition.

Global arms trading in the *Longue Durée*

In the unfolding narrative of the Indian Ocean arms trade, we uncover historical issues that resonate with contemporary interest. In some cases, there is enhanced appreciation of how current problems are connected to—if not caused more directly by—the 'waves', 'layers', and 'cycles' of long-term processes. The possibility that European soldiers or civilians might be slain by the very weapons sold by European companies still raises questions about the problematic nature of arms procurement and proliferation across porous borders. This example of 'bad' globalization is intensified by the asymmetries and ambiguities of power in the contemporary world, where indigenous communities are still alienated and polarized in opposition to Western neo-colonialism. The provision of Western arms to potential adversaries—either directly or by proxy—still risks reducing military advantage and undermining security, implying with the transfer of new technologies that the cost of going to war with a client will be that much higher.

Various forms of 'low-intensity' conflict and 'limited' warfare, backed by firepower, have long served as a means to secure a new symbiosis between increasingly interconnected, interdependent Eastern and Western hemispheres. Small wonder that the volatility and violence spiralling from globalized arms trafficking should still bedevil the Indian Ocean arena in the early twenty-first century. In the hands of gun-runners, insurgents, pirates, and terrorists, weapons of Western manufacture or design continue to make their presence felt with astonishing regularity across the African interior and Somali coast, the Arabian Peninsula and Gulf region, Afghanistan, Pakistan, and Kashmir; or further eastwards around the Malacca-Singapore Straits, the Indonesian and Philippine Archipelagos, and the South China Sea.[111]

In the radically altered post-Cold War scenario, weapons obtained from Europe, the United States, or the former Soviet Empire have flowed from regional arms bazaars in Cambodia and Thailand to other parts of Southeast Asia. In further resonances between past and present events, arms smuggling routes have again extended from Penang, through the Malacca Strait to Aceh, or across the porous border area between East Malaysia and the Philippines. Singapore also regained something of its past reputation as a trading entrepôt for the redistribution of illegal weapons, with shipments apparently reaching states as distant as Iran or Iraq. Other illicit consignments have ended up arming pirates or insurgents, or indigenous populations having to protect themselves against such 'non-state actors'. In the post-9/11 climate, and the shifting frontlines of an international war against terrorism, revamped security measures have likewise increased surveillance and restrictions. Recent terrorist attacks around the world— from the United States to Bali, Istanbul, Madrid, Jakarta, London, and Bombay—have altogether sounded a call to the global community to curtail the illicit spread of conventional weapons, along with weapons of mass destruction (WMD).

In many ways, these recent trends and patterns represent the culmination of long-term cycles of production and procurement in the evolving international arms transfer system. The dynamics of military supply and demand, as characterized by one strategic analyst, will sound strangely familiar:

The world's wealthier states appear to procure newly produced weapons at regular intervals. By contrast, the world's poorer states procure on an ad hoc basis. Discounting transfers of surplus stocks to states in conflict or those recovering from conflict, these states probably procure more frequently, but they acquire fewer weapons in any one initiative. Poorer state acquisitions are very often of older surplus weapons and this is one further reason why these states have far higher stock ages than their rich counterparts.[112]

Arms transfers have always occurred in the context of military innovation, production, and diffusion, flowing from a dynamic relationship between 'metropolitan' centres of arms production and 'peripheral' regions of arms consumption. Expressed emphatically in intensifying cycles of profit and violence, and manifested globally for the first time during the nineteenth century, it is that relationship which shall be explored over the course of the following chapters.

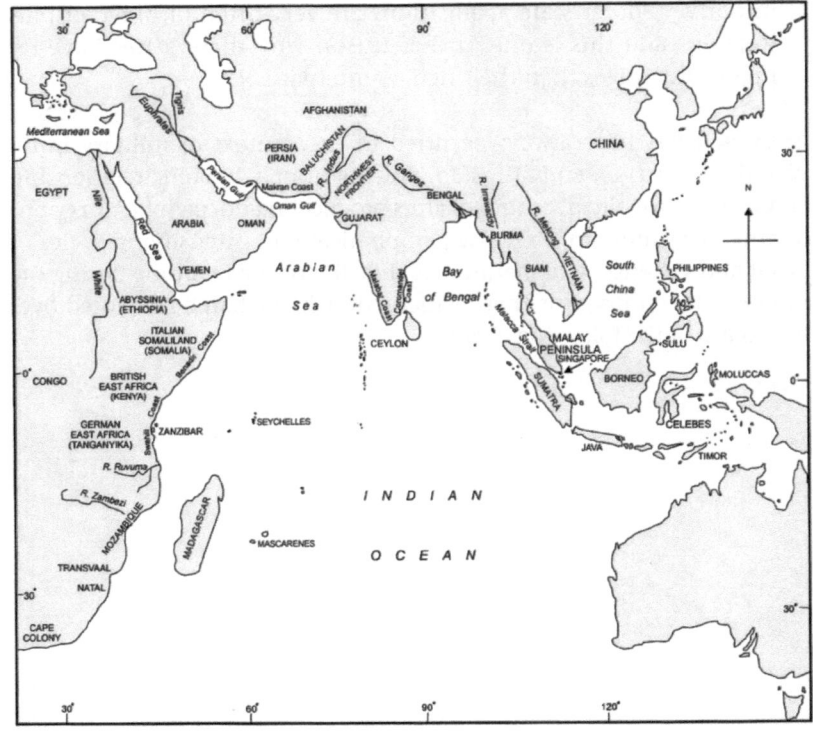

The Indian Ocean World

2
The Arms Trade in the Metropolis

Now thrive the armourers, and honour's thought
Reigns solely in the breast of every man:
They sell the pasture now to buy the horse.
William Shakespeare, *Henry V*, Act II, Prologue

New patterns of centralization, competition, and conflict in early modern Europe called for the establishment of an equally novel military-fiscal basis, remarkable in its ability to extend long-term credit for public finance of the war machine. Such were the 'sinews of power' in the Western metropole, which helped to set the pace and direction of change in both military technology and manufacturing capability. Progressive waves of technological innovation and industrial production thereby supplied the requisite armament to drive forward, if not secure, Europe's frontiers of economic, political, and military expansion.[1] How then was the supply side of the international arms trade sustained by the emerging European states system—with its growing panoply of socio-economic, political, and military institutions? How did the cycles of military innovation, production, and diffusion either reflect or reinforce the pursuit of wealth, power, and victory in warfare by those states?

From the sixteenth century, the European dynastic states found themselves developing a more advanced military economy to meet the rising tempo of inter-state rivalry and overseas expansion. While great Muslim dynasties like the Ottomans and the Mughals were constructing large 'gunpowder empires' in Asia, the very nature of European inter-state rivalry—with a growing sense of national sovereignty underpinning the patchwork of states—rather precluded the formation of similar empires in the West. Instead, a primitive type of arms race among Europe's fractious city-states and kingdoms engendered an armaments spiral that could only be sustained by enhanced military-fiscal structures in each rival nation, through mutually reinforcing cycles of profit and violence.[2]

At the most basic military-strategic level, burgeoning nation-states in early modern Europe needed a system of official patronage and funding that would effectively finance the war machine and keep weapons-manufacturing capabilities at the cutting edge. Within this collaborative framework of government sponsorship and free enterprise, bankers, artisans, and arms dealers would become essential, not peripheral, members of society. According to C. A. Bayly:

> The influence of war permeated most aspects of government and society: its frequency encouraged technological innovation; its huge expense and the need for ready cash in wartime meant the establishment of close links between governments and financiers; and its escalating costs saw a search for new revenues with or without the consent of the nation represented in parliament.[3]

Scavenging ideas and techniques from its continental competitors, England as the 'magpie of early modern Europe' would come from behind to overtake them in political, economic, and military terms. Through a series of institutional transformations—a 'revolution in government' under the Tudors, a 'revolution in parliamentary authority' under the later Stuarts, and the 'financial revolution' that accompanied the founding of the Bank of England in 1694—this 'sceptred isle' would develop the requisite fiscal sinews to sustain the exercise of political authority, economic strength, naval power, and military prowess.[4]

Sponsored by the state as part of this developing military economy, and driven by the imperatives of mercantilism, it was between the sixteenth and eighteenth centuries that the nascent arms trade of England grew to match, if not outclass, every continental counterpart. The arms-producing centres of London and Birmingham proved admirably responsive to pressures and uncertainties arising from the fluctuating patterns of military supply and demand. Ultimately, they succeeded in maintaining their advantage over external competitors by capitalizing on a highly variegated but concentrated proto-industrial base, which comprised workshops and cottage industries with an extraordinary capacity for hard work and industrious enterprise.

Between 1780 and 1914, however, the struggle for mastery in industrial Europe would translate into imperial rivalry and warfare at a global level. The fiscal-military state in Europe was galvanized by the exigencies of international conflict; the augmented war machine and the accelerating manufacture of weapons, in turn, enabled the Europeans to project their power and secure resources along with markets on other continents. Dramatic surges in the cycles of profit and violence would thereby stimulate new patterns of arms production and procurement, and new trends in military modernization and technological transformation. French arms were thus sought and supplied as far afield as North America, Africa, and South Asia, even as France in revolutionary vein became the trendsetter in European military affairs from 1789 until the end of the Napoleonic Wars in 1815. A resurgent Germany and Austria would become, by the 1890s, major exporters of precision-arms to the less developed periphery of Europe and to the non-European world.[5]

Still, for much of the long nineteenth century, it was again the British example that impressed consistently. Operating dynamically at the various levels of weapons production, from replication through adaptation to innovation, Britain's evolving arms trade enabled the state to play a unique global role that transcended the struggle for mastery in Europe. The challenge of Napoleonic warfare further stimulated British military-fiscalism, from areas of domestic taxation to imperial revenue gathering. In the vanguard of 'industrious revolutions' and industrialization were private entrepreneurs and inventors whose contributions significantly enhanced traditional proto-industry through the first half of the nineteenth century, creating further opportunities for military technological innovation, weapons production, and arms transfers. Official conservatism in military technology, exemplified by the attitudes of Wellington and Napoleon, did not prevent the range and accuracy of infantry fire from increasing at least fourfold, nor stem the development of armament engineering after Waterloo.[6]

And yet, when the pressures of the Crimean War compelled the British Government to assume a more direct role in spearheading the manufacture of military firearms, the stimulus for the vital technological breakthrough would only come from across the Atlantic, from a former colony of Britain. Free of the colonial yoke since 1783, the United States had inherited most of its gunmaking traditions from Britain, notably in the craft origins of small arms production. But where drastic shortages of skilled gunsmiths had become apparent during the War of 1812, America soon attempted to apply new machine methods of mass production to the manufacture of military firearms. Long-term government contracts had already encouraged private firms to collaborate with the arsenals at Springfield and Harpers Ferry in designing machines capable of turning out vast quantities of almost identical parts with sufficient accuracy to allow unskilled operators to assemble them into workable small arms.

Much of the skilled filing and fitting that had traditionally characterized the gunmaker's craft became unnecessary, and older quantitative limits on production were suddenly lifted. When the deficiencies of British military arms production became obvious during the Crimean War, the stage was set for a radical new experiment. The 'American system of manufacture' was by 1856 transplanted to the new Enfield Arsenal on the outskirts of London, resulting in revolutionary repercussions not only for Britain, but for the rest of Europe and the wider world. Other European powers would follow suit by the 1870s, heralding a new era of great power rivalry persisting right through to the cataclysm of the First World War.[7]

When the supply side of the international arms transfer system is surveyed from this broader historical perspective, the British small arms trade dominates the 'view from the metropolis'. By providing a crucial interface between pre-industrial techniques and industrial technologies, the arms manufacturing centres at London and Birmingham played a vital role in facilitating interregional flows of military innovation, production, and diffusion. Certainly, the manner in which the British chose to finance and conduct their arms trade was shaped by particular preconditions and circumstances. An island realm off the European Continent with distinctive craft and military traditions, the United Kingdom had become by the nineteenth century an established 'blue water' naval and industrial power with extensive imperial interests. As we chart the course of that singularly historic transformation, we also hear echoes of contemporary debates about military industrial policy and the supply side of the arms transfer system. How did Britain—as a leading industrial and imperial power of the Western metropole—then decide which weapons to produce or procure, and which ones to discontinue, dismantle, or dump? Under what conditions were earlier forms of joint private-public production or civil-military integration pursued in order to keep strategic military industries viable in the long run? Under what circumstances were the historical forerunners of 'dual-use' technology permitted to enable the conversion of equipment from wartime to peacetime applications?

This chapter examines the long-term development of the British small arms trade in the metropolitan 'core' of the arms transfer system: in particular, the craft origins of small arms production in England, and the shift in private arms production from London to Birmingham during Britain's age of industrial transformation; the curiously ambivalent relationship between state and private arms production in Britain; and British responses to the twin challenges of the new technology and foreign competition.

England's small arms trade and Britain's industrial transformation

Beautifully illuminated manuscripts depict the use—and possible manufacture—of small arms in England during the fourteenth century. If that were indeed the case, it would put the English (and eventually British) small arms trade ahead of continental counterparts on the scale of technological innovation. But what remains obscure is the earliest pattern of supply and demand. These fragmentary accounts scarcely constitute a comprehensive overview of the incipient gun trade's organizational and manufacturing capabilities. They simply hint at a loose combination of state initiative and private enterprise in the earliest decades of small arms production, geared sporadically towards domestic use in local or national defence.[8]

The picture clarifies somewhat during the fifteenth century, when a gun trade in its infancy started to take shape. Against the backdrop of memorable English victories in the Hundred Years' War—such as Crécy (1346) or Agincourt (1415)—it was unsurprising that domestic opinion had initially gone against the deployment of firearms in established military tactics. While English cannon had been fired in field battle at Crécy, Englishmen could not fathom the growing popularity overseas of the 'miserably ineffective handgun', unless that weapon could be promoted more consistently or proven more convincingly against the traditional longbow. Such preconditions were obtained internally over the three decades of sustained dynastic strife known as the 'Wars of the Roses' (1455–85), when four classes of craftsmen—artillers, gunners, smiths, and founders—were engaged in firearms production in London, the largest manufacturing centre in the kingdom.[9] When Henry VII finally took power in 1485, he organized the Yeomen of the Guard and equipped half of them with arquebuses. The introduction of small arms as official weapons of the Royal Guard would gradually raise their profile in English society.[10]

England's small arms trade was consolidated through the 'proto-imperialism' of Henry VIII (r.1509–47), whose great divorce from the Roman Catholic Church had led him to reinvent the English church and state as an imperium with revised military-strategic imperatives. As a matter of internal security, Henrician acts and proclamations broadly discouraged the populace from owning firearms, and English gunmakers found it correspondingly difficult to get organized as a distinct trade.[11] Yet the King grasped the importance of cultivating a domestic firearms industry, for he recruited professional gunsmiths from the Continent and had in regular service by 1545 a group of Hainaulters, skilled in the manufacture, usage, and repair of the arquebus.[12] The main impetus for small arms production came not from the need for internal pacification, but from intensifying external threats. The demands of war against France in the 1540s, and the fear of Spanish invasion during the 1560s–80s, brought together native and foreign gunmakers in an informal association revolving around the Tower of London. By the reign of Elizabeth I (r.1558–1603), there was a nucleus of some 37 accredited gunsmiths plying their trade in the Minories of the Tower of London.[13]

The rising profile of the London gun trade would lead to the incorporation of the London Company of Gunmakers by royal charter in 1637. It was intended, perhaps primarily, to protect and profit the gunmaker in the tradition of the older trade guilds. The incorporation of the new company also marked the introduction of professional proof into England. Although it is uncertain from the charter whether private proof houses or a trade proof house were to be used, the proving of guns for the public good was clearly stipulated. After the disruptions of the English Civil War (1642–51), the gunmakers' own proof house was built in 1657. A second charter was granted in 1672, enabling the gunmakers to enforce proof: to 'search for, prove and mark all manner of hand-guns great and small, dags and pistols, and every part thereof, whether made in London or the suburbs, or within ten miles thereof, or imported from foreign parts, or otherwise brought thither for sale'.[14]

By the second half of the seventeenth century, firearms were also required to serve Britain's imperial interests beyond Europe. Apart from the Board of Ordnance, the London gunmakers' major customers were the great joint-stock companies that dominated the mercantilist scene: the East India Company (chartered, 1600); the Hudson's Bay Company (chartered, 1670); and the Royal African Company (chartered, 1672). Contracts for military arms issued by the Board of Ordnance were the mainstay of the trade. But the English East India Company had started ordering its own arms in 1664 from individual makers, the most important of whom were William Bolton (from 1670–83), John Williams (from 1711–26), Thomas Gregory (from 1717–32), and John Bumford & Co. (from 1745–73). Diverse weaponry reached the Company's forts in India; in addition to standard military muskets and pistols, there were products as highly differentiated as 'blunderbusses', 'musquetoons', 'fuzees', 'buccaneer guns', 'fowling pieces', and 'grenadier guns'. Throughout its early history, the East India Company's small arms were not far short of the high standard of weapons employed by the British Army and Navy. Indeed, the Ordnance Department commandeered the Company's stock of muskets during the Napoleonic Wars, and then produced its own India Pattern musket of similar design. Having achieved renown for good workmanship, London gunmakers remained the main suppliers, while all the arms manufactured were proved at the Gunmakers' Company Proof House.[15]

The London gun trade's connection with the Hudson's Bay Company and the Royal African Company proved comparatively less successful, as London gunmakers always struggled to produce custom-made weapons for the cheap end of the trade. For commercial exchanges with the Amerindians who gathered at the Company's forts and trading posts, the Company needed cheap hunting guns of a type that became known as the 'Northwest Gun'.[16] Even cheaper trade-guns were sought for the African trade; by the 1740s, James Farmer of Birmingham had become the main supplier.[17] With the rising competition from the Dutch and Belgian arms production centres, and Birmingham gunmakers also entering the market as mass-producers, the London gunmakers no longer bothered to compete. By the 1850s, they had virtually committed themselves to the production of highly priced, individually crafted sporting-guns and rifles.[18]

For all its prestige as the capital city of a great power, or its importance as a financial and business hub, London was even in the best of times an 'artificial' centre of arms production,[19] It was Birmingham, at the heart of the English Midlands, which developed into the natural epicentre of the British small arms industry. The Birmingham gun trade benefited from superior factor endowments, enjoying from the start various geographical and infrastructural advantages over its older London counterpart. In the long run, it was better positioned to meet the requirements of the state and the great trading companies for large quantities of arms, even at relatively short notice. By the dawn of the nineteenth century, Birmingham's ascent to pre-eminence as the principal manufacturing centre of guns and gun parts for the rest of the country's gunmakers was all but assured. Its key advantages included the proximity to raw materials and fuel; a larger pool of labour trained for metalworking in 'proto-industrial' conditions; and a subdivision of that labour. Beech-wood from Gloucestershire and Herefordshire, or imported Italian and German walnut, could be combined readily with various grades of iron and steel from the Birmingham district, and thereby fashioned into firearms. Given the abundance of coal and river-based commercial activity in the area, steam and watermills would become the chief sources of power for Birmingham's gun trade. Finally, the availability of gunsmiths—the first gunmakers to be concentrated in what would become Birmingham's 'Gun Quarter'—proved that the industry could be followed more economically there than in London, where the necessary materials and manpower had to be conveyed from afar.[20]

From medieval times into the early modern period, Birmingham earned its reputation for exceptional craft-based production. It had been a village of manufacturers in the England of Edward III (r.1327–77), for 'its location in the centre of a region rich in iron ore and coal deposits made natural the development of such trades as nail-making and iron-ware manufacturing'.[21] In 1538, the Tudor chronicler John Leland reported the presence of many 'smiths, lorimers, naylors, and cutlers' in Birmingham; and, though he does not specify, there were probably a number of gunsmiths among them. In 1643, a master gunner named Nye remarked that small arms were being fashioned at Bromsgrove, a town situated between Birmingham and Worcester.[22] Whereas the English Civil War had merely stymied progress in the case of the London gun trade, that crisis acted as a catalyst to Birmingham's development as a centre of small arms production. In 1638, the local historian John Morfitt noted that Birmingham's metal products were mainly practical and plain articles of iron—nails, hinges, kitchen utensils, and farming implements—manufactured from the abundance of coal and iron mines in the locality. Given the highly charged atmosphere of the 'English Revolution', however, instruments of peace were easily refashioned into tools of war, with ploughshares converted into swords and gun barrels. Birmingham opposed the Royalists during the Civil War, incurring the wrath of Prince Rupert in 1643 when it not only supplied swords but muskets and pistols to the forces of Parliament. As conflict escalated in the ensuing 'revolt of the provinces', country squires found the local smiths better placed to repair their London-made weapons. Such patronage contributed towards the further consolidation of small arms production in the Midlands.[23]

Before long, the Birmingham trade was given more impetus by a series of conflicts with even wider dimensions. These domestic and external crises included the 'Glorious Revolution' and accession of William of Orange (1688–89), and the 'Grand Alliance' against Louis XIV's France in the run-up to the War of the Spanish Succession (1701–14). A local tradition maintains that William III was soon lamenting that 'guns were not manufactured in his dominions, but that he was obliged to procure them from Holland at a great expense and a greater difficulty'.[24] Strictly speaking, muskets and pistols were being made in London at the time; and Birmingham also, though not supplied domestically on any regular basis. But London gunmakers could not sustain the King's demand for military arms, a requirement made more urgent by the latest technological transition from the old pattern matchlock to the flintlock musket. For someone who was simultaneously the Stadtholder of the Netherlands (r.1672–1702) and King of England (r.1689–1702), the fact that the English were then at war with the Dutch presented even 'greater difficulty' in procuring weapons for the one side from the other. At any rate, in response to the royal lamentations, Sir Richard Newdigate, a Member of Parliament for Warwickshire, managed to persuade a group of Birmingham contractors to supply firelock muskets to the Board of Ordnance from 1689.[25]

Once it became obvious that the Birmingham firelock was superior to the old matchlock, approaching the standard of the newly introduced flintlock, the finished product and its competitive pricing both earned the approval of the Board. In 1692, the Birmingham gunmakers obtained a trial order from the Government.[26] Much chagrined, the London Company soon complained to Parliament and, in an ironic twist, Parliament left it to the Board to 'compose the matter in dispute'. In recognition of the service rendered by Birmingham at a critical hour, the Board arranged a further contract in 1693, whereby certain Birmingham gunmakers pledged themselves to supply 2,400 more firelocks, at the rate of 200 per month. These were to be 'proved at Birmingham according to the Tower proof', duly inspected and marked by an Ordnance official, and then financed by the Government 'after the rate of seventeen shillings per piece, ready money by way of debenture, within one week after the delivery thereof into their Majesties' stores in the Tower of London or any other place within this kingdom, as the Board shall order and direct'. The gunmakers were also to be 'paid and allowed three shillings for the carriage of every one hundred weight from Birmingham to the Tower and so proportionally to any other place ... the money paid to them without any charge or trouble'.[27] A third order, placed sometime between the trial order and the contract of 1693, is verified by the existence of an Ordnance Department warrant, to 'pay John Smith for Thomas Hadley and the rest of the Birmingham gunsmiths, one debenture of four score and sixteen pounds and eighteen shillings'.[28]

These interlocking events signified the start of a major transition that saw the Birmingham gun trade evolve into an industry. From the structural and technological angles, the type of arms production was beginning to shift from mere replication through adaptation to innovation. From an organizational perspective, it marked the movement from an informal trade, run by a loose fraternity of Birmingham gunsmiths, to 'The Company of Gunmakers in Birmingham', a quasi-corporate entity nearly as exclusive as the London Company. This newfound sense of corporate identity is underscored in a letter of thanks from the Birmingham gunmakers to their parliamentary patron in November 1692. Nevertheless, the finances of the Birmingham Company were far from secure at this stage. Owing to a 'scarcity and want of money', Sir Richard

Newdigate had to extend his personal credit lines to tide them over until the Board of Ordnance paid up.[29] Furthermore, the rising star of the Birmingham gunmakers continued to be challenged by the rivalry from London. In February 1707, some 400 Birmingham gunmakers petitioned Parliament to end the persecution of the London guild, which if allowed to persist, would force the petitioners to seek employment overseas. The Birmingham gunmakers alleged that the London Company, having monopoly of proof, had exerted pressure on the petitioners, forced down their prices, and obstructed their trade to the plantations in the West Indies.[30]

Once again, it took a sustained cycle of profit and violence to place the Birmingham gun trade on a more secure commercial footing. As the eighteenth century wore on, and the long nineteenth century began in earnest, a recurrence of protracted warfare virtually guaranteed the production of military firearms. Many of the conflicts were fought on an even bigger international stage, which thus widened the scope of arms procurement and galvanized the basis of arms production. The War of the Austrian Succession (1740–48), the Jacobite Rebellion (1745), the Seven Years War (1756–63), the American War of Independence (1775–83), and the French Revolutionary and Napoleonic Wars (1792–1815) all drew massive orders from the Board of Ordnance. The pressure to increase small arms production was immense, given the persistence of pitched battles and the importance of infantry formations in these extended confrontations.

The period of global warfare between 1780 and 1815 was a watershed in the history of British small arms production. During the 1790s, it is estimated on average that 300,000 muskets were made each year in Birmingham for the Board of Ordnance. At 25 shillings apiece, an approximate £370,500 was turned over every year for government orders alone. These figures exclude the weapons manufactured for private use and those supplied to the East India Company, which would not have fallen far short of the number provided for the Army. At the height of the Napoleonic struggle, from 1804 to 1815, Birmingham's production of muskets, rifles, carbines, and pistols averaged over 400,000 weapons per year, over half of which comprised military firearms ordered by the Board of Ordnance.[31] In contrast, France's ten state-controlled manufacturing establishments of the same period combined to yield only an annual average output of 190,000 small arms.[32]

Most would agree, however, that the Birmingham small arms industry reached its apogee in the middle decades of the nineteenth century, which witnessed the Crimean War (1853–56), the American Civil War

(1861–65), and the Franco-Prussian War (1870–71). In terms of 'quantitative' growth, manufacturing capacity was substantially enlarged to meet the vastly augmented demand for military weapons. The annual average quantity of barrels proved in Birmingham between 1855 and 1864 numbered 327,781 for the private trade and foreign government orders, and 97,824 for the Board of Ordnance. London was at this time proving an annual average of only 135,513 for the private trade and foreign consumption, while Enfield between 1858 and 1864 was proving a mere annual average of 72,157 for the Board of Ordnance. During the first four years of the American Civil War, Birmingham was supplying both the Union and the Confederate States with an annual average of 184,350 rifles, as compared to London's annual average of 86,200. Even compared to the combined small arms production of the technically sophisticated American state arsenals and private armouries, turning out an annual average of around 365,000 rifles from 1861–64, Birmingham's annual average output alone amounted to 469,464 small arms, of which 195,850 were military rifles.[33] Despite a severe temporary slump in the mid-1860s, the decade from 1862 to 1872 witnessed the highest production figures ever achieved in the trade. The annual average for these 11 years was 676,758, with the peak years attaining figures as high as 766,893 (1867), 961,459 (1868), and 891,228 (1871), the numbers swelled by extraordinary French orders for Chassepot and other military rifles in the run-up to the Franco-Prussian War.[34]

The expansion of the trade is also evident from the growing number of skilled artisans employed in the variegated processes of small arms production. Sketchley's *Directory of Birmingham* recorded in 1767 that 'from this place all the kingdom are supplied with barrels and locks, and the consumption is very great'; the supply was made possible by at least 62 workers involved in the various branches of the Birmingham gun trade: gun and pistol makers (35); gun-barrel makers and filers (8); gun-barrel polishers and finishers (5); gun-lock makers, forgers, finishers, and filers (11); and gun-swivel makers and stockers (3).[35] At the time of the Great Exhibition in 1851, Birmingham—and not London—contributed 5,167 to a total of 7,731 British gunsmiths and workers.[36] By 1855, the numbers in Birmingham had grown to no fewer than 7,340 'material makers' and 'setters-up', subdivided into specialist categories that had sprung up with the accelerating pace of technological change. These included gun-barrel makers and filers (700); gun-barrel polishers and finishers (1,350); gun-lock makers, forgers, finishers, and filers (1,350); gun-stock makers and stockers (1,100); furniture, rod, bayonet, band, and odd-work makers (1,420); 'machiners' and 'percussioners' (250); barrel borers and riflers

(100); as well as screwers, strippers, sighters, and sight-adjusters (1,070).[37] From start to finish, a single piece might pass through as many as 50 hands, since many of the gunmaking processes were not dignified with separate categories by the directory compilers. Only after the introduction of machinery from the 1850s were these branches cut back and streamlined significantly; and even then, in years of extraordinary demand, numbers of gunmakers could still prove substantial. In 1861, the first year of 'free trade' in Europe, there were 8,459 Birmingham gunmakers out of a national total of 11,873.[38] A newspaper article of 1863, written at the height of the American Civil War, estimated a figure of 7,000 gunmakers and outworkers employed in Birmingham's military gun trade alone, each worker with his personal boy-assistant. In 1871, 5,931 out of 7,362 British gunmakers were reported to have come from Birmingham.[39]

Additionally, there were major 'qualitative' developments within the industry, reflected primarily in the changing rules of proof. Although London had legal possession of the first public proof house in England, the passage of new legislation—the Proof Acts of 1813, 1815, 1855, and 1868—acknowledged the growing strategic importance of Birmingham's gun industry. As with so many other 'family-based' provincial gunmakers throughout England and Wales, the system of private proof operating inside the Birmingham trade was too informal, and hence increasingly inadequate, to handle the phenomenal expansion of orders. Given Birmingham's unparalleled production rate, it is unsurprising that large numbers of substandard firearms also entered into circulation. Such weapons were below par in terms of material and workmanship, turned out by unscrupulous makers who failed to submit their wares to official proof. Most notorious perhaps were the cheap trade-guns made for barter in African markets. In response, Westminster introduced landmark legislation to tighten the rules on quality control, which accorded special prominence to Birmingham:

> Whereas serious injuries are frequently sustained by persons using guns, fowling pieces, blunderbusses, pistols, and other Firearms, from the bursting thereof, in consequence of the barrels of such guns not having been sufficiently proved, and it is therefore expedient that the Manufacturers of Firearms should be compelled to prove the same, at some place appropriated for that purpose as a public Proof House:
> And whereas great quantities of Firearms and barrels for Firearms are manufactured in the Town of *Birmingham* and the vicinity thereof, and it would tend to the safety and security of the public if a Proof

House for Firearms, under proper superintendence and inspection, were to be established in or near the said Town.[40]

Proofing was made mandatory in 1813 through the first Proof Act and the establishment of the Birmingham Proof House.[41]

The passage of proof legislation from 1813 further demonstrated the rising influence of the Birmingham gunmakers' lobby. Frustrated by the Birmingham trade's growing preponderance, the London gunmakers' lobby introduced a bill in the House of Commons in 1813, requiring all manufacturers of firearms to assign real names and addresses to their products. But such was the clout of the Birmingham trade that they were able to quash the bill, indicating that they now supplied most of the parts for the London gunmakers, so that the only difference between the respective products was that some were assembled and signed in London. Birmingham's gunmakers were determined— and had the connections—to capitalize on London's relatively greater prestige at this time. Although the 1813 Proof Act stipulated that every barrel made in England must be proved either at the London Proof House or the new Birmingham Proof House, many Birmingham gunmakers managed to evade the provisions of the Act by indiscriminately 'passing' the barrels of guns belonging to the cheaper commercial categories. Attesting to the increasing popularity, if not prestige, of Birmingham firearms is the fact that Liège gunmakers soon began to forge Birmingham proof-marks.[42] A new Act was passed in 1815, framed upon lines that were generally favourable to the Birmingham trade. In 1855, the Gun Barrel Proof Act made further substantial revisions to the rules of proof to accommodate the gunmakers' production of breech-loading firearms. Fine-tuning was accomplished through the Proof Act of 1868—with minor amendments in 1887, 1896, and 1904—under which the administration of the Birmingham Proof House would continue, essentially unchanged, until the end of the twentieth century.[43] Of the London trade, only a reputation remained; the reality was increasingly to be found in Birmingham's position as the principal centre of British small arms production.

In retrospect, the period 1780–1914 saw Britain's arms manufacturing capacity enlarged if not stretched to meet the demands of crises at home and abroad. Was this a case of industrial overstretch? The exigencies of international warfare, extending from Britain's continental involvement to its global involvement, served to fuel and finance an unprecedented expansion of the Birmingham gun trade. Such wartime expansion brought about a shift in the bulk of private production from London to Birmingham. But we ought to determine the extent to which peacetime industrialization also contributed to that expansionary process. While it makes good sense to set the developmental cycles of Birmingham's gun trade against the soaring trajectories of Britain's Industrial Revolution, it cannot be assumed that the rapid and widespread industrial changes sweeping across the country from the second half of the eighteenth century must have automatically 'revolutionized' the trade.

Indeed, until the 1850s, the trade remained *par excellence* a workshop industry, reaping the traditional benefits of economic concentration. Given the need for rapid communication and transport between the many individuals engaged with the production process, Birmingham gunmaking was by the beginning of the eighteenth century situated mainly in the Digbeth area, though the movement towards the St. Mary's district had started by the 1740s. By 1829, three-fifths of the trade was concentrated in the so-called 'Gun Quarter' centred on St. Mary's Church, with regional proto-industry revolving around it and extending outward to encompass the nearby Black Country towns of Staffordshire. Lock-making, for instance, was located in Darlaston, Wednesbury, Willenhall, and Wolverhampton. Barrel-making—requiring water and, in due course, steam power—centred in Aston and Deritend, and Smethwick and West Bromwich.[44]

In the context of Britain's emerging industrial economy, as De Witt Bailey has rightly indicated, 'the English gun trade, including military manufacture, was one of the last trades to be substantially affected by the Industrial Revolution'.[45] Only during subsequent decades was the gun trade transformed by a second Industrial Revolution, the machine revolution of the mid-nineteenth century. Only through the introduction of mechanized production from the 1850s did the small arms manufacturers of Birmingham acquire, in stages, the capability to mass-produce breech-loading precision weapons for later conflicts, such as the Second Anglo-Boer War (1899–1902).

In the case of Birmingham, it might even be argued that the trade could afford to be a late developer, given its strong proto-industrial base and its astounding capacity for industrious enterprise. A key reason for this later industrial transition was that the gun trade was already an advanced craft at the time of the first Industrial Revolution. Until the 1850s at least, the well-established yet flexible proto-industrial framework had been able to handle, more or less, the pressures of domestic and overseas demand. Fundamentally, however, there were other factors embedded in the curious relationship between public and private sector arms production, and the wider challenges of competing systems of trade and technology.

State or private enterprise?

Combined royal patronage and government supervision had permitted the gradual organization of both the private gun trade and the state armouries, rendering them accountable by law to the Board of Ordnance. The first reference to any such direction by a government department dates back to 1572.[46] Spurred by both military-strategic objectives on the world stage and monopolistic tendencies within the mercantilist system, the state would expand its role in arms procurement and production over the course of the next two centuries.

Under the administrative reorganization of 1683, the Board of Ordnance was revamped and assumed responsibilities for the provision of munitions: from small arms to heavy armaments, ammunition, and other 'warlike stores' for the Army and Navy. Following the establishment of the Bank of England in 1694, there would be a considerable extension of credit and comprehensive military-fiscal structures in place to support the state's pursuit of wealth, power, and victory in warfare.[47] The new system of military arms production for the British Government, set in motion in 1715, came to be known as 'the Ordnance system of manufacture'. Rather than simply purchasing weapons from numerous private contractors in England, Ireland, and the Netherlands, the result of which was 'a confused mass of non-standardized weapons of indifferent quality', the new system was expected to ensure a reasonably steady flow of standard parts, the manufacture and finishing of which could be more closely supervised, inspected, and controlled by agents of the Board.[48] The Board superintended, even sponsored, the invention and development of standard weapons, either arranging for their manufacture in its own manufacturing departments or buying them from the civilian contractors. Having acquired them, the Board managed their storage and maintenance until properly authorized to issue them for service. Thereafter, it monitored the transport and operation of the arms.

Several Ordnance departments were tasked specifically to make and maintain the nation's munitions, ranging from heavy armaments to small arms and gunpowder. The Woolwich Arsenal (renamed 'Royal Arsenal' in 1805) was first used for the proof of ordnance in 1651, the year that Parliamentary forces triumphed in the English Civil War. Storage premises at the Tower of London were acquired in 1670, nearly a decade after the restoration of the monarchy. Manufacturing commenced when the Laboratory (later the Royal Laboratory) was set up in 1696, even as England geared up for its confrontation with the Bourbons. The Royal Brass Foundry for the production of brass ordnance was established in

1716 and the Royal Carriage Department added in 1803. Gunpowder production began with the establishment of the Royal Powder Mill at Faversham in 1759, although this would be largely superseded by the Royal Gunpowder Factory set up at Waltham Abbey in 1787. Gunpowder was stored at Greenwich (until 1763, when the magazine there was moved to Purfleet), and also at Upnor Castle (later the Upnor Armaments Supply Depot).

Compared with heavy armaments and gunpowder, the state-controlled production of small arms had a relatively late start. It began at the Tower of London from 1804, at the height of the Napoleonic struggle, and spread in 1807 to the Royal Manufactory of Small Arms at Lewisham. In 1811, a marshy site at Enfield Lock was acquired for a new manufacturing establishment, where operations would commence five years later. In 1818, state facilities for small arms production were divided up between the Tower of London and the Royal Small Arms Factory at Enfield. Each of the Ordnance manufacturing departments underwent subsequent changes in administrative control during the nineteenth century. In 1855, the Royal Arsenal at Woolwich, the Royal Gunpowder Factory at Waltham Abbey, and the Royal Small Arms Factory at Enfield were transferred from the Board of Ordnance to the War Office. In 1887, they were called 'Royal Ordnance Factories' and placed momentarily under the jurisdiction of the Financial Secretary's Civil Department, before being returned to the Board of Ordnance in 1899.[49]

The pace and direction of state-controlled weapons production under the Ordnance system of manufacture may be explained largely by the Government's relationship with the private sector. Throughout the period 1780–1914, that public-private sector relationship remained flexible, at best, and fraught with contradictions, at worst. State-led enterprise in arms production, spearheaded by the government departments in London and the Royal Ordnance Factories, moved in tandem with the consolidation of military-fiscal policies and the expanding role of the domestic state. For much of the nineteenth century, however, the Government generally 'maintain[ed] the Royal Ordnance Factories in relatively steady employment while forcing the private trade to carry the full burden of the "armament cycle"'.[50] This prevented the public sector from having to be so massive as to supply the reserve production power during major crises, while lying idle at other times. The role of the Royal Ordnance Factories, most notably Enfield in the case of small arms production, was to supply quality and price control; the private sector was expected to provide volume. Paradoxically, the sudden stresses and steady strain of the armament cycle—or the 'gun trade cycle', in the case of small arms—often proved debilitating for the smaller private firms, ultimately putting many out of business.

In the production of heavy armaments, the system was to work well for the state through the contributions of the more resourceful, big private companies. As European armaments industries developed over the second half of the nineteenth century, M. J. Bastable has observed how 'weapons technology evolved mostly from the research and development efforts of private entrepreneurs', while 'governments contributed relatively little to the major innovations'. Governments 'encouraged their arsenals to make cheaper versions of those innovations which could not be ignored and were content to play the role of customer to entrepreneurs and took no interest in the plight of private manufacturers whose production costs grew while government orders fluctuated according to the oscillation of international relations'.[51] But from the 1880s, the costs of developing and producing modern armaments escalated upon entering a new era of high demand. Between 1880 and 1914, as industrial processes gained in complexity and the concept of deterrence shaped the policy-makers' thinking on international and strategic relations, governments—the British state included—found themselves compelled to redefine their relationship with private industrialists to maintain the facilities needed for conducting a total war.[52] State-owned arsenals and shipyards were incorporated into an expanded armaments industry, led by private companies whose welfare acquired national importance. In Britain, the Royal Commission on Warlike Stores, chaired by the Earl of Morley in 1887, represented the first of several parliamentary enquiries into the subject of public versus private arms production.[53] By 1888, the Government was encouraging Vickers to enter the armaments business, thus posing a challenge to Armstrong's dominant position. Compared to the vigour of Armstrong and Vickers, the royal arsenals had become outmoded in all areas by 1914, no longer preserving their monopoly of armaments production.

It was often argued, in the case of armaments, that the export trade maintained the technological cutting edge and innovative superiority of British arms manufacturing. It did so by having to respond to the specifications of different governments and other potential clientele, while competing with other arms manufacturers. The trade in the larger, more expensive weapons systems could also reap suitably handsome financial rewards to tide over the industry in the long run. Small arms, by and large, enjoyed neither such a cosmopolitan market of governments nor such hefty financial remuneration. The trade in light weaponry depended much more on the direction of the domestic state, or the demand generated at the periphery, handled collectively by private traders, company agents, and other intermediaries. This tension between short-term unpredictability and the quest for long-term stability imparted greater volatility to the fortunes of the Birmingham small arms industry, while simultaneously reinforcing conservative tendencies within its craft tradition.

For small arms production, the British Government resorted to both its state arsenals and the private manufacturers in times of war, yet left the latter largely to their own devices in peacetime. In 1793, following the outbreak of the French Revolutionary Wars, Britain's arsenals were relatively empty—60,000 muskets in the Tower of London and other state armouries, compared to France's 700,000—and the English gunmakers were hardly expected to meet government requirements in a flash. Instead, an Ordnance official was despatched to procure 293,000 muskets from the Continent. Where government small arms had been previously made in Birmingham and later assembled in London, it was soon appreciated that quicker processing methods were needed. A state-owned proofing establishment known as 'the Tower' (not to be confused with the ancient Tower of London) was opened in 1798 at Bagot Street, Birmingham, for the purpose of viewing and stamping all new government arms with a 'Tower' mark. When war resumed after the end of the Peace of Amiens (1802), the British Government was again in urgent need of weapons. Supplies were once more sought from abroad, but the key difference this time round was that after March 1804 the Board of Ordnance was finally able to engage the services of Birmingham's gunmakers on a full-time basis, over and above the output of the Tower of London.[54] It was an arrangement capable of remarkable results, for the period 1804–15 proved to be the longest stretch of sustained military production experienced by the Birmingham trade, supplying the British Government and its continental allies with a grand total of 7,660,229 completed arms and components.[55]

The Birmingham gun trade was until the mid-nineteenth century an advanced craft characterized by 'wonderful elasticity'.[56] Viewing this attribute in more positive light, its artisans, working in their crowded network of workshops,

> were able to produce surprising quantities of arms in a pre-technological, non-machine oriented complex through the device of extreme specialization of labour. ... The highly specialized and fragmented nature of the production processes gave the trade a great deal of elasticity, which enabled the workforce to adapt to changes in demand. When there was little work about many would work in other allied fields, some

returned when demand for guns rose. Given the British Government's habit of not restocking its supplies of military arms until the outbreak of a new war such sudden demands were not unlikely.[57]

The trade was also lucrative in seasons of high demand, such as during 1870, when wartime demand for Chassepots and Sniders drove prices up threefold over a four-month period, from £3 to £9 per gun.[58] Of course, prices and profits fluctuated a great deal, according to the potential demand, and this could lead to either the stretching or shrinkage of business. In the absence of sustained demand or state direction in peacetime, the private trade could scarcely afford to maintain in employment the larger volume of workers required for military arms production. Once the fighting ceased, the trade automatically contracted, and many workmen were discharged or drifted into other trades. This happened during the years 1784–93, 1802–4, 1816–29, 1857–60 and, intermittently, from 1864–70.

The volatile character of the gun trade cycle, intensified over a century of tumultuous change and conflict in Europe, could emphasize either the trade's seasonal profitability or its structural vulnerability. Its implications and consequences were understood by A. Wylie, President of the Birmingham Gunmakers' and Inventors' Club, whose observations were made precisely during the boom-bust cycle of the Franco-Prussian War:

> There are peculiar circumstances connected with the gun manufacture that call for especial tact and vigilance in those who conduct it. First of all, the trade, especially the military branch of it, is, in its nature, exceedingly spasmodic and irregular; at one time utterly stagnant, at another in a perfect fever of activity. During the period of slackness men take to other branches, sometimes to totally different pursuits. When the trade suddenly revives, and the transition is always sudden, these men return to their former posts, but of course not so efficient as if they had remained in it all along. This is one cause of the indifferent work that is always turned out when any sudden demand arises. ...

> Some of the men who have given up the trade cannot or do not return to it, and their place is filled up by inferior hands, and, as the work must be done at railway speed, a host of outsiders, attracted by the high rate of wages, which at such a time necessarily prevails, swarm into the trade, and although not in any way fitted for it, either by intelligence, education, or mechanical training, their services cannot

well be dispensed with, and they soon come to know it, and have a peculiar weakness for showing their independence. ...

When the inevitable collapse comes, when the work falls off, to remain perhaps in a stagnant condition for years to come, then the difficult problem presents itself how, whilst keeping production to the lowest point, to maintain the skeleton of a complete establishment ready to be expanded at a moment's notice into full working condition. In truth, the manager of an efficient factory has then to do on a small scale what the Government of the country has to do (or, at least, ought to do) with its military establishment. ... [T]he manufacture of arms is no mean handicraft ... it requires in those whose part it is to conduct it good taste, varied knowledge, and intellectual and moral qualifications of a high order.[59]

Unfortunately, where skill and experience were prerequisites—especially in the barrel-making and lock-forging processes—this loss of skilled labour during peacetime contractions amounted to a critical weakness of the system in the long run:

When such people left the trade in order to earn a living during lean times, there was always a problem in enticing them back into the trade when demand required their skills. The result, as far as the all-important production of military arms was concerned, was that until this skilled labour force could be recruited the Government had to buy arms abroad.[60]

The resultant loss of productive capacity, compounded by the typical bureaucratic delay of the Board of Ordnance in placing orders before the nation was actually at war, compelled the Government to purchase significant quantities of foreign weapons as a stop-gap measure at the onset of almost every nineteenth-century war, at least until the momentum and means of domestic small arms production could be regained. Not until the Anglo-Boer War, at the turn of the century, had the state's expectation of the private trade matured sufficiently, and only then because new public-private sector arrangements and the major technological adjustments wrought by the 'Breech-loader Revolution' seemed at last to satisfy government requirements. Even so, as we shall see from the perspective of the private trade, the biblical injunction not to put one's trust in princes remained just as relevant, while the need to diversify the industry became more pressing than ever.

If the trade's elastic response to the exigencies of the state had meant expansion, then that same elasticity in periods of declension meant either dealing with other clients and diversifying the industry, or facing the certain prospect of depression and decay. The more prudent, enter-prising gunmakers (like the Galtons of Birmingham) appreciated the need to simultaneously capitalize on international conflicts and invest in other financial concerns (like banking and the slave trade) to offset the violent fluctuations of the cycle and tide themselves over the 'lean years' of peace.[61] An alternative was to supply foreign customers, as indicated by a letter of 1830 from the first agent of the Birmingham branch to the Bank of England, London:

> It may not be uninteresting to the Governor [of the Bank of England] to learn that there has within the last few days been an Agent of the French Government in Birmingham endeavouring to contract for 300,000 Stand of Arms—The matter has been referred to Government, and there is now a Deputation of the Manufacturers in London on the subject—They wish Government to supply the arms from the Tower, and they will engage to replace them within three years—This I appre-hend will hardly be complied with. 150,000 Stand of Arms passed the Proof House here last year, and the means do not at present exist of more than doubling that quantity. It is understood that the Spanish Government has likewise applied for a supply of Arms—I am told 200,000.[62]

Responding to sudden, international demand for British arms, several Birmingham gunmakers journeyed to London to obtain government permission to supply these foreign clients from out of 'the Tower', the state-run small arms establishment on Bagot Street in Birmingham. By undertaking to replace government stocks, the gunmakers were also hoping to contract some private business, a matter to be underwritten by the Birmingham branch of the Bank of England. But as the vicissitudes of the gun trade cycle grew more pronounced, and the state's attitude toward the private trade remained just as ambivalent, even such survival tactics were found wanting.

It is from this critical mid-point of the nineteenth century that an informative trade magazine, *The Ironmonger*, sheds more light on the tensions between the private sector (at Birmingham or London) and public sector (at Enfield, under the Board of Ordnance and War Office).[63] *The Ironmonger* suggests that the curious partnership was forced rapidly through an unstable period of transition, subject to the pressures of profound industrial and technological changes over the second half of the nineteenth century, and hostage to dramatic fluctuations not only of the domestic gun trade cycle but also of the wider international arms trade cycle (see Appendix B). At this point in the historical narrative, there are distinct echoes of more contemporary metropolitan debates about the formulation of military industrial policy. At what stage did the state decide which weapons to produce or procure, and which ones to discontinue or dispose of? In what ways did early forms of joint private-public production or civil-military integration maintain the productive capacity and cutting edge of the arms industry in the long run? To what extent did historical examples of 'dual-use' technology enable the conversion of equipment from military applications in times of war to civilian applications in times of peace?

By the mid-nineteenth century, the mutual dissatisfaction between the British Government and the private gunmakers had reached a state of ferment. This crisis of confidence became the subject of a major parliamentary enquiry, highlighting the conspicuous lack of continuity in government orders:

> The complaint of the contractors, that the orders of the Government for Small Arms have not been continuous, raises a question of great importance. There is no doubt that a large demand for Small Arms, spread over a number of years, would attract to the gun trade a supply of hands sufficient to meet the demand. ... On the other hand, it would not be judicious on the part of the Board of Ordnance to pledge themselves to large and continuous orders, for in this age of rapid invention, such a course might be attended with very inconvenient consequences. For instance, the pattern of 1853 had been substituted for that of 1851; therefore, if large orders for the pattern of 1851 existed, they would be orders for a now obsolete arm. Some new pattern may soon supply the place of that which is now ordered. The contractors, however, state that the continuous employment which they desire need not depend upon the identity of the pattern.[64]

Whatever the views of the private trade, the Government's opinion was that a system of commercial contracts which had hitherto met the needs of the nation in crisis had finally defaulted once too often—a perception of structural inadequacy perhaps justified given the delays caused by later strike action within the Birmingham trade. The failure of the commercial contractors to make punctual delivery of a large batch of Minié rifled-muskets by March 1851 soon triggered the despatch of two commissions to the United States—in 1853 and 1854—to study the American system of small arms production.[65] They inspected the Springfield National Armory in Massachusetts, together with several commercial small arms factories. The findings of the first commission informed the conclusions of the Parliamentary Select Committee on Small Arms in May 1854:

> While Your Committee recommend the system of contracting for the supply of Small Arms should not be discontinued, they are, nevertheless of opinion that a manufactory of Small Arms under the Board of Ordnance should be tried to a limited extent. This manufactory would serve as an experiment of the advantages to be derived from the more extensive application of machinery, as a check upon the price of contractors, and as a resource in times of emergency, and it should be arranged with a view to economical working. ...

> If the Board of Ordnance were to construct a factory on the extensive plan originally proposed to Parliament, the manufacturers would foresee that their employment as contractors must soon cease. The public service might thus lose the advantage of the skill, capital, and enterprise of private traders at the time when their assistance is needed, and the further experiment of providing arms by contract would not be fairly tried. ...

> The Government manufactory at Enfield should be continued, since it appears that it is well adapted, and can be made available for the execution of the plan, and that it is capable of any extension that can be required at far less cost, and much less delay than would be inevitable in the construction of a new establishment.[66]

The Parliamentary Select Committee therefore recommended the modest reconstruction of a British arsenal along the lines of the American system. A mechanized plant was installed at the Royal Small Arms Factory at Enfield, while the post of superintendent was offered to James Burton, formerly of Harpers Ferry National Armory, Virginia, the other major American arms-producing centre.

The Crimean War supplied the context for vastly increased military orders at this pivotal moment, proving decisive in persuading the Board of Ordnance to revise plans for a small-scale armoury in favour of a much larger one. War demand lasting through 1856 generated orders for interchangeable rifles that Enfield could not supply initially, such that the

Ordnance Department had to settle for mostly non-interchangeable rifled-muskets from the hand contractors in Birmingham. Signifying the transition from 'old' to 'new' technologies, these rifled-muskets were the output of an intermediate phase of small arms production, comprising converted 'Brown Bess' flintlock muskets and later 'Enfield pattern' percussion muskets, all rifled for greater accuracy of fire. To supplement domestic small arms production and ease the technological transition in these experimental stages, the British Government again resorted to procurement from foreign sources: 20,000 rifled-muskets from Liège (October 1854); 25,000 from Robbins & Lawrence of Windsor, Vermont (February 1855); and 20,000 from St. Etienne (August 1855).[67]

Notwithstanding the vagaries of the trial period, the end of the Crimean War and the beginning of Enfield factory production had serious implications for the future of Britain's small arms industry. These overlapping events signalled, in effect, the advent of 'the American system of manufacture' in Britain. As the era of non-interchangeable, craft-based workshop production drew to its close, a new epoch of interchangeable, machine-based factory production was about to begin. When overall demand for the traditional 'Brown Bess' flintlock musket declined, and the Enfield rifled-musket was mass-produced with its percussion mechanism for the first time, the British Government indicated that it would not place any more orders for non-interchangeable military firearms after 1858. This decision, above all, drove a wedge right through the centre of the private trade: the bigger private manufacturers embarking on a bold course of action that involved state assistance; the smaller gunmakers essentially picking up the pieces, scraping out a living on the basis of far more modest traditional resources.

The bigger manufacturers realized that any further contracts would depend upon the construction of a plant capable of meeting interchangeable standards.[68] In 1860–61, 15 of the largest Birmingham contractors, members of the former Birmingham Small Arms Trade Association, were amalgamated into the Birmingham Small Arms Company (BSA) under the auspices of the War Office. This official connection provided access to state facilities, models, gauges, and technical drawings at Enfield, which had by this time adopted the new system of interchangeable-parts production. The BSA directorate was drawn from a small circle of local businessmen: BSA's first chairman was John Dent Goodman, President of the Birmingham Gun Trade; the rest of the directorate would comprise 24 gunmakers—all of whom had been for many years engaged in the supply of small arms to the domestic and foreign governments—and later, two bankers, from the Midland Bank. The directors were joined by

Joseph Whitworth and the Manchester Ordnance and Rifle Company, as shareholders, backed up by an agreement that allowed BSA to benefit from Whitworth's 'high reputation and influence' plus the manufacture of his rifles. Some 3,400 shares (collectively worth £85,000) were taken by the original promoters of the company, while the directors sought to issue a further 3,000 shares at £25 each—representing, in total, an initial capital of £160,000.[69]

Even so, everyone at the fledgling company soon realized that the Government's fluctuating orders, as in the past, made business precarious. In 1866, it was reported that business was 'flat':

> The military small arms business has been notoriously overdone, and but for a few orders occasionally dropping in for fowling-pieces, the Birmingham gun trade would be an entire blank.[70]

A month later, there was general cause for celebration, yet also specific cause for concern:

> The gun trade is improving, but there are great complaints of the system of trafficking in 'job lots', whenever a foreign order for military small arms reaches any of our principal outposts.[71]

The following month, it seemed even more likely that Enfield would secure the lion's share of business at Birmingham's expense:

> Birmingham profits little, if at all, by the tardy decision of the War Office, in regard to the conversion of the army Enfields into breech-loaders. At present, orders are given for the conversion of about 20,000 of these rifles, and it is expected that double that number will be converted during the present year. None of this work comes to Birmingham, but is to be done at the Government factory at Enfield, on the principle known as Snider's patent.[72]

The appointment of Lord Dufferin to the War Office (1866–68) proved fortunate for a time. Under his influence, the Government decided not to depend wholly on Enfield for the conversion of its standard firearms, but to involve BSA in that process. The new Under-Secretary for War even offered monetary prizes as an incentive for Birmingham gunmakers to develop the best model of military firearm that met government specifications. Lord Dufferin promised 'to take care that no ingenious novelty produced in answer to this invitation shall be adopted into the service without proper acknowledgment, and that the name of the individual with whom it originates shall be recorded in connection with it'. The rifle that won the first prize of £1,000 would not only be adopted into the service, but also bear the inventor's name.[73]

Still, in the new era of interchangeable-parts production, such state-led initiatives were few and far between, more sporadic than sustained. The first BSA-War Office contract was not agreed until 1868, while BSA's factory at Small Heath was shut for the year 1879 through lack of work. Notwithstanding the growing involvement of the state in the armaments industry, in addition to the crucial innovation and state-sponsored adoption of a succession of light weapons (the Lee-Metford and Lee-Enfield magazine rifles, the Maxim machine-gun, and cordite, and smokeless cartridges) from the 1880s–90s, BSA was forced onto the path of industrial diversification into more stable civilian markets—bicycles (1880), tricycles (1885), bicycle parts (1893), and motorcycles (1895)—with a corresponding withdrawal from dependence on government arms orders. The trade journals reassure us that BSA did survive financially in this period, although military small arms comprised only 21.3 per cent of BSA's turnover by 1910, and only 33 per cent in 1913, on the eve of the First World War.[74]

Other 'makers of interchangeable small arms', likewise equipped with American gunmaking machinery, appeared in the 1850s. They included the London Armoury Company, superseded in 1867 by the London Small Arms Company (LSA), and the National Arms and Ammunition Company (NAA).[75] Again, success was limited and fleeting, for their military arms factories received orders too inadequate and intermittent to keep the whole plant occupied. To strengthen their hand collectively, BSA, LSA, and NAA had actually formed an agreement for dealing with the British Government, allocating to each of the firms a fixed proportion of the output required by the state: BSA was to supply 40 per cent; LSA, 27 per cent; and NAA, 33 per cent. But this pact drew fiery protest from the War Office, thus proving unsustainable.[76] BSA found some consolation in industrial diversification; and, when pressed, remained capable of making 30,000–40,000 weapons per year in the late 1880s–90s. LSA could still turn out 20,000–30,000 weapons per year over this same period, but NAA was altogether less successful than either partner.[77] NAA folded in 1882, its large factory at Sparkbrook abandoned following various experiments with tricycle manufacture, and then acquired by the Government in 1885–87 for the repair of rifles. Between 1887 and 1894, even 'the Tower'—one of the last vestiges of direct state interest in Birmingham—was relocated to Sparkbrook, and not without sudden and serious reductions in output, employment, and wages. While Enfield itself suffered a 25 per cent drop in government finance in 1892–93 due to state budgetary policy, from £160,000 to £120,000, Sparkbrook's reduction was greater still, a 44 per cent cut from £62,000 to £35,000.[78]

If the big manufacturers themselves faced such difficulties in the state-led transition to interchangeable-parts production, then the situation facing the smaller manufacturers was bleaker yet. The British Government's decision to buy only interchangeable military rifles after 1858 severely circumscribed the business of civilian gunmakers who made traditional military muskets in addition to sporting-arms and trade-guns. Essentially, they could no longer compete for British military contracts when the Government required interchangeable-parts standards they could not meet. Conversely, the state failed to provide any alternative means of remuneration, the vital economic safety net and financial security needed for a small firm's survival during the downturn of each production cycle.

In the afterglow of Enfield's early success as a centre of interchangeable small arms production, *The Ironmonger* reports in 1861 the visit of a high-level delegation involving leading Birmingham and London gunmakers and the MPs for Warwickshire and Birmingham to Lord Herbert of Lea, the Secretary of State for War, 'to present a memorial on matters affecting the interests of this trade', namely, 'the unfair position in which the small arms trade is placed in regard to the Government manufactory at Enfield'.[79] The deputation contended that the whole reduction for 1860–61 in the state subsidy for the production of small arms was made to fall on the private trade. Meanwhile, Enfield continued to work at full capacity, with an additional £35,000 grant to purchase fresh machinery that would enable the government establishment to monopolize a branch of trade previously in private hands. It was argued that this would effectively reduce orders given to the trade by at least 50 per cent, rendering it incapable of meeting government demands in an emergency. Even if higher rates were charged to provide some compensation, workers were likely to be dismissed, entailing a further loss of skilled labour. Collectively, the maintenance of Enfield on its present scale, the tempting monetary inducements for workmen to leave Birmingham, and the reduced capacity of masters to meet sudden government demand would all prove injurious to the industry in the long run. The deputation urged the Government to allow the reduction of orders to be borne equally by both Enfield and the private trade, since the matter was of such regional importance for Birmingham and its neighbouring districts: a reduction from 80,000 to 40,000 military firearms in one financial year would entail a loss of £150,000 to the community. But there is no indication that the Government acted to ease their plight.

On another occasion, the projected closure of 'the Tower' had, as far back as the 1860s, convinced Birmingham's military gun trade that 'such a course would seriously affect thousands of hand-made gun-operatives in the town', scattering and driving into other trades thousands of skilled artisans. Several trade representatives had therefore sent a 'memorial' to the Earl de Grey and Ripon at the War Office, highlighting the long-standing relationship between the state and the private sector. They claimed that the Government had relied on them in previous conflicts such as the Crimean War and, on the threshold of interchangeable-parts production, had even been supplied with the new Enfield pattern rifled-musket of 1853, hand-made yet mass-produced in Birmingham at the rate of a thousand per week. The gunmakers suggested that 'the Tower' might be used as a repairing establishment for furnishing the weapons to volunteers and local militia.[80] Again, this appeal seemed to fall on deaf ears. By 1866, the old gun trade was deriving no further direct benefit from British Government orders.[81]

Emphasizing the extent to which private enterprise was threatened by state-driven requirements for interchangeable-parts production is a memorandum sent by the BSA gunmakers at Small Heath to Edward Cardwell at the War Office in 1869. From the biggest private manufac-turers of interchangeable small arms, we read how the once-flourishing trade of Birmingham had 'gradually sunk into a state of heavy depres-sion' as a consequence of Enfield's growing stranglehold on domestic production:

> Firstly, the Royal manufacturing establishment of small arms at Enfield, originally intended by the Government as a model factory, has gradually extended its operations so as, in a great measure, to absorb the orders formerly given to the contractors for small arms in Birmingham. ... [T]he existence of such an establishment on so large a scale is incompatible with true economy and opposed to the true interests of the country, by discouraging the private manufac-turer, thus preventing his being prepared to give assistance when required in a period of emergency.

> Secondly, the manufacture at Enfield of the arms required for the mili-tary service in India and the colonies. A few years ago, the arms for these services were obtained exclusively from the trade, and the demand not only encouraged skilled labour in the gun trade, but mitigated the con-

sequences occasioned by the fluctuating character of the orders given out by the Home Government. ... [T]he distress existing in the trade at the present time is marked by a severity and length of duration unexampled for many years.[82]

The gunmakers feared that such extensive state-controlled production would also erode the prestige and high reputation which had traditionally belonged to the private trade, with a wider dampening effect on orders from abroad. Foreign governments, seeing that the British Government would not employ the Birmingham gunmakers, would surely sense that there was 'something wrong', and that 'Birmingham men could not be relied upon'.[83]

Only war demand in Europe, combining with the wider dynamic of crisis in indigenous societies beyond Europe, would allow the 'new' private producers of interchangeable arms and the 'old' private non-interchangeable producers to sell their output of military firearms to foreign clientele. Trade journal reports, proof returns, and export records from the 1850s all suggest that it normally took such dire exigencies to effect a shift in military buying preference from 'quality' (interchangeable weapons made in machine factories with a long lead time) to 'quantity' (non-interchangeable weapons turned out rapidly by all makers).[84] Even then, under circumstances which in previous generations would have ordinarily provided work for the private trade, the fate of this sector was gradually sealed by the systematic adoption of interchangeable-parts techniques in other countries: first in Europe, and then elsewhere, as increasingly sophisticated firearms were dumped by stages in Asia and Africa. Foreign governments were by the late 1870s manufacturing their own precision weaponry to arm their troops; American gunmakers competed to capture the higher end of the market from Birmingham, while Belgian gunmakers threatened the cheaper end.

Under most circumstances, the cultivation of foreign clientele remained vital to the Birmingham gun trade. Where overseas markets had always been important for the survival of the industry in periods of declension, Birmingham gunmakers—big and small—stepped up their supply of firearms and explosives to a host of foreign customers in the later decades of the nineteenth century. This included several major contracts for the new French Chassepot rifle (1867), involving BSA and other big firms; Snider rifles and 10,000 barrels of gunpowder for the Turkish Government (1871); 150,000 improved Mauser rifles and a million cartridges for the Prussian Government (1872), involving Westley-Richards Small Arms Company; large weekly shipments of gun barrels for the Providence Tool Company,

Rhode Island, to be completed as Martini-Henry rifles for Turkish Government contracts (1877–78); and numerous large orders for ammunition in the late 1870s and 1880s, involving Eley Brothers, Westley-Richards, and Kynoch & Co. Wartime demand ameliorated the overall condition of the arms industry and its adjunct trades, such as during the Franco-Prussian War: 'The war has brought some relief to the poorly paid gun-lock filers of Darlaston. They have plenty of work, at an enormous increase in prices.'[85] Orders for ammunition included a Romanian contract for eight million cartridges (1879) and a Chinese contract for 11 million cartridges (1885), both supplied by Kynoch.[86]

Nonetheless, the fighting of foreign or colonial wars could not always guarantee growth in the industry, particularly from the mid-1870s. The outbreak of war in Afghanistan and South Africa in the late 1870s had no marked effect on the slack overall demand for military small arms. Apart from small government orders from C. G. Bonehill of Birmingham for 2,000 Sniders, the large factories of BSA and NAA were at a standstill for lack of employment—an estimated capital investment of some £500,000 lying idle.[87] By 1883, the National Company had to be liquidated amid forecasts of doom and gloom, now sounding increasingly prophetic: 'It is just possible that the consequent restriction of competition may enable the other local producers of military rifles to prolong their existence until something turns up in the shape of a big war to stimulate this flagging branch; but, so far as Birmingham is concerned, the military gun trade is evidently played out.'[88]

The clearest evidence for this trend would be the overall decline in British rifle exports from the 1870s to 1914. During the American Civil War, British rifle exports had constituted 76 per cent of the total value of small arms shipped abroad, while in the years 1880–81, rifles comprised only 30 per cent. The actual value in current pounds of rifles exported dropped by more than two-thirds over the same period. Whereas government military arms exports were sustained to meet the demands of empire, the remaining rifles exported in the 1880s were probably mostly civilian models made by sporting-arms makers.[89] Between 1872 and 1910, Birmingham trade production figures reflecting the consolidated output of the smaller gunmakers also declined steadily, notably during the last two decades. From 1873 to 1880, the annual average number of barrels proved was 594,191, and for 1881–90 it was 578,897. But for 1891–1900 it had fallen to 378,979, and for 1901–10 it was 363,103, indicating that the bulk of military work had been either absorbed by the state and private factories, or had shifted overseas. Even when both production and employment figures for Enfield and the big private rifle makers—BSA,

LSA, and the Henry Rifle Company—soared to meet Boer War demand across the years 1899–1902, production and employment figures for the old gun trade remained sunk in depression.[90] Finally, the depression faced by all branches of the trade devastated the remnants of regional proto-industry at least as old as Britain's first Industrial Age: the last lock-makers of Darlaston and Wednesbury abandoned their trade mostly for the factory production of nuts, bolts, and tubes; and the gun trade disappeared from Wednesbury, even as the last firm—Spittle Brothers—sold its machinery and spluttered out of existence by the late 1880s.[91]

In a major polemic of 1907 concerning the 'decaying' British small arms industry centred upon Birmingham, two leading gunmakers writing under the pseudonyms 'Artifex' and 'Opifex' criticized the Government's policies with unalloyed disgust:

[T]he Government attempted to transplant the industry from the provinces to the capital. The Lewisham factory was a costly failure, and the choice of Enfield for the small arms factory a mistake. Birmingham found the skilled men for both factories, and their removal to London was a loss to the Birmingham industry without being any gain to the country. If the State factories had been established at Birmingham the Government would have had the advantage of being in the centre of the industry, and the Birmingham manufacturers and artisans would have improved their processes and products through being in closer touch with the State factories. ...

Isolation is equivalent to stagnation. It is nearly a century since the Government broached its Lewisham scheme for a State small arms factory, and its long decades of experience in arms manufacture have resulted in no advance ahead of improvements in the firearms industry. ... [T]here are men still living who made flintlock, 'Brown Bess' muskets for the British Army, and made them at a date when continental armies possessed breech-loading rifles. The Government has taken the expanding bullet, the Snider, the Whitworth, the Lancaster, the Westley Richards, the Henry, the Martini, the Lee, and inventions too many to enumerate, from the trade, and meantime what novelties or improvements have originated at Lewisham or Enfield? ...

Now a Government benefiting by the experience gained, and therefore wiser than all its predecessors, relinquishes the Birmingham factory and concentrates small arms manufacture at Enfield, an act which excites the trade to ribald laughter, whilst it moves the taxpayer to sighs and tears.[92]

Considering the comparative success enjoyed by Enfield *vis-à-vis* interchangeable-parts manufacture, the partisan views of Artifex and Opifex may come across as a simple case of sour grapes. Yet the overall record of state initiatives in small arms production does support those misgivings about the Government's increasingly 'baroque' arsenals at Lewisham and Enfield, in addition to the state's oddly indifferent attitude towards the private trade. Although the Government appreciated the importance of maintaining the private gun trade in a state of efficiency, and Birmingham could produce guns fully 20 per cent cheaper than Enfield, the Government was still making plans to increase the latter's productive capacity to 8,000 barrels per week (or 1,000 rifles per day) by the late 1870s. As one commentator observed, 'The government establishment at Enfield is very busy on Martinis [Martini-Henry rifles], on which the staff are working overtime, but this circumstance is equally difficult to reconcile with good faith and public policy, ministers having repeatedly recognized the wisdom of encouraging private enterprise in the military gun trade, so as to maintain a source of independent supply against times of emergency, and having led the small arms companies to understand that about one-third of the state requirements would be obtained from them.'[93]

Whereas state policy from the mid-nineteenth century tended to privilege public-sector small arms production, the whole tenor of state legislation seemed to undermine the private sector:

The history of the firearms industry in this country and abroad, shows that both on the Continent and in England those who followed the industry were protected by powerful patrons, by sovereign rulers, influential guilds, and, ultimately, by the state. ...

No manufacturing industry, such being an artificial and not a natural product, can originate or progress without some sort of protection. ... No country has such natural and organized facilities for the rapid production of firearms as England possesses, and in no foreign centre is the industry so heavily handicapped by legislation as it is at Birmingham.[94]

From the perspective of metropolitan gunmaking, this would encompass special legislation affecting production, including the competition of the state factories for the supply of military weapons, as well as more general domestic legislation. There was legislation limiting the sale and use of firearms, including the Game Laws, the Gun Licence Act, the Explosives Act, the Pistols Act, and the Irish Arms Act; and legislation regulating the export of British manufactures, such as the Foreign Enlistment Act, the Customs Act, and Orders-in-Council that restricted the export trade to foreign countries. Even the Patent Laws operated in such a way 'that the new inventions of foreigners are, or can be, protected everywhere throughout the British Empire, but may be produced in any other country, whilst the goods invented by British subjects, in order to have the monopoly a patent confers of the Belgian, French, German, or other foreign market, must be manufactured in Belgium, France, Germany, and other countries respectively'.[95] Such legislation could only dampen private-sector production and innovation in Britain, just as it elevated the importance of foreign manufacturing centres.

One example was the 'Prohibition of Arms Act' of 1900. Given the urgency of international crises such as the Anglo-Boer War and the Boxer Uprising in China, this law progressed rapidly through Parliament as an emergency measure, controlled by royal proclamation. Ironically, the legal limits imposed on the supply of British-made weapons could serve to boost foreign competition, rival suppliers from within the metropolitan core of the arms transfer system:

> The Act itself merely gives powers for prohibiting the exportation to countries which may be specified of arms, ammunition, and such military or naval stores as may be used directly or indirectly against her Majesty's subjects or forces. Such powers as are granted under the Act may be of great importance to the country if intelligently used, but, on the other hand, they may create many injustices. ...

> China has been a large buyer of arms and ammunition in recent years, and obviously such munitions are likely to be used against her Majesty's forces and subjects. On the other hand, the stoppage of English supplies of these commodities would not, in the smallest way, check the Chinese Government in acquiring from other countries all they require. Our English Government would serve no useful purpose in stopping such traffic, and our manufacturers would be the only sufferers. ...

> Krupp and Waffenfabrik Mauser may smile at the Act as removing one or two of their most powerful rivals. Let us hope that the use

made of the Act by our Foreign Office may not result in granting a trade monopoly to our foreign rivals.[96]

Artifex and Opifex were simply echoing a common refrain of industrialists at the time; the strategic imperatives of Britain's gun laws tended to harm the special interests of its gun trade, while simultaneously encouraging foreign competitors by offering them a free market for their manufactured goods: 'The chief cause of decay is the burden imposed on the productive industry by legislative enactments, by laws of which the administration is invariably costly. If these were removed it is apparent that productive industry would be relieved, and might then support other burdens, foreign or domestic ... Legislation may hamper and hinder commerce; a different kind of legislation may encourage production and develop trade.'[97]

Over the six decades between the outbreak of the Crimean War and the First World War, the traditional elasticity of the Birmingham gun trade was eroded steadily by its ambiguous relationship with the British state. This was manifested in the priority and privilege accorded to government establishments like Enfield, as well as the pattern of government legislation. Clearly, the interdependence between private and public sectors required mutual accommodation and many readjustments after the 1850s, often more than either was able or willing to offer. Finally, though, it is worthwhile exploring two distinct but related external developments that reinforced the internal pressures on the arms industry.

New technology and new imperial competition

The first of these twin challenges was technological, arising from the American system of manufacture. From 1783 until the aftermath of the War of 1812, the craft origins of small arms production in the United States had been significantly indebted to 'the guilds of Birmingham'.[98] Between 1815 and the early 1850s, however, the shortage of skilled artisans and the high cost of labour had necessitated the progressive, concerted application of labour-saving devices, such as the Blanchard gun-stock lathe and barrel-welding machinery. This mechanization of small arms production on the interchangeable-parts principle came to be known as 'the American system of manufacture'. Though not without its problems—not least the conservative resistance of the rural community at Harpers Ferry—the American system exemplified by the dynamism of the Springfield Arsenal became conspicuous for its efficiency, high standards, and low costs of production.[99]

Conversely, even when military technological innovation in Britain seemed to be blessed by years of fertile invention, such individual achievements—occurring within a framework of handcrafted-parts production—could scarcely compete with America's overall systemic changes. Forsyth's fulminating powder had, for example, led to the development of the percussion cap (1807–16). Other noteworthy British innovations included Osborne's barrel-welding machinery (1817); Richards' improvements in gun-locks (1820s); and Greener's expanding bullet (1836).[100] But in failing to harness them on any consistent, systematic basis, the British state proved unable to really capitalize on such innovations. By the mid-nineteenth century, it was the United States that was supplying Britain with the stimulus for the next phase of development in military technology and small arms production.

The Great Exhibition of 1851, as several American historians have emphasized, marked the onset of a technological sea-change in the British small arms industry. For the first time, British Ordnance officials saw the American system exhibited in large quantities. As a result of overlapping events that flowed from this display—the production of American guns in Britain on a trial basis, the investigation of American machinery by a British commission, and a detailed British parliamentary enquiry into the nature of small arms manufacture—British military gun-making was transformed almost overnight.[101] Under the American Ordnance Department's policy of awarding contracts, the producer benefited from interchangeable manufacturing, since any cost savings in the production process accrued to him. The British Board of Ordnance, on the other hand, had always issued separate contracts for parts and assembly. The parts-production contractors could not therefore reap the economic benefits of low-cost assembly of interchangeable parts, and they had no economic incentive to initiate the new system of interchangeability. British contracting policy, rather than stimulating innovation, tended to reinforce the traditional craft structure of the industry.[102]

By 1854, the American system would have to be imposed from above by the British Board of Ordnance, the consumer of military arms, rather than coming from the contractors. The British state's justification for importing the new system was spelt out in the report of the Parliamentary Select Committee on Small Arms:

> The general gun trade of this country is said to be increasing. Merchants and gunmakers have affirmed that they export more extensively than formerly finished guns and parts of guns, locks especially, and that even in the United States, where machinery is stated to be so

advantageously employed in the gun manufacture, the general gun trade, nevertheless, derive their materials from this country.

It is to be observed that the increase has been in the inferior class of arms, or else in guns far superior to those usually employed for warlike purposes.

It appears that the manufacturers of this country have supplied the East India Company with all the classes of small arms which they have hitherto required. They have also, during many years, supplied foreign governments, and it is said, that they can successfully compete with manufacturers in foreign countries. But it also appears that they have not succeeded in satisfactorily supplying the Board of Ordnance with the quantity of the superior class of arms required in the last three years for the use of the British Army.[103]

The 'new' technology was formally introduced to Britain from 1854, and it varied primarily according to the *type* of firearm produced. In order to produce that 'superior class' of 1851 Minié and 1853 Enfield rifled-muskets, military arms manufacturers were forced to adopt—lock, stock, and barrel—American technology and the American system of manufacture. Shotgun producers adopted some American machines but not the American system. African trade-musket makers seem to have hardly changed their methods at all. At one end of this spectrum, the military gun trade was restructured and expanded to meet the massive, exacting requirements of government forces engaged in the Crimean War and subsequent conflicts. As we have seen, this left a trail of economic casualties along the way. At the other end, trade-gun models and production techniques tended to remain fixed by the cycles of fluctuating orders and availability of skilled labour in the metropolis, or by colonial laws and monopolistic conditions at the periphery.

The introduction of the new technology and the implementation of the new system would have two key consequences for metropolitan gunmaking in Britain. Under the new system, Enfield would prove how capable it could be in satisfying the peacetime demand of the services and the domestic state. Conversely, the adoption of large-scale mechanical production would fragment forever the traditional structure of the Birmingham trade. Under the old system, the gunmakers could turn to supplying sporting-arms and trade-gun demand in times of slack military trade, and then rise to the challenge of military production to supply government wartime demand. The new system deprived the gunmakers of a large clientele, while the factories of the big interchangeable producers, after being hugely expanded during periods of active trade, were likewise bereft of any means of employing their plant once military demand ebbed.[104] During slack periods, crippling overhead charges were incurred because of the difficulty in adapting the specialized tools and machinery to the production of other manufactured goods; simultaneously, much skilled labour was made redundant.

Whereas the flood of new orders generated by the American Civil War brought respite to the Birmingham gun trade in the 1860s, it is equally true that the trade was passing through an industrial sea-change. In 1867, we read about the briskness of commercial activity, but also notice the shifting landscape:

> The erection of another large factory in the vicinity of the town for the production of guns and pistols by machinery has been instanced as an indication of prosperity in the trade; but although the times look promising enough for the manufacturers of warlike weapons, the real significance of the building of new factories is that the gun trade is entering upon a new phase. The old system of attic and small workshop production is passing gradually away, and giving place to greater establishments, where firearms are manufactured on what ought to be now pretty well known, even to the general public, as the 'interchangeable' plan.[105]

Insofar as the Birmingham gun trade succeeded in readjusting its manufacturing output to the more exacting industrial specifications of the new technology, so too did its subsidiary branches in neighbouring districts, at least superficially: 'At Wednesbury and Darlaston, the gunlock filers are busy, in consequence of the continued activity in the gun trade of Birmingham.'[106] Nevertheless, hardly any firms were able to keep their workers fully employed, smaller masters were driven out of the trade altogether, and a significant proportion of traditional gunmakers were drafted into other branches of industry which offered better employment prospects. For the gun trade, the dawning of a new age of mechanized production meant that a considerable pool of skilled labour remained unemployed due to the difficulty of getting workmen to 'turn their hand' to something of a different order, which 'although looked upon as kindred to that which they have been brought up, is practically another and wholly different trade'.[107]

Given the organically rooted craft traditions of rural England, a conservative reaction to the mere perception of newfangled organization seemed only natural, and the recruitment of suitable labour for the factories proved very difficult. For instance, a number of Darlaston lock and spring filers engaged by BSA could not be persuaded by their foremen to give up old ways:

> These men still followed the method of a hundred years previously. ... They still resorted to 'fiddle-drilling' (i.e. bow and breast drilling) when, by going a few yards, they could use power machinery. They still used tallow dip candles (purchased by themselves) when tempering springs, although the Company had offered to supply them with best Russian tallow, free. They would not do tempering after 10 o'clock in the morning, owing to their superstitious belief that springs tempered after that hour would break.[108]

If the 1860s appeared to mark the apogee of the British small arms industry, both in terms of state and private production, then doubts were raised rapidly as to its continuing prosperity. This unstable cyclical pattern would form a recurrent theme throughout the later nineteenth century as the machines took over: rapid technological transformation within the British small arms industry, in general; and vicissitudes in the largely declining fortunes of the Birmingham gun trade, in particular.

Upon closer inspection of the 'boom' years of the Birmingham trade, there is indication that American Civil War demand was something of a false bonanza. For over two years, it had provided the basis for a vast expansion of the British trade, but was then abruptly terminated in September 1863 when American factories began to meet Federal demands. Export to the Confederate States was hampered progressively by the capture of ports into which British-made weapons had been delivered. Most harmful of all was the formation of even more machine-equipped United States factories producing breech-loaders and ammunition. Many of these folded at the war's end, but those that survived formed the nucleus of a new American gun trade, which for the first time in that nation's history was able to supply its own internal demand for firearms. Only breech-loading shotguns could not be produced cheaply by factory methods; and while these continued to be imported from Britain, even more came from Belgium where costs (and therefore prices) were lower.[109] In a public debate that raged in the *Birmingham Daily Post* between 1864 and 1869, the thesis of 'decay' posited by Artifex and Opifex in 1907 was foreshadowed ominously. Was blame for the relative decline of the Birmingham gun trade to be apportioned to the opportunism of government policies and legislation; or to the conservatism of masters and merchants, manufacturing, and manpower organization; or to the dynamism of competing foreign machines, manufacturers, and markets?[110]

Against the changing complexion of technological development and industrial transformation at home, the innovation of breech-loading firearms and the partial adoption of the American system of manufacture would have revolutionary repercussions on the entire arms transfer system. Within four years of Britain's two fact-finding missions and purchase of $105,000 worth of American machinery in 1854, several European governments had sent similar investigatory teams to the United States.[111] At the metropolitan core of the arms transfer system, something of a Pandora's Box had been opened, furnishing the tools and machines to mass-produce batches of the latest, most lethal breech-loading weapons: from single-shot rifles to precision-arms of repeating and rapid-firing capability in the later phases. Whether exported as war surplus or obsolete equipment, these modern firearms would make their presence felt within decades, even at the most remote frontiers. The proponents of the new technology, having sown the wind of technological change in Europe, would shortly reap the whirlwind of destruction in the quest for profits and spread of violence across Asia and Africa.

The second twin challenge to the British small arms industry took the form of trade competition from Britain's continental rivals. Notwithstanding the pronounced fluctuations of the trade cycle, the Birmingham gun trade was until the mid-nineteenth century unsurpassed in terms of the general cost of production, in addition to the overall quantity and quality of its products; and Britain was, until around 1870, the metropolitan centre for over 50 per cent of the global small arms trade.[112] Yet the interruption of British small arms sales to the Continent and the steep inflation of prices during the American Civil War finally induced other Europeans to commence the mechanized production of their own breech-loaders. Given the widespread adoption of the American system of manufacture by 1876, the leading continental powers—France, Germany, and Austria-Hungary—would also acquire factories capable of satisfying both domestic and foreign requirements for precision weaponry. In the case of France, the 'new' trade of St. Etienne began to flourish and spread. In the case of Germany, the trade dags found ready markets for its military merchandise inside the German Customs Union (*Zollverein*), and further afield in Austria-Hungary and the Russian Empire. German rifles from Waffenfabrik Mauser were exported in large quantities to clientele both within and outside Europe. Austria, too, developed an exceptional small arms industry, the most noteworthy firm being the Mannlicher Company at Steyer.[113]

Most persistent and perplexing, however, was Belgian competition centred upon Liège, starting initially at the cheaper end but spreading gradually across the entire range of production. Birmingham and Liège shared many similarities in terms of organization and output. Like Birmingham, Liège had some renown as a gunmaking centre since the first introduction of firearms in Europe. Raw materials were abundant locally, or imported from nearby regions, such as Westphalia. Its earliest gunmakers were makers of nails who also forged gun barrels; the barrel-makers belonged to the guild of smiths, and the stocks were fitted by members of the carpenters' guild. Gun-proofing was made mandatory by the state as far back as 1672, though gun production remained a monopoly of the trade guilds until the French Revolution. Among the firms that survived into the twentieth century were Renkin et fils (founded 1772) and seven other civilian gunmakers established between 1801 and the independence of Belgium in 1831. In addition, there seems to have been a degree of interdependency between the entrepreneurs of Birmingham and Liège. In the course of the nineteenth century, Birmingham sold to Liège large quantities of gun castings and components, especially gun furniture, which could be made better and more cheaply in Birmingham, while probably buying back from Liège to meet some of its export orders.[114]

The challenge from Belgium was making a clear impression by the mid-1840s. William Greener, a prominent Birmingham gunmaker and a patriarchal figure in the British small arms industry, noted that 'Birmingham makes considerably *less* guns than Liège ... the gun-manufacturers of Belgium are making rapid improvement, while [the Birmingham trade] are retrograding.'[115] Belgian gunmakers had adopted the model and the style of finish from Birmingham, the extent of which was described by John Goodman, another distinguished Birmingham gunmaker, as 'the baneful practice of counterfeiting English trade marks'.[116] Liège only used two qualities of iron and two proofs, so even the commonest guns were safe; and these, too, were bored well. Liège gunmakers were adept at discerning the tastes of different markets, and adaptable in production. Above all, the primary strength of the Belgian trade lay in the cheapness of its labour; whereas in the late 1850s the weekly wages of Birmingham gunmakers, outworkers, and boy-assistants ranged from 30 shillings to £6, 15 to 25 shillings, and 5 to 10 shillings respectively, Liège gunmakers were paid the equivalent of just 12 to 17 shillings. With its bearing on prices and profits, this would in turn enable Liège to capture a larger share of the market at opportune moments, beginning with the cheaper end of the small arms trade.[117]

The African trade was the first to give way. From the late 1830s, as Birmingham gunmakers busied themselves with government military work and East India Company orders at the expense of African trade-musket production, many merchants and their agents turned to Liège for a more reliable, cheaper supply of trade-guns and associated component parts. The number of African trade-muskets proved at Liège shows a marked upward trend from 1837 onwards, with an annual average of 14,803 in the 1840s, and 29,591 in the 1850s. Moreover, while Birmingham was still exporting an annual average of 100,000–150,000 African trade-muskets between the 1860s and 1890s, barrels made for the African trade were sent increasingly to Liège for setting-up.[118] In 1890, Birmingham produced 176,000 African barrels, of which 100,000 were duly proof-marked and then sold to be finished in Belgium with Belgian locks and stocks.[119] The production of trade-guns in Birmingham had been rendered virtually impossible by the introduction of machinery and the increased cost of labour. The competition amongst Birmingham entrepreneurs to secure orders and the raising of prices at the slightest rumour of large contracts by otherwise slack workmen, combining with the large-scale removal of government work to the Enfield factory, all contrasted unfavourably with the abundant cheap labour of Liège, the apparent industry of its gunmakers, and the general absence of expensive machine-equipped factories.[120]

The loss of Birmingham's African trade to Belgium was followed rapidly by the loss of the sporting-gun trade. From the 1860s, as Birmingham gunmakers neglected sporting-arms manufacture for military production, Liège once again stepped into the breach and began to supply sporting-gun demand at relatively low prices. Between 1870 and 1890, Belgian gunmakers successfully undercut and secured much of the trade for which Birmingham had become famous internationally, especially in the United States. The American market, along with the African trade, was Birmingham's biggest customer; its loss, confirmed by the McKinley Tariff of 1890, was irreplaceable. Over a quarter-century, Birmingham's exports to the United States declined in value from $1,169,214 (1882), to $348,839 (1890), $82,029 (1893), and $19,552 (1905); it represented a reduction from an 80 per cent share of the total value of America's arms imports to a mere 3 per cent.[121] Meanwhile, by the early 1890s, Liège-made guns and revolvers were being sold in Birmingham itself at a lower price than those turned out by the Birmingham factories. By the turn of the century, Britain was exporting a meagre annual average of 269 shot-guns and rifles to Belgium over the period 1901–5 (worth £2,311, or £8.59 per gun), whereas Belgium was exporting the annual average equivalent of 24,411 weapons to Britain (worth £30,653, or £1.26 per gun), in addition to barrels and other semi-finished parts.[122] By 1913, Britain was importing £16,954 worth of Belgian shot-guns, carbines, and rifles.[123]

Compared to its British counterpart, the small arms trade of Belgium was making an arguably more successful transition to the mixed industry that balanced the best of both manual- and machine-operated worlds. At the mid-point of the nineteenth century, few of Belgium's 160 existing firms had attempted to produce 'high-class weapons'; and many were content at that stage to make 'muskets, shot-guns, and rifles of the most ordinary type'.[124] Yet Artifex and Opifex were utterly convinced that 'English, German, French, and American ingenuity' plus foreign capital were being utilized to such an extent that Liège tended 'more and more to become the manufacturing annexe, not only of free trade England, but of protective Germany and France'.[125] Clearly, the patronage of the Belgian state and the promotional work of Belgian consuls abroad had something to do with it; the British commissioners were already observing, following the end of the Crimean War in 1856, that 'thanks to the efforts of their Government, the Belgian gun manufacturers made great progress and would become formidable rivals in all foreign markets'.[126]

Writing 50 years on, in 1906, Britain's Consul-General in Belgium Sir Cecil Hertslet would confirm much the same thing in comments about the highly specialized nature of Belgium's small arms industry. Even at the turn of the century, Liège continued to operate with undiminished vigour on the traditional craft basis of 'great division of labour'.[127] In many instances, however, the worker was supplemented—not superseded—by machinery, the two systems operating in conjunction with each other.[128] Recognizing that manual labour could not be displaced, a union of manufacturers had established the Liège School of Armoury for the instruction of young men seeking to take up the trade. Whereas the Birmingham School of Gunmaking had failed to attract sufficient pupils for want of favourable prospects in the declining industry, the Liège School drew many applications for every vacancy, expanding its training capacity in due course.[129] At the same time, the introduction of mechanical means of production enabled manufacturers to accept and execute orders which, under the old system, they would have had to refuse as a result of inability to deliver the products within the specified time. Under the new system, Liège gunmakers—already renowned for the diversity and originality of their products—supplied an even greater demand for firearms of various descriptions and models. As a foreign diplomatic observer, Hertslet could see how the Belgian industry owed a large part of its success to this facility of its manufacturers to vary continually the type of production, and to offer trade-guns 'thoroughly up-to-date', responding to every specification.[130]

A brief statistical survey comparing the production, exportation, and importation figures of several states in the Western hemisphere will confirm a relative decline in the overall performance of Britain's small arms industry between the 1850s and 1914. The official figures for the decade 1855–64 set the annual average value of British small arms exports at £504,205 (out of a total production value of £1,044,399) with the biggest rival—Belgian small arms exports—attaining £551,095 (out of an estimated £821,071). By 1900, however, the declared value of British small arms exports (£193,838) was far below Belgium's £719,518. It was clear that Britain had failed to secure any share of the additional demand for small arms. Even considering the fact that countless guns were exported from Britain without being entered at the Customs House as guns, Artifex and Opifex estimated that Britain's share of over 50 per cent of the world's trade in firearms in 1870 had shrunk to less than one-tenth of the remaining foreign trade, excluding foreign arms produced for domestic consumption. By the turn of the century, Belgium had about 65 per cent of the foreign demand for firearms; the United States nearly 11 per cent; Germany 6 per cent; France 3 per cent; Austria less than 2 per cent; Spain and Italy less than 1 per cent each; and other manufacturing countries producing 'but very little more than needed for their own home markets'.[131]

Furthermore, while there was evidence of lessening output from the British small arms industry, there was increased output at other metropolitan centres of production. For the decade 1865–74, Birmingham proof figures reached their highest annual average of 752,263, compared to Liège's 744,718. By 1895–1904, the annual average number of Birmingham proofs had fallen steadily to 373,877, while Liège proofs were averaging an astonishing 2,270,279 per year.[132] The situation was summed up by one representative of the British gun trade in 1904: 'England thus has the name and Belgium the turnover.'[133] And while British exports of small arms were declining in number and value, British imports were increasing. Whereas the numerical strength of foreign small arms industries was rising generally, the number of British small arms manufacturers and manufacturing establishments was on a downward trend between 1870 and 1900. What increased was the number of British retailers and merchants dealing in foreign weapons. Foreign-made barrels, such as those supplied by Belgium, were sent over to Britain 'in the rough' and given preference by British manufacturers owing to their greater uniformity of quality than British barrels. The subsequent call for marks of identification to be placed on foreign-made guns and gun parts—ostensibly to safeguard the reputation of the British small arms trade—was simply the plaintive cry of an industry in retreat.[134] By the 1900s, certain branches of the old gun trade had all but withered away, and the industry had become far less prominent or profitable than ever before.[135]

Yet, in the last analysis, our study of the metropolitan dimension of Britain's small arms trade does not leave us at a dead end. When it is set within the expansive drama of British industrial history as a whole, it aids our understanding of the competitive decline afflicting many other British industries in this era, and engages with wider debates about the nature of Britain's economic 'retardation' between the mid-1870s and 1914. At least until the 1850s, the largely Birmingham-based small arms trade had been a successful workshop industry with strong proto-industrial connections and a remarkable talent for industrious enterprise. After the mid-nineteenth century, however, the trade was transformed progressively by the second Industrial Revolution and its corollary of sweeping technological and organizational changes, which yielded mixed results for the industry. After the 1870s, in particular, the British small arms industry seemed to be affected significantly by comparative technological and organizational lag, especially since industrialization had by this time proceeded more vigorously in the United States and Germany—featuring more thorough mechanization and automation, specialization of labour and output, and standardization of products with interchangeable components—against the well-established craft traditions of gun production in Britain. Conversely, in comparison with the small arms trade of Liège, the Birmingham gun trade appeared relatively less successful in making the transition to the mixed industry that could reap the benefits of both craft- and machine-based production. This general perception of the motors of progress being held back by the weight of tradition and retrograde tendencies was largely true of British industry as a whole.[136]

While there were outstanding exceptions to the rule, the British small arms industry was after the mid-nineteenth increasingly pervaded by a sense of cultural conservatism and institutional rigidity that characterized both the gunmakers and the workers. This profile of a broadly tradition-bound industry blends into an even bigger picture, incorporating the self-confident yet complacent 'gentlemanly' capitalist elites of late Victorian and Edwardian society, and a larger working class cautious in its response to the march of scientific and technological progress.[137] Having been the world's first industrial giant, here was institutional 'ageing' writ large:

> Britain's distinctiveness derived less from the conservatism of its cultural values per se than from a matrix of rigid institutional struc-

tures that reinforced these values and obstructed individualistic as well as collective efforts at economic renovation. ...

[Unlike the United States, Germany, and later, Japan] Britain was impeded from adopting these modern technological and organizational innovations by the institutional legacy associated with the atomistic, nineteenth-century organization.[138]

These institutional rigidities affected the small arms industry in ways similar to industries such as cotton or iron and steel. Being hugely dependent on international export markets, the small arms industry—like the cotton industry—proved vulnerable to the invasion of open markets by 'cheap labour' products and import substitution strategies in less technologically-advanced countries. Being heavily reliant on traditional technologies, it ultimately possessed neither the incentive nor the ability to undertake internal reorganization. Like the iron and steel industry, the small arms industry remained at the mercy of Britain's free trade policies, which left it exposed to adverse competitive dynamics, leading ultimately to competitive decline.[139]

Viewed in such light, the gun trade becomes another instance of British deceleration in the decades of the late Victorian 'Great Depression'. This was the period of British social and economic history characterized from the mid-1870s by general economic malaise, pronounced slumps, the decline of specific industries, and depression in particular regions (especially when unaccompanied by commensurate shifts of labour to new and advanced industries); a period that witnessed both the emergence of new imperial and economic competitors, and the tightening grip of trade protectionism (and later, arms control measures) in foreign markets.[140] Addressing the assembled gunmakers of Birmingham in 1894, their chairman Charles Playfair would sound a warning that summed up the twin challenges of the new technology and foreign competition: 'Modern guns, unfortunately for us at present, are being so largely turned out by machinery, and of such excellent quality and workmanship, that those gifts which gave our workmen the supremacy will be more and more at a discount, unless applied in new directions.'[141] The messenger, Cassandra-like, was doomed not to be heeded by the representatives of the private trade or the domestic state until it was too late. Against the cumulative burdens of cyclical depression and competitive decline in the metropolis, Joseph Chamberlain's picturesque metaphor for Britain's overall sense of firmamental exhaustion would seem especially apt at the turn of the century: 'The weary Titan staggers under the too vast orb of its fate.'[142]

One shining exception to 'weary Titan' syndrome was, however, the superior competitive performance of most other sectors of the arms industry—explosives, heavy guns, warships—which did not suffer this decline. The most significant variable is almost certainly the different impact of the British state on the different sectors, and the divergent market conditions which this created. M. J. Bastable goes so far as to claim that, by 1914, the business of Armstrong's armaments industry at Elswick and the business of the British state had become 'virtually indistinguishable'.[143] Yet, as we have noted previously, we would do well to look further afield for answers: first, these heavier armaments and explosives industries enjoyed the steadier, cosmopolitan custom of foreign governments within and outside Europe; and, second, the immediate context of Chamberlain's words was, in fact, the *Colonial* Conference of 1902. Against the vicissitudes of the armament cycle, British warship constructors and the makers of large weaponry—such as Vickers, Armstrong, John Brown, and Coventry Ordnance, or the explosives manufacturers of the Anglo-German Nobel Dynamite Trust—remained world leaders, fully able to compete with Krupp, Schneider-Creusot, Skoda, or Dupont. But their markets were very different; the British 'armament kings' could draw upon Britain's needs as a primarily maritime power, and they were kept up to the competitive mark by the demanding specifications of foreign governments that were pursuing their share of maritime power.[144]

The metropolitan dimension of the small arms industry, with its fluctuating cycles of production, must be considered alongside those markets at the periphery that functioned as a lifeline or dynamo in otherwise lean times. In the case of Britain, Birmingham's export trade was gradually built up over the long nineteenth century to extend to all parts of the world:

> [T]he savage in Africa exchanged his gold dust, his ivory, and his spices for Birmingham muskets. The Boer of the Cape shot elephants with a gun expressly made for that purpose by the Birmingham manufacturers. The Army, Navy, and the East India Company's services all drew from Birmingham their principal supplies of the weapons of destruction ... the sword, the pistol, and the musket. The riflemen of the backwoods of Canada and the Hudson's Bay Territory would have been deprived for a while of the means of trade or sport if the Birmingham trade had ceased to fabricate gun-barrels and locks; and all the tribe of sportsmen, whether frequenting the

jungle, the moor, the mountain, or the lake, carried on their recreations by the aid of Birmingham.[145]

Similarly, even with its comparative success at interchangeable-parts manufacture in metropolitan circles, Enfield's production of small arms after the mid-nineteenth century came to rely increasingly on the deployment of military rifles in foreign fields and their distribution for sale in overseas markets. The earlier Snider, Martini-Henry, and Lee-Metford rifles were approved for production by Enfield; and they were successfully re-adapted and released to both domestic and international markets as 'original' Snider-Enfield, Martini-Enfield, and Lee-Enfield rifles. The Lee-Enfield, in particular, came to be adopted as the British Army's standard bolt-action, magazine-fed, repeating rifle from 1895 onwards. In various incarnations, it was used by Britain's colonies and continued to see official service in a number of British Commonwealth countries even beyond the twentieth century. Total production of all Lee-Enfields has been estimated at over 17 million rifles, making it the one of the most numerous military rifles ever manufactured, second only to the Russian Mosin-Nagant, which was a contemporary firearm introduced in Czarist Russia and later deployed throughout the Soviet Empire. That the Lee-Enfield family of rifles remains the oldest rifle design in official service attests to the durability of the original Lee-Enfield design, as well as the reach of the British Empire.[146]

The international and imperial aspects of Britain's small arms trade were, however, riddled with ambiguities of their own. The increasingly vexed issue of arms transfers to the periphery would provoke new questions about production and procurement, and engender new problems related to proliferation and prohibition. There is little doubt that the new weaponry would continue to prove instrumental in the establishment and enforcement of British colonial authority. But from the mid-nineteenth century, the new technology—with its greater potential for destruction—would raise new hopes and fears in the wider colonial sphere. Between the 1870s and 1914, the French, the Germans, and the Belgians were also expanding their colonial interests across large sections of Africa and Asia. Whether in the Western metropole or the wider periphery, the acquisition of the new technology was now augmented by the acceleration of this 'new imperialism'. As the scramble for colonies intensified, the serious commercial competition from Britain's continental rivals also gained momentum. The Belgian arms industry, centred upon Liège, became a bugbear to the British small arms industry in ever-increasing circles. Not only in European and

American markets, but also in colonial markets and regional arms bazaars across Africa and Asia, Birmingham and London found themselves being upstaged by Liège in terms of turnover and cheapness of production, if not in reputation.[147]

In the emerging metropolitan debates about military industrial policy and associated arms transfers, it was the global dimension of empire that polarized views on the small arms industry. From the 1870s, the 'dumping' of obsolete or surplus munitions by the British Government—as 'old stores' at colonial outposts—became increasingly commonplace. There were even official advertisements appearing in various provincial newspapers offering large lots of ex-service rifles for sale at specified low prices in municipal centres such as Sheffield, Manchester, Liverpool, and Glasgow. Totally indignant, *The Ironmonger* declared in 1882 that 'the system of selling rifles indiscriminately has been a direct incitement to their acquisition'.[148] One Birmingham correspondent found sufficient grounds to launch a stinging diatribe against the Government:

> The recent disaster to the British arms in South Africa has given fresh and painful emphasis to the complaint, more than once expressed, of the conduct of the Government in the disposal of what are termed obsolete small-arms. ...

> Government, instead of befriending the private gunmakers, upon whose assistance it must rely in time of war, was constantly engaged in undermining their trade by the wholesale flooding of the market with surplus or obsolete guns. These are straightway exported by the dealers to the principal foreign and colonial markets, in some of which there can be no doubt they are used for the arming of native tribes, with a view to their employment against British troops. If a fair price were realized for these surplus stocks, and any appreciable benefit accrued to the nation for their sale, there would be less room for complaint; but the guns are mostly sold at the price of firewood, and more money is believed to be spent on freight and packing than the goods realize. ...

> [T]he recent sale of some 120,000 rifles at Weedon fetched, in some instances, only 1s. 4d. to 1s. 6d. each, for guns which were never made under 50s., and which, for all practical purposes, were as good as the day they were made; and ... there was reason to believe that Government had spent as much as 2s. per gun in preparing them for the market. ... The estimate of 2s. per gun purports to be no

more than an estimate, though it is made by an experienced gun-maker, who is himself a large purchaser of these surplus stocks, ... [and] has had the assurance of officials at India House that his accepted tender of 1s. 6d. per gun did not cover the cost of preparing the arms for market. ...

The economical objection to these sacrificial sales, however, is not only or the most serious one. They constitute a danger to the nation, owing to the facilities they afford to unscrupulous traders for arming against us the aborigines of Africa, India, and the Fiji Islands. There can be no doubt that the arming of Cetewayo's 40,000 warriors was largely facilitated by the selling at nominal prices thousands of efficient small-arms—rifles as well as smooth-bore muskets—and that the recent calamity at Rorke's Drift, therefore, is traceable in a measure to this penny-wise pound foolish policy. It is no secret that in London some gentlemen of the Hebrew persuasion have been doing a brisk business for some time past in the export of obsolete Government guns to the African coast, as well as to the Fiji Islands, and no surprise therefore need be felt at the futility of the restrictions imposed upon the import of guns into Cape Colony to prevent the arming of the Zulus and other native tribes. If Government would be at the trouble to break up the guns before selling them, they would realize as much, if not more, for the locks and barrels alone than they now get for the entire guns, and they would place an effectual obstacle in the way of the shipment of such dangerous commodities to foreign parts.[149]

Nevertheless, the private trade could not always expect to occupy the high moral ground. From the government perspective, it was obvious that munitions shipments handled by private traders often went surreptitiously under the cover of 'hardware' labels. Such ingenuity would explain why so much contraband escaped detection by official agencies, slipping through the net of Customs officers and Board of Trade records on innumerable occasions, from ports such as Liverpool, Bristol, and London.[150]

By the turn of the century, the metropolitan debate on the British small arms trade hinged upon diametrically opposed interests of imperial security and imperial political economy. 'Herein lies the paradox of empire,' declared one British official, 'Government maintains warships in the Gulf to preserve peace, and the Birmingham manufacturers and London exporters ship out whole cargoes of rifles and cartridges which can only destroy peace.'[151] On the other side of the debate were sounded the alarms of Artifex and Opifex:

'It is evident that the industry is decaying at Birmingham, and as Birmingham is the only producing centre in the British Empire, a total population of 400,000,000 British subjects should depend upon the Birmingham industry, not only for their sporting firearms, but may have to rely upon this centre for all the weapons of war needed to defend the Empire. Thus the importance of the industry is not easily overestimated, and its decay is a matter of national concern.'[152]

For all the high controversy surrounding the funding and organization of arms transfers from across the production centres of the Western metropole, it remains vitally important to consider the trade's various components through the kaleidoscope of interactions at the periphery. Many indigenous societies in receipt (or on the receiving end) of Western firearms belonged to turbulent regions of the 'non-West', with altogether more volatile political traditions and military cultures. The dynamics of the arms trade would be shaped by indigenous traditions of power-broking, profit, and plunder, along with indigenous modes of warfare, military organization, and weapons manufacture. In that broader international context of frontier encounters and exchanges, the trade of Birmingham and other arms-producing metropolitan centres would proliferate as never before. Ultimately, it would be impossible to comprehend the complexities of a global arms transfer system, without careful examination of how the small arms traffic from Europe intersected the trade and politics of indigenous societies across monsoon Africa and Asia.

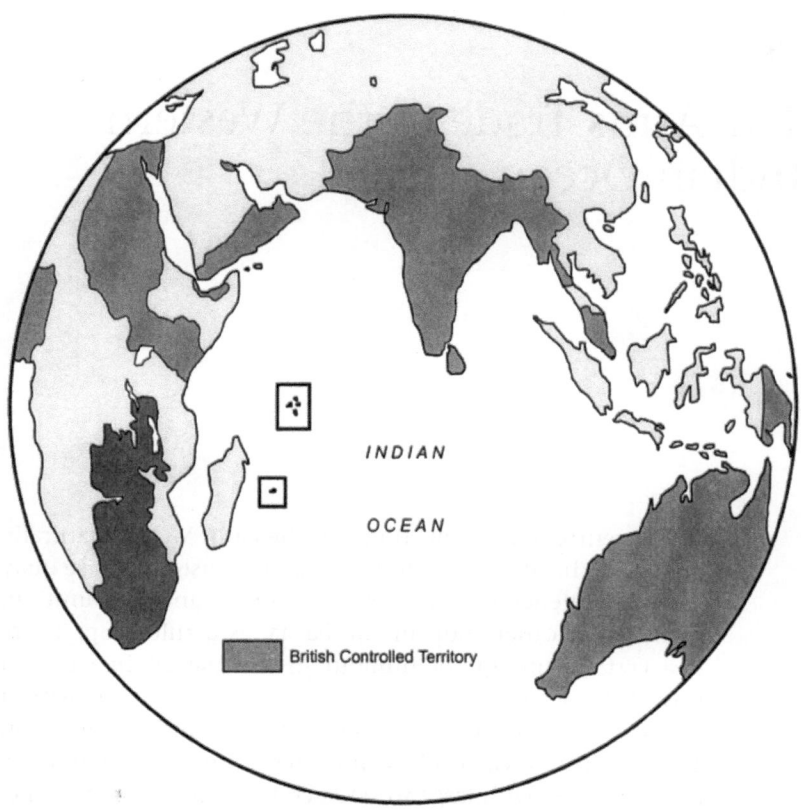

The Indian Ocean as a 'British Lake' during the Nineteenth Century

3
The Arms Trade in the Western Indian Ocean

As the centre of the arms traffic in the Gulf, Muscat naturally bristles with rifle depots and stores. The Custom House quay is seldom unencumbered with cases of rifles and ammunition, while every other shop in the bazaar is a rifle-shop. There is a certain amount of humour in a situation in which a British cruiser is actually at anchor in the harbour, with a dhow loaded to the water-line with rifles and ammunition almost within a cable's length of her, a Custom House quay fairly crowded with large wooden crates containing the same, and a bazaar simply bursting with arms and ammunition.

Arnold Keppel, 1911[1]

By the closing decades of the long nineteenth century, the port-cities of Muscat and Zanzibar had achieved notoriety as principal gateways for arms traffic in the western Indian Ocean. Even as munitions made in Europe inundated their harbours and warehouses, there were European navies mounting surveillance and interdiction operations in nearby tropical waters. By 1880, it was estimated that firearms comprised more than one-third of Zanzibar's total imports.[2] Breech-loaders had begun to flow into East Africa within a year of their being superseded by repeaters in Europe, so that as early as 1886, numerous modern rifles were found in the arsenal of a Chagga chief.[3] In 1889, when the Welsh journalist and explorer Henry Morton Stanley 'rescued' Emin Pasha in the Lake Albert area, he found a section of the latter's native troops equipped with breech-loaders.[4]

The nature of the problem facing colonial authorities at the periphery was elucidated in a consular report of 1888, from Colonel Charles Euan-Smith, the British Consul-General at Zanzibar, to the Earl of Rosebery, a leading Liberal Imperialist who was the first to conceive of the Empire as a 'Commonwealth of Nations':

> The great question is that regarding the import of arms and ammunition into East Africa. This trade had now assumed proportions of which your Lordship may possibly be unaware. Formerly the arms so imported were cheap and worthless weapons manufactured to last for a maximum period of some two or three years and after that time becoming useless and worn out. Now, however, arms of precision and breech-loading rifles and ammunition are being imported in very large quantities and are rapidly taking the place of the flintlock and muzzle-loading cheap muskets. ...

> Unless some steps are taken to check this immense import of arms into East Africa, the development and pacification of this great continent will have to be carried out in the face of an enormous population, the majority of whom will probably be armed with first-class breech-loading rifles.[5]

According to Euan-Smith's calculations, between 80,000 and 100,000 firearms were entering Africa each year via the East African ports; and the returns of the Zanzibar Custom House tend to corroborate this figure.[6]

Euan-Smith's alarm in the official sphere echoed the anxieties of Alexander Murdoch Mackay, a Scottish Presbyterian missionary from the Church Missionary Society (CMS), stationed much further inland in the East Central African kingdom of Buganda. In a private letter to Euan-Smith, Mackay had highlightedthe arrival of a notorious gun-runner named Charles Henry Stokes. The gun-runner's consignment of 100 breech-loaders had included some of the latest Winchester repeaters, introduced to Europe from the United States only the previous decade, plus 20,000 rounds of ammunition. Mackay remarked how 'short-sighted' it was 'to place these arms into the hands of a vain young princelet' such as Buganda's 20-year old Kabaka, Mwanga II:

> [T]hese arms will be used in raids for women and slaves. The fact that almost all the European Powers being at present about to adopt magazine rifles, will not be without its effect on East Africa. Discarded Martini-Henry, Mauser, Gras and other breech-loaders will

now be poured into the Zanzibar market and unless prompt measures are taken to the contrary these will soon be in the hands of all the tribes in the interior.[7]

Already responsible for the martyrdom of James Hannington, the newly ordained Anglican Bishop of Eastern Equatorial Africa, along with 45 royal pages who were Christian converts, Mwanga's ruthless ambition and deployment of the new weaponry to resist Western 'civilizing' influences would eventually seal his fate as independent Buganda's last monarch.

For two decades after the Berlin Conference of 1884, approximately one million firearms, over four million pounds (lbs) of gunpowder, and many million percussion caps and rounds of ammunition deluged the British and German spheres in East Africa. These munitions exacerbated conflicts among local communities, and empowered anti-colonial protest and small wars, thus necessitating the Anglo-German and Italian coastal blockade from the Tana to the Ruvuma (1888–89) and the Brussels Conference Act of 1890. In addition to checking the slave trade, the Brussels Act attempted by international agreement to curb small arms proliferation throughout tropical Africa.[8] Yet it is hardly surprising that the world's first major attempt at international arms control only managed, ultimately, to alter the direction and distribution of arms traffic.

When the stipulations of the Act began to be enforced by naval patrols, the torrents of contraband were diverted from Zanzibar to Muscat, a trading entrepôt situated outside the zone of prohibition. From Muscat, the munitions were redistributed with relative ease to other destinations around the region. Arms imports at Muscat were assuming formidable proportions by the mid-1890s, and rising sharply: in 1895, 4,350 rifles and 604,600 cartridges; and, in 1896, 20,000 rifles with 'a proportionate number of cartridges'.[9] By the turn of the century, the Government of India was complaining that small arms and ammunition were again 'filtering into the hands of the hill tribes round the Tirah Valley', despite all the seizures, searches, and surveillance to pin-point possible channels of supply. Embarrassingly enough, most of the weapons were of British manufacture and a source of 'considerable trouble' in the region.[10] In 1900 alone, an estimated 25,000 rifles and 2.25 million rounds of ammunition entered the Persian Gulf.[11] In 1906, 45,000 rifles and a million rounds of ammunition were imported and then re-exported via Muscat. In 1908, 80,000 rifles passed through Muscat and the arms trade amounted to nearly 43 per cent of the port-city's total imports. Of this, the Indian Chief of Staff estimated that some 30,000 'Muscat Martinis', along with other small arms, were reaching the Northwest Frontier and Afghanistan each year. Nothing illustrated the gravity of the situation more than the price of rifles on the Frontier, which had fallen 50 per cent over the course of a year.[12] All of a sudden, nightmare visions of a hundred thousand tribal warriors—fiercely independent, formidably armed—appeared neither imponderable nor improbable. Sooner or later, a fearsome indigenous resistance significantly enhanced by the new weaponry could confront British colonial authority in the rugged frontier terrain. It was a concern that multiplied in direct proportion to the increase of the trade.

Yet the coalescence of arms trading interests in the western Indian Ocean over the period 1880–1914, with its phenomenal statistics and frightening repercussions, actually had a far lengthier gestation. For at least a hundred years previously, changes in the nature of European imperialism as well as the interaction between indigenous state formation and tribal politics had escalated a spiral of profitability and plunder, more dehumanizing and violent than ever. The advent of more assertive forms of European capitalism and colonial authority provoked the outbreak of various forms of anti-colonial resistance and ethno-religious strife in local communities undergoing their own complex transformation. In the long run, this growing web of cross-cultural interactions would militarize geopolitical frontiers extending from East Africa through the Gulfs of Oman and Persia to India's Northwest Frontier. By drawing together the strands of regional commerce and indigenous crisis, these interweaving patterns of collaboration and conflict would supply the necessary preconditions for the emergence of an international arms bazaar. For the first time, policy-makers and law-enforcers would find themselves grappling with the vexed issue of what could be done, on some multilateral basis, to eradicate small arms proliferation along clandestine routes while enhancing security across porous border areas.

European expansion into the Western zone

Centuries before those terrifying images of tribesmen armed and ready on the East African and Southwest Asian frontiers, Europeans had first rounded the Cape of Good Hope on tentative voyages of exploration. Generally, they had entered the Indian Ocean arena in hope of obtaining commercial profits, cultural prestige, and Christian converts through various forms of contact with indigenous societies. More specifically, they had wanted the 'little nut tree' memorialized in a traditional nursery rhyme—replete with its 'silver nutmeg' and 'golden pear'—signifying the sources of the spice trade from the fabled 'Indies'.[13] Every successive empire that the Europeans built in monsoon Asia, as Carl Trocki has suggested, were modified versions of that initial venture:

> The exact 'spice' varied over time, but there was always one exotic chemical or another that became the object of European trading monopolies—first it was nutmeg and cloves; later it was pepper, coffee, cacao, and sugar; still later it was tea and finally it was opium. In most cases, the aim was to monopolize the delivery of the desired substance to the European market.[14]

The trade balance between the Eastern and Western hemispheres was maintained for much of the early modern period by the reverse flow of precious metals from Europe to Asia; or, more accurately, from the Americas, via European agency, to Asia. Initially, there was little that Europe could actually offer in exchange for the exotic produce of monsoon Asia, apart from gold and guns, silver and saltpetre. Under the constraints of the prevailing mercantilist ethic, the supply of 'spices' could be either secured by force of arms, or procured through flows of arms in addition to hefty silver bullion.

During the early phases of European expansion, the agents of the trading companies vying for any such strategic or commercial advantage would help to pioneer the firearms trade in the western Indian Ocean. Back in the sixteenth century, the Portuguese impact had been felt through force of arms and the establishment of fortified trading bases they called 'factories' (*feitoria*). Portuguese factories were established at Aden and Hormuz, controlling the commercial traffic of the Red Sea and Persian Gulf; Goa, on the west coast of India; and Sofala, Mombasa, and Mozambique in coastal East Africa.[15] Portuguese soldiers carried muskets up the Zambezi, the southern frontier of East Central Africa, although they were generally reluctant arms traders.[16] The release of Spanish silver from the New World into the political economy of the Indian Ocean went further, paving the way for general crisis through the creation of wealth and power in the outlying provinces of the Safavid and Mughal Empires, largely at the expense of the centre. The growing volatility of that climate would, in turn, create markets for military equipment and all manner of strategic goods. From the early seventeenth century, the European presence tended to become more restrictive and destructive as the Dutch and English fought to monopolize and redirect trade. After the start of white settlement at the Cape in 1652, the arming of Dutch and English colonists contributed to the formation of a 'gun society' across the southwestern frontier of the Indian Ocean, a development that would cast long shadows and come to haunt the British in later years.[17]

But the era of sustained arms transfers in the region only really commenced with the phase of intensifying Anglo-French rivalry from the middle decades of the eighteenth century. This was a duel for commercial and strategic control of resources and routes, the products and productivity of native lands and peoples. Anglo-French expansion from maritime and coastal areas into the interior was characterized by an extension of the 'rule of law', with uniform notions of sovereignty and progress imposed upon societies distant from one another. From the 1780s, even more assertive forms of colonial authority and commercial activity were manifested: legitimated by more domineering ideologies of civilizing mission; reinforced by cross-cultural stereotyping; and evidenced by growing numbers of soldiers, settlers, merchants, missionaries, and administrators. Wherever these Europeans traded weapons for the produce of East Africa or South Asia, they armed native populations and militarized local communities. Even where they did not engage directly in arms trading, their activities so altered or disrupted the western Indian Ocean zone that they paved the way for larger arms transfers by the second half of the nineteenth century.

Commercial linkage between the supply of firearms and slaves was also established. French slaving interests had been mostly satisfied from the 1730s by supplies of human cargo from Madagascar and, intermittently, Mozambique and the East African coast. From the 1770s, however, the development of sugar plantations on the French Mascarene islands of Ile de France (Mauritius) and Bourbon (Réunion) gathered pace as an extension of similar developments in the West Indies that fed into the massive commercial nexus of the Atlantic gun-slave trade.[18] French influence at Zanzibar was exercised through commercial treaties and agreements with its Arab rulers concerning the trade in slaves and ivory, which were exchanged for muskets, gunpowder, and shot. A French agent took up residence at Muscat in 1807, as an extension of Napoleonic grand strategy, when it had become even more obvious that Oman's strategic position commanded the entrance to the Persian Gulf, the overland route between the Mediterranean and India, and the sea route from Suez to Bombay.[19] Growing French involvement in the region meant that France was able to secure a commercial treaty with Oman in 1844, expressly permitting the unrestricted import and export of all kinds of merchandise—including munitions—in the hands of French firms. These arrangements would eventually affect the stability of the region and trouble the British in the long term.[20]

Britain, for its part, was generally wary of direct involvement in the western Indian Ocean. Backed by the political power and military might of the nation-state, rivalries between the chartered trading companies of Britain and France had erupted into open warfare during the second half of the eighteenth century. The disintegration of the great Islamic land empires had exposed the struggle for political mastery in India and the Indian Ocean, thus also allowing the British and the French to contest the subcontinent between 1740 and 1815 as a 'globalized' extension of their European conflict. The East India Companies of both countries shipped massive quantities of weapons to indigenous allies. In the ensuing chaos, the English East India Company further acquired vast land revenues that enabled it to build and finance a huge native army, supplied from its own arsenals.[21] Having contended successfully for control of South Asia's military economy, Britain's consuming interest would remain the trade and territorial revenues of India, revolving around territorial annexation and military-fiscal consolidation on the subcontinent.[22]

Initially, it took the threat of Revolutionary and Napoleonic France, together with the ambitions of Governor-General Richard Wellesley and his protégé Captain John Malcolm, to give some tangible basis to British influence in the wider region. By emerging as the dominant European power in India, the British would come to regard the defence of their new imperial domain (and eventually its external arteries of trade) as a full-time preoccupation. The Persian Gulf and the Red Sea were two such arteries; and, for both of these, Muscat was a principal port-of-call. British involvement did not extend beyond engagements contracted with the states of Oman between 1798 and 1820, but arms and ammunition made in Birmingham or London were brokered as standard stock-in-trade in the bargaining process, proving useful for strengthening friendly native states against hostile powers.

Any reorientation of Britain's official strategy and policy toward the western Indian Ocean came only gradually. It was conceded that the Sultan of Zanzibar, as ruler of Oman, exercised considerable local influence on both the maritime and overland trade routes to India. Alongside pre-existing concerns about French interests in the Indian Ocean, experts in London and Calcutta grew perplexed over the possibility of a Russian invasion of India, progressing from Russian expansionism in Central Asia and Russian adventurism in the Ottoman Empire and Persia. A series of dramatic events, including the outbreak of the

Turco-Egyptian war (1831–48) in which Czarist Russia realigned itself with the Ottoman Empire, jolted Lord Palmerston at the Foreign Office into rethinking the status of Asia in the international order. With the politicians in the metropolis and the agents at the periphery tending to interpret events within the same broad speculative framework, it followed that the more dynamic would be tempted to exercise British power and influence over the lands and waterways of the Indian Ocean. Mounting threat perceptions made them willing to explore the whole range of policy options, from gun-running to gunboat diplomacy.[23]

Apart from direct hostilities in the Crimean War, the 'Great Game' was staged by Britain and Russia with varying degrees of intrigue and intensity between 1828 and 1907. It was a nineteenth-century cold war in which the two great powers sought to counter each other's expansion through subversion, coercion, and other means of informal influence—including the supply of arms to friendly native powers. Anxious to eradicate alien influences from the lines of communication to India, Britain endeavoured not only to preserve Afghanistan and the Northwest Frontier region as a buffer zone between the Russian Empire and British India, but also to maintain paramount authority over the Persian Gulf and the coastal areas flanking it to the north, the south, and the west.[24] British representatives were stationed permanently at Zanzibar or Muscat after 1830. Following Britain's implementation of the Trucial System (1835–53) and intervention in the Oman-Zanzibar succession crisis with the Canning Award (1861), the three decades from 1861 to 1891 witnessed a progressive extension of British influence over Oman and the Trucial Coast. By the terms of the Award, India henceforth designated the respective new rulers of Oman and Zanzibar, setting a precedent that required British recognition of the successive rulers of Oman in return for the payment of British subsidies. With the pacification of the more quarrelsome Gulf sheikhdoms and the opening of the Suez Canal in 1869, Britons saw the Indian Ocean increasingly as a 'British Lake' connecting British interests in Asia and Africa—a strategic theatre that had to be secured at all costs against old rivals and new competitors.[25]

The advent of a 'new imperialism' in international affairs from the 1870s would mark the start of the most strident phase of European expansion. It would also lead to widespread small arms proliferation throughout the frontiers of the western Indian Ocean. Formidable indigenous armies and arms races presented serious resistance to the establishment of British rule on the Indian subcontinent, but even those would be dwarfed by the sheer magnitude of arms traffic in East Africa and South Asia's wider periphery between 1870 and 1914. In that high imperial renaissance, British statesmen irrespective of political ideology—from conservatives such as

Benjamin Disraeli and the Marquess of Salisbury to liberals like William Ewart Gladstone and the Earl of Rosebery—all had a hand in shaping Britain's imperial destiny, as did financiers like Sir William Mackinnon, founder of the Imperial British East Africa Company. As 'Iron Chancellor' of an ascendant Prussia-led Germany, Otto von Bismarck projected the *realpolitik* of Europe's balance of power into the global arena, but his successors Leo von Caprivi and Bernhard von Bülow charted a 'new course' that led to a more aggressive *Weltpolitik*, advancing the ambitions of Kaiser Wilhelm II by pursuing the Reich's 'place in the sun'. Leopold II, King of the Belgians, promoted his own private colonial project—the Congo Free State—in the heart of the Dark Continent. New French republican governments joined the fray alongside their Italian rivals, each pursuing an unashamedly nationalistic colonial agenda. In the rising crescendo of profit and violence, arms transfers from Europe to East Africa, the Gulf region, Afghanistan, and India's Northwest Frontier—largely unregulated till the late nineteenth century—coalesced into one of the most pressing problems facing European statesmen, administrators, missionaries, explorers, and traders.

Great power rivalry at the Berlin Conference of 1884 was prelude to a mad scramble for Africa and a race for empire in the western Indian Ocean. Initially, the statesmen of Europe and America had met in Berlin to smooth the path to an orderly partition of Africa. Unfortunately, guided by a combination of profit-seeking and colonial strategy, and then a quest for stable revenues and frontiers underpinned by law and paramount authority, every state pursued its own ambitions in the oft-tangled realities of African geopolitics and society. Between 1894 and 1914, the arrival of Germany, Italy, Portugal, and the United States in Gulf affairs, alongside France, Russia, and the Ottoman Empire, was altogether viewed by Britain as an unwelcome intrusion into its private lake, for each new wave of imperial competition was calculated to stir the waters and generate wider 'ripple effects'.

One notable manifestation of the growth of British interests and involvement in the western Indian Ocean zone was the expanding British diplomatic presence. Partly to contain (if not counteract) the competitors, and partly to extend the official 'gaze' of the colonial state over issues of mounting concern, such as the unrestricted trade in arms and ammunition, formal British Protectorates were proclaimed over Zanzibar in 1890; Uganda (incorporating Buganda, Bunyoro, Toro, Ankole, and Busoga) in 1894–96; and the remaining East African territory between Uganda and Mombasa in 1895.[27] In the early 1890s, the principal British officer in the Gulf was the Political Resident in the Persian Gulf and Consul-General for Fars and Khuzistan. In 1894, he was assisted by just one subordinate officer, stationed at Muscat. By the outbreak of the First World War, British officers were positioned at Bahrein, Kuwait, and Bandar Abbas as well, with two additional assistants assigned to the Resident at Bushire. Another Resident was posted in Baghdad, assisted by a subordinate Consul in Basra.[28]

By the 1890s, this expanding geopolitical and strategic involvement was not confined to Britain. France, an ally of Britain's Russian opponent, posed a direct challenge to British influence by stationing a Consul at Oman in 1894. Germany appointed a Consul to Bushire in 1897. Russia followed suit with a consul-generalship there in 1901, with subordinate consulates at Mohammera, Shiraz, and Bandar Abbas.[29] 'The essential problem was simple,' as B. C. Busch would argue, 'the "Lake" was no lake at all, but an international waterway of steadily increasing importance in an age of imperial rivalries, diplomatic flux, and sizable dangers to international peace of mind in the cycles of decay and revolutionary activity in the Ottoman and Persian states.'[30]

At this section of the periphery, the competing claims of the European great powers would culminate in a series of distinct but overlapping crises. The dynamics of the ensuing confrontation would be shaped more by the subjects and institutions of European governance at the periphery—soldier-administrators, tax-collectors and law-enforcement agencies—than the policy-makers in the metropolis, whom they mirrored. In the case of Britain, colonial competition and crisis often highlighted the differing attitudes of the metropole and the Raj toward the periphery. From India's perspective, issues as problematic as the arms trade further exposed the vulnerability of British interests in the western Indian Ocean to meddlesome foreign powers like France, Russia, or Germany. International crises triggered by incidents related to arms trafficking would emphasize the dangers of having a conflicting foreign policy, pulled in all sorts of directions by the complex hierarchy of control involving Whitehall government ministries (the War Office, the Foreign Office, the Colonial Office, the India Office, and the Admiralty) as well as the overextended administrative apparatus of the Government of India, pretentious agents-on-the-spot, along with aggressive armies and navies.

In this new global epoch of cross-cultural enterprise and mission, the importance of 'personality' and 'caste' in the top officials was crucial, especially where the men on the spot were concerned. Many of them, gentlemanly capitalists or high imperialists of the pre-1914 order, were motivated by a peculiar kind of worldview. In the words of one soldier-administrator: 'Before the Great War my generation served men who believed in the righteousness of the vocation to which they were called, and we shared their belief. They were the priests, and we the acolytes, of a cult—*Pax Britannica*—for which we worked happily and, if need be, died gladly. Curzon, at his best, was our spokesman and Kipling, at his noblest, our inspiration.'[31] Driven by these expansive ideas and neo-religious convictions, the interventions of the colonial state in the western Indian Ocean zone seemed destined to create a tangle of colonial intrigues, spark off cycles of indigenous warfare and resistance, and, in so doing, galvanize the arms trade. The established colonial regimes and the chartered trading companies from Europe, with their agents and armed forces, would strive to execute the aims of far-off policy-makers; the aggressive, provocative secondary imperialism that often ensued—'sub-imperialism'—was an amalgamation of metropolitan policy, colonial interests, personal ambitions, and local adventurism.

From this angle of sub-imperialism, the vastly expanded small arms supply in East Africa from the late 1880s was not simply due to the unprecedented numbers of obsolete weapons thrown suddenly on the market by the rearming of Europe. It was also the result of the changing impact of colonial authority upon the region; and, more specifically, the northward shift of a trade that had been conducted for over a decade in southern Africa. Arms transfers had been initiated by Portuguese agents who sold muskets to the Zulus and various tribal communities in southern Africa; later Boer demand for arms against the British provided a good market. Notwithstanding the customs regulations, registration and stamping, thousands of guns and hundreds of barrels of gunpowder had made their way into the interior, chiefly via the Delagoa Bay route over which the British had no control.[32] Where the old 'Tower' muskets had been for years the regular native wage in the diamond fields, escalating arms transfers would become a significant flashpoint between the British and the Boers in the Transvaal. Vigorously opposed to British laxity in allowing the arming of the indigenous population, the Boers had armed themselves so as to take matters into their own hands. In 1873, a group of Boers had ambushed a large party of natives returning from the diamond fields with guns they had earned there through hard labour. The Boers shot down the entire party in cold blood, upon their refusal to surrender these weapons.[33] By the 1880s, however, the market for gun-running in South Africa was reduced. The Transvaal had been handed back to the Boers by Gladstone's Liberal Government. The British South Africa Company was beginning to be active further in the interior, and a settled administration made gun-running more difficult, tending instead to divert arms traffic from the Delagoa Bay route northwards to virgin markets in East Africa.[34] Thus, by the time East Africa was divided into British and German spheres of influence under the Agreement of 1886, rival Imperial British and German East Africa Companies faced an indiscriminate arms trade that was fast becoming a security conundrum.

Within the German sphere, the German East Africa Company (*Deutsch-Ostafrikanische Gesellschaft*) found itself under considerable pressure. Not only did the Company have to please its metropolitan masters and pursue its sub-imperialistic programme to maximize profits, it also had to pacify its newly acquired native 'charges' while simultaneously placating its colonial competitors. In 1888, with the outbreak of a major Arab-Swahili rebellion arising from clear German provocation, the Company came under fire in the Reichstag and in the German newspapers for its ineptitude in handling the Reich's East African affairs. Back in Berlin, there was even speculation that the Kaiser's brother, Prince Henry, was to command a naval expedition to pacify the coast.[35] The designs of metropolitan policy-makers were, however, rapidly overtaken by events on the ground, with an increasing likelihood that other Europeans would be drawn into the dynamics of armed confrontation at the unstable colonial frontier.

Needless to say, the British were among the first to be affected by developments in the German sphere. The Arab insurgents were subjects of the Sultan of Zanzibar, who was virtually under British protection, and anti-colonial sentiment threatened to spread to the British Company's territory. Considering this danger, Euan-Smith, the British Consul-General at Zanzibar, had more than once remonstrated how important it was to curtail the arms trade, and he discussed the feasibility of implementing countermeasures with the Director of the German East Africa Company. Further discussion, involving the Italians who had recently acquired an interest in the Somali coast, resulted in the imposition of a tripartite blockade (commencing 30 November 1888) on the import of munitions along the east coast of Africa, from the Tana to the Ruvuma.

As an exercise in damage limitation, the blockade was to prove ineffective. The stretch of coastline covered by the blockade scarcely extended far enough; it merely re-diverted arms traffic southwards to the Delagoa Bay route, so that by early 1889, massive quantities of firearms and gunpowder were reported as passing through Ibo and up the River Ruvuma. Furthermore, the naval authorities were hampered by their limited right of search, with only British, German, and Italian flags being subject to it. The Germans found the blockade costly, in terms of the deteriorating health of the German naval crew, as well as the ivory that was sacrificed in consequence of being diverted either westward to the Congo route, or northward to British territory, where the Imperial British East Africa Company (IBEAC) contemplated sending agents inland to guide ivory caravans to Mombasa. The blockade proved disruptive to local shipping, commerce, and relations with the Sultan of Zanzibar, who suffered a loss of revenue. It also temporarily disrupted the expeditionary plans of explorers such as Dr Carl Peters and Count Sámuel Teleki. Missionaries argued that it jeopardized the survival of civilized tribes along the coast, who were denied defensive weapons and left at the mercy of marauding tribes from the interior, such as the Masai or the Makua.[36] Even among British colonial officials, there was little sympathy for the blockade. Sir John Kirk, the influential former British Consul at Zanzibar, had condemned it from the start as impractical and expensive. British participation had apparently stemmed from a policy of cooperation with (or at least, non-opposition to) German colonial aspirations.

When the blockade was lifted on 1 October 1889, various arms control regulations were instituted in the British and German spheres.[37] In the German sphere, a decree was issued in January 1890, stipulating that all guns must be stamped or confiscated. Breech-loaders were banned and had to be surrendered in exchange for their equivalent in money or muzzle-loaders; and no new breech-loaders could be imported.[38] Under British pressure, the Sultan of Zanzibar had agreed in February 1889 to a regulation prohibiting the unlicensed importation of arms and ammunition into Zanzibar and Pemba.[39] On the mainland, the British and the German Companies both permitted caravans to carry only up to 100 lbs of gunpowder per 100 men, while breech-loaders were to remain in the hands of Europeans or those under European supervision.[40] The British Company was also empowered to issue licences for game-hunting and the use of sporting weapons.

Given the pressing need for a multilateral approach to arms control, such tentative measures were superseded by the Brussels Conference Act of 1890. The Conference aimed to 'draw the sting' from wars between Africans through a combination of international arbitration and prohibition, relating to all aspects of trafficking in arms, alcohol, and slaves. Under the provisions of its General Act, the signatory powers agreed to ban all further importation of firearms, gunpowder, balls, and cartridges, within a defined zone, with exceptions only under certain conditions. The zone of restriction covered a wider area than had been attempted previously by any European government: essentially, all of tropical Africa as far south as the Limpopo, encompassing territories between the 20th parallel of north latitude and the 22nd parallel of south latitude, from the Atlantic to the Indian Ocean, and including islands 100 miles from shore. All imported 'improved' weaponry—the more lethal 'arms of precision' and their ammunition—had to be deposited in public warehouses under government control. Private warehouses were allowed to store only unrifled flintlock guns and gunpowder, which could be withdrawn for sale, but local authorities would then determine where they could be sold specifically, always excluding slave-trading areas. The only exceptions to the rule were cases where arms were needed by individuals whose position was a sufficient guarantee they would not use them improperly, or by travellers for their own protection. The signatory powers also agreed to exchange information with one another and prevent arms transfers between territories. The arms control system would remain in force for 12 years, with the possibility of renewal.[41]

The Brussels Act would shape colonial authority and the arms trade in East Africa with mixed results, owing to inconsistencies in application as well as loopholes and limitations of scope. Arms control proved to be an uneven process, with metropolitan, colonial, local, and personal factors all influencing the pace and depth of enforcement. In the German sphere, all arms and ammunition were kept under government control in accordance with the Act, and were obtained by private individuals only after registration and licensing. But the sale of these articles became a monopoly of the colonial administration, which was often prepared to throw caution to the winds because the trade was so lucrative.[42] By 1891, local German administrators were clearly dealing in these articles and reaping substantial profits. Travellers were even encouraged to purchase gunpowder; as far west as Tanganyika and as far south to the German station of Langenburg on the Nyasa, trade powder was being sold at not more than 4 rupees per 5 lb keg compared to the previous price of 6 rupees per keg asked by an independent trader.[43] Explorers in the German sphere commented on the large volume of the gunpowder trade. Dr Joseph Moloney, the British medical officer of the Stairs Expedition (1891–92) that annexed Katanga to Leopold II's Congo Free State, noted how the Germans were using this powder to obtain ivory in the Tanganyika region.[44] French missionaries blamed their sale of arms and powder for the continuation of the slave trade in the vicinity of Lake Tanganyika.[45]

In the British sphere, Captain Frederick Lugard and the IBEAC officers pondered the source of the inexhaustible supplies entering Uganda. Finally, they ascertained that the munitions were coming from German territory and concluded that German officials were guilty of complicity in trafficking. In September 1891, 70 loads of gunpowder were seized. In

October 1892, Lugard's deputy Captain W. H. Williams wrote to the IBEAC headquarters at Mombasa, informing them of the serious implications of this arms traffic from German East Africa:

> [T]he result being that the greater portion of Uganda and Unyoro ivory finds its way into German territory, so that not only do the British lose the ordinary profits of trade but the armed strength of the natives is gradually increased. ... It is easy to tell when there is much being sold, as the Kampala purchases fall at once, and men are found to have quietly gone off to Baziba country with their tusk or two.[46]

The arms trade via western Uganda persisted until the close of the decade, playing a critical role in fuelling indigenous resistance against British colonial authority. It was one of the decisive factors enabling Kabarega, the Omukama of Bunyoro, to wage a protracted campaign against the British (1894–99) that was only terminated by his capture and deportation to the Seychelles.[47]

The IBEAC seemed far more determined than its German counterpart to uphold the Brussels Act. Being responsible for the administration of Uganda (1890–93) and the East Africa Protectorate (until 1895), the British Company proved steadfast in attempting to control arms traffic within the sphere under its charge. Lugard, in particular, was much exercised about the illicit trade during his time with the Company:

> The curse of Africa is the muzzle-loader, ammunition for the breech-loader is harder to procure and its importation more easily controlled but trade powder is not only smuggled into Africa to any extent, but enormous stores of it, I believe, exist in the country already.[48]

Lugard's remedial proposal was to introduce a new type of breech-loader with a special bore. Unlike the Germans who were issuing muzzle-loaders in return for existing breech-loaders, Lugard's intention was that all existing muzzle-loaders—an estimated 6,000 in Uganda alone—would be exchanged for his modified weapons, at the rate of 3–4 muzzle-loaders per new breech-loader. The IBEAC alone would oversee the stamping, registration and licensing of these new weapons, while keeping strict account of their disposal. Every breech-loader would be issued along with a special cartridge so constructed that the bullet could not be extracted or the powder removed without destroying the whole. The ammunition would only be issued to reliable individuals.[49] The one loophole in Lugard's scheme lay in the assumption that the registration and licensing of the weapons would establish control over their use, when it could merely provide a record of the traders who disposed of them, so there was no actual check once the weapons left the traders' hands.

Paradoxically, Lugard had himself sought to extend British colonial authority in East Africa through the use of firearms. Yoked by the 'white man's burden' in the pacification of 'anarchic' Buganda, Lugard equipped the Kabaka's Christian subjects with modern rifles and deployed his own Maxim guns freely, 'for the purpose of causing a moral effect in the country, and striking terror into those who are disposed to think they are a match for us'.[50] On less elevated moral grounds and at various times in the IBEAC's history, it was also tainted by accusations of selling arms to the natives, especially in the German sphere. In the Kilimanjaro district, the Germans seized rifles carrying the IBEAC stamp and the letter 'T'. These rifles belonged previously to an expedition that had travelled up the River Tana and experienced a large-scale desertion of porters with their weapons; the rifles reappeared in the Chagga region within two years.[51]

It has been argued that a more settled colonial administration in East Africa was better equipped to curb the arms trade in the region.[52] In Uganda, where the British colonial government had employed former IBEAC officers, the rise of colonial surveillance meant that arms transfers could be checked: the registration of small arms, for instance, revealed that in Kampala and Buddu alone, there were 8,020 guns in 1898; in Toro district, there were 1,050 guns in 1899, of which 150 were breech-loaders; and in Busoga, there were 3,500 guns in 1900, of which 100 were breech-loaders.[53] Under the Uganda Agreement of 1900, the institution of a gun tax of 3 rupees per firearm ensured the surrender of many obsolete weapons not deemed worth the payment of such a tax.[54] In the German sphere, the arms trade also appeared to diminish by the late 1890s as colonial rule was consolidated. Governor von Wissman introduced numerous game regulations, including by 1897 severe restrictions on the use of firearms. New German customs stations and forts further tightened control, extending from Mahenge and Songea in the south to the fortress of Langenburg at the north end of Lake Nyasa (where a steamer patrolled the lake), and then northwards through Ujiji, Kitega, and Kigali, up to Mwanza and Bukoba on Lake Victoria.[55]

Yet the extension of European colonial authority and the pacification of East Africa actually had a wider redistributive effect on the arms trade. If anything, trafficking activities were pushed even further northward. In northern Uganda, the arms trade persisted well into the 1900s as weapons entered via the northeastern route from Ethiopia and Somaliland, skirting the north of Mount Elgon, and cutting across the Lake Rudolf area. Even after his sources of supply from German East Africa were eliminated, Kabarega continued to obtain weapons via the northeast until his capture and deportation in 1899. By 1890, large supplies of arms were already entering the Benadir ports on the Somali coast from Jibuti in French Somaliland, while large shipments of arms bound for the interior were being diverted from Zanzibar to Muscat and other ports in the Gulfs of Oman and Persia.[56] Muscat, in particular, was located beyond the restricted zone specified in the Brussels Act, so there was nothing the Zanzibar authorities could do to halt arms shipments there, except urge British Residents in the Persian Gulf, Berbera, and Jibuti to monitor the weapons as they arrived. Reports from the Residents indicate that a large proportion of the arms did eventually return to East Africa via the Somali coast.[57] Jibuti, the terminus of the Jibuti-Addis Ababa Railway launched in 1896, became the entry-point for thousands of rifles intended for the Ethiopian Emperor Menelik II, whom the French were arming against their colonial rival, Italy. Menelik managed to procure some 100,000 rifles, which his forces deployed to inflict crushing defeat on the Italians at Adowa in 1896. Jibuti also became the gateway for other weapons that filtered down the East African coast, much to the dismay of the British.

Fuelled by these arms transfers, the old Anglo-French rivalry seemed to revive from the 1890s until the eve of the First World War. With more colonial competitors in play, the French adopted a more subtle strategy by conniving at—if not actually engaging in—arms trafficking. To prove French complicity, British Residents in the Gulf of Aden had to furnish as evidence confiscated French rifles of the Gras pattern. Originally manufactured at St. Etienne in 1874, these breech-loaders were stamped for Jibuti and seemed destined for East African markets.[58] Captain Dugmore of the Freeland expedition up the River Tana would observe, upon his return in 1894, the extreme facility with which such weapons could be obtained from an agency in Aden, which shipped arms in dhows flying the French flag: 'I have heard it remarked that "there can never be any difficulty in getting arms into the country so long as there is a French man-of-war at Zanzibar".'[59] The truth was that native vessels under protection of the French flag could not be searched at sea, since France had refused to accept this procedure at the Brussels Conference.[60]

Notwithstanding calls by the British Foreign Office to cooperate in the sharing of information, as required under the Brussels Act, the

French authorities defaulted on several counts. Even with possession of the requisite information, French officials failed to notify British naval patrols whenever dhows were clearing Jibuti with arms; they did not apprehend the arms dealers who did business through Arab brokers; and they issued dhows with clearance papers made out to Yemen when it was definite knowledge that their destination was the coast of Somali and East Africa. The provisions of the Brussels Act pertaining to the storage and re-importation of firearms—especially 'arms of precision'— were flouted repeatedly at Jibuti, yet the demeanour of French colonial officialdom remained one of calculated indifference.[61]

The volume of small arms traffic at Jibuti offers some indication of the scale of the problem facing East Africa and the wider region by the end of the nineteenth century. There was no general crisis or great conflict in metropolitan Europe to justify the trade, as had been the case during the Napoleonic Wars. Yet arms imports at Jibuti reached astonishing proportions: in one month alone (August to September 1902), five steamers (three French, one British, one Belgian) docked at Jibuti, carrying 985 cases of arms (with no fewer than 20 rifles per case); 625 cases of cartridges (typically 1,600 cartridges per case); and 50 pigs of lead.[62] Arms exports from Jibuti were a matter of equal concern: within a span of nine days (16 to 25 August 1902), ten dhows left Jibuti harbour, each carrying an average of five or six cases of rifles (80 to 100 weapons); 8,000 to 10,000 rounds of ammunition; 20 barrels of sulfur (each containing 1 cwt or 100 lbs of powder); and a quantity of lead.[63]

The volume of small arms traffic at Muscat provides another index. The trade in the Gulf had revolved initially around Persian Bushire, although there were other redistribution centres, including Muscat. In the late 1890s, official reports put the consolidated figure of arms reaching the Persian Gulf at some 17,000 per annum.[64] By the turn of the century, however, Muscat was unquestionably the main arms entre- pôt of the Gulf region, and it continued as such for at least another decade. In the first quarter of 1902 alone, some 5,000 rifles and 327,000 cartridges could be transhipped from Muscat to the coast of East Africa.[65] Muscat also became the centre for the re-export of arms to Afghanistan and the Northwest Frontier, via the ports and coastal areas of the Persian Gulf. Major Grey, a British agent at Muscat, calculated that 200 rifles a week crossed over to Makran in 1907.[66] Statistics provided by other British agents, travelling in disguise along the Makran coast, suggest the likelihood of over 30,000 rifles of different patterns and three million rounds of ammunition landing during the cold season of 1908–9, most of which would have reached the Northwest Frontier tribes. These figures would be corroborated by reports from officials of the Indo-European Telegraph Department who were employed in Makran and the interior.[67]

Back in 1897, the Government of India had refused to believe that a significant proportion of munitions imported into the Gulf were reaching the Frontier. 'India is not, perhaps, directly interested from a military point of view,' noted Viceroy Lord Elgin, 'but Imperial interests are involved in questions which might arise from the trade.'[68] Sir William Lee-Warner, a seasoned administrator and analyst of Indian affairs, reached a different conclusion. He was convinced that arms were indeed arriving in Afghanistan via Bandar Abbas, Chahbar, and other ports opposite Muscat, and 'the tribes in revolt on the Indian frontier were obtaining ammunition and arms from the Persian Gulf'.[69] Elgin, however, dismissed this assessment as 'more than improbable'; Lee-Warner was merely 'hunting a shadow'.[70]

The men on the spot knew better. Colonel Malcolm Meade, Britain's newly appointed Resident in the Persian Gulf, wrote privately that modern breech-loaders were definitely reaching the Frontier, such that 'the blood of our poor fellow lies at the door of those who have carried this traffic'.[71] As the intelligence was gathered, these supplies began to loom larger in the Indian Government's threat perception, and it would eventually acknowledge that the best policy was to take action at Muscat to 'curtail or extinguish a trade which is fraught with such menace to our future peace in the North-West'.[72] The freshly-minted Baron Curzon of Kedleston, appointed Viceroy upon the Earl of Elgin's retirement in 1899, appreciated the danger on the Frontier and determined to take all necessary steps to control the trade.[73] The main obstacle was Muscat, and at Muscat, the French: no lasting arms control could be achieved without the cooperation of France, whose treaty rights since 1844 had included an article that forbade extra duties or prohibited items.

Anglo-French competition in the Gulf of Oman came to a head at the turn of the century. The colonial contest between two longstanding rivals climaxed in sharp Anglo-French disagreements over the French attempt to establish a coal depot (1899); the perennial issue of French flags to protect local shipping (1900–5); and, despite the Anglo-French Entente achieved on the metropolitan stage in 1904, the ongoing question of growing French participation in arms traffic at the periphery, settled by mutual agreement only months before the outbreak of war in 1914.[74] As the coal depot crisis peaked, the use of French flags and papers was already established and the arms trade a flourishing aspect of commercial enterprise in the Gulf. When the trade began at Muscat, arms imports were almost exclusively of British manufacture and predominantly in

British hands, though a proportion had Belgian provenance. By the turn of the century, however, it was clear that the French share of the trade was rising sharply: from 14 per cent in 1900, to 40 per cent in 1905, and 49 per cent (compared to 27 per cent Omani and 24 per cent British) in 1908.[75]

Over the two decades leading up to 1914, it was indeed ironic that the augmentation and acceleration of arms transfers from Europe would encourage European colonial regimes to expand their scope of authority and exercise stricter control at the periphery. In the respective colonial spheres, the situation prompted further preventive measures: naval patrols along the coastline of notorious gun-running areas; new colonial legislation; and novel methods of surveillance and disarmament. British and Italian men-of-war stepped up patrols off the East African, Somali and Gulf coasts, in anticipation of seasonal changes in local maritime traffic as determined by the monsoon: from southwest to northeast between April and October, and from northeast to southwest between November and March.[76] Despite their vigilance, however, it was apparent that only a small percentage of dhows were actually intercepted, while the trade continued to expand. As one official paper put it, 'The attitude of the British Government was, during the period 1898 to 1905, one of vigilant attention without the power to intervene directly or effectually. ... Whenever Masqat harbour was left without a British man-of-war a general exodus followed of native craft carrying contraband arms.'[77] Commander Pears of the British patrol noted that 'in the suppression of the arms traffic the occasional cruises of a man-of-war were costly, temporary and merely restrictive'.[78] His solution was to experiment with small, fast cruisers rigged to look like native dhows, and cheaper to maintain. These decoys proved successful, but only in diverting the arms traffic from the closely monitored British coast to the more unguarded Italian sphere.[79]

Legislation was also introduced to compel all trading dhows to proceed directly to ports where customs stations had been established, such as Zeyla, Berbera, Bulah, Hais, and Karam. At Aden, proper port clearances were more strictly enforced, while at the Arabian ports of Ras Irmean, Balhaf and Macullah, police posts were installed to monitor Mijjertein dhows in harbour.[80] The Italians were keener on intercepting the arms after they had been unloaded; monitoring waterholes, checking caravan routes, and apprehending the Somalis who carried the weapons inland. But these efforts were markedly less successful, a number of skirmishes ending with rifles falling into the hands of Somalis who wiped out bands of Wagosha hired by the Italians as patrols.[81] By the completion of the

Jibuti-Harrar Railway at the end of 1902, when it was again apparent that arms were penetrating the British and Italian spheres via Harrar and infiltrating into the interior, the colonial authorities braced themselves for the implementation of a more draconian policy of native disarmament. District officers were employed full-time to disarm entire districts; some of these, it was said, were 'bristling' with weapons obtained in exchange for ivory and livestock from the Ethiopians and Somalis— and even from Karamoja poachers—all of whom derived their arms supplies from the same original source.[82]

The vastly expanded Gulf trade called for even more comprehensive measures backed by concerted efforts at law-enforcement, nothing short of a reconvention of the Brussels Conference in 1908. Where the original conference had marked out areas in which the arms traffic would be controlled, it was painfully obvious that one glaring omission—the Gulf— could no longer be overlooked. Yet it was equally apparent that the colonial powers could not see eye-to-eye on a range of crucial issues: the scope and methods of control; the implementation of those measures; and the territories or treaty rights that had to be sacrificed in order to make arms control a reality. Nearly all parties—Britain, Russia, Germany, the Netherlands, Portugal, and the United States—agreed that the zone of restriction should be extended to encompass the Gulf. To deal with its troubles in Somaliland, Italy also pressed for the inclusion of the Red Sea. The Ottoman Empire and Persia were both amenable to the overtures of the Europeans.[83] From Britain's perspective, however, it was argued that the extension of the zone would tend to increase the chance of other powers meddling in Oman and the Gulf, with the increased likelihood of one European power after another—Italy, followed by Russia and Germany—demanding consular representation at Muscat and intelligence agencies on the spot. Commodore Sir George Warrender, Commander-in-Chief, East Indies Station, devised practical methods to tackle the actual gun-running aspect of the trade: an augmented blockade consisting of at least seven ships, supported by small military posts along the Makran coast, all utilizing the recently established telephone line to form an early warning system pin-pointing arms caravans and contraband.[84]

Nearly all powers having relations with Muscat were in broad agreement over the termination of the trade, but it was evident that one power actually stood in the way of its suppression: France. The arms trade was immensely profitable to French firms trading at Muscat, and it was clear that the stakes were even higher. The second Brussels Conference became a long drawn-out affair largely in consequence of protracted Anglo-

French bargaining over the exchange of territories for treaty rights: namely, Britain's possible concession of Gambia, which controlled an important outlet for a large area of French territory in West Africa; and frontier changes in British India favourable to remaining French enclaves at Chandernagor and Pondicherry, in return for France's acquiescence at Oman, the key to the Gulf.[85] After much diplomatic haggling and departmental wrangling in both the metropolis and the periphery, no territories were finally exchanged and the Brussels Conference was officially dissolved for the last time on 30 December 1909.

Before the end of 1909, however, the augmented blockade had been instituted along with other countermeasures, proving their effectiveness through the increased captures and dislocation of the trade. Afghan arms purchasers travelling to the Gulf on steamers from Bombay and Karachi were outlawed, and a Gulf intelligence system was created to trace any new approaches to Muscat and arms transfers across the Gulf. With the construction of a telegraph network along the Persian coast, Bushire, Bahrein, and the Trucial Coast, the gun-running could be eradicated more systematically between 1909 and 1914. The telegraph stations at Jask and Chahbar, along the Makran coast, were especially instrumental in the efforts to thwart local gun-runners. Assisted by the officers and crew of ships such as H.M.S. *Fox*, H.M.S. *Hyacinth*, and H.M.S. *Hardinge*, and backed by British Indian troops with the cooperation of local chiefs, they achieved some measure of success in capturing consignments of Snider, Gras, and other rifles, curtailing the activities of numerous Afghan and Baluch arms traders.[86] Collectively, these developments marked the start of real control. By the end of 1913, the bargaining with France was over. After lengthy negotiations stretching into the early part of 1914, and with war looming in Europe, the French Government agreed to surrender France's paper privileges and recognize the arms limitation arrangements at Muscat after the British authorities had paid £64,495 in compensation for the existing stocks and predicted profits of two French firms, Dieu and Goguyer.[87] Summing up the entire episode, B. C. Busch concluded that 'French influence vanished with the last rifle'.[88]

In the long term, these transfers of small arms and other strategic goods had a definite impact on how colonial authority in the western Indian Ocean zone came to be exercised, whether at the regional level or within the microcosm of a local community. They elicited further incursions of European colonial sovereignty into the geopolitics of autonomous states, often undermining indigenous systems of rule and provoking anti-colonial backlash. In so doing, they were a catalyst of a golden age of sub-imperialism: British intervention spearheaded at the periphery by agents on the spot, and then sanctioned, after the fact, by the metropolis.

The 'Dubai Incident' of 1910 provides a classic example of the 'microcosmic' sub-imperialism that could be triggered by offences related to arms trafficking. While patrolling the Trucial Coast on 27 December 1910, H.M.S. *Hyacinth* received reports of an arms cache at Dubai and took the initiative to investigate the area. The crew's rapid search for arms in the town provoked local Arab resistance, resulting in four sailors dead and ten wounded or missing, and 37 Arab deaths. The British Resident, Major Cox, and the Commander-in-Chief, East Indies Station, Admiral Slade, both arrived on the spot and delivered an ultimatum to the somewhat bewildered Sheikh: they demanded compensation in the form of 50,000 rupees; the return of some 400 rifles; and approval for the installation of a British agent, a small guard, a sub-post office as well as a radio transmission station at Dubai, should the British Government require any of these rights.[89] In the end, the Sheikh complained to the U.S. Consul in Muscat, handed over the money and the rifles to the British, yet refused to accept an agent, which he regarded as a major infringement on his personal authority. The British themselves—the press in London and Delhi, home government, colonial officials, and naval commanders—all agreed that the affair was unfortunate, but were divided in their opinion of the final settlement.[90] The British could not retreat from their position at the periphery without a loss of national honour and imperial prestige, so Cox was simply informed by the India Office that such forceful action must in future receive prior approval, while the remaining points of the ultimatum were left in abeyance.[91]

There were also numerous instances of high-handed interference with local customs, traditions, and institutions, which were either dismantled or diminished in conjunction with the arms trade. Particularly emotive was the issue of slavery perpetuated by weaponry trafficked and deployed in the enslavement, which thereby provided important justification for British attempts at promoting free enterprise and responsible government in the western Indian Ocean. The Foreign Office saw in Zanzibar and its Sultan a promising entrepôt and agent, such that by 1845 they secured a treaty outlawing the slave trade, thereby anticipating the expansion of 'legitimate' commerce. Yet Britain's suppression of the slave trade and the gradual dislocation of the slave economy would disrupt the indigenous political economy as a whole, especially during the 1870s–90s. Both of Zanzibar's main exports, ivory and cloves, responded to the growth of external demand, but neither could provide the basis for long-term export development. Ivory depended on essentially finite resources, which began to decline from the mid-1870s, whereas the production of cloves was confined to coastal estates that continued to employ slave labour.[92]

As with Atlantic slave emancipation, abolitionist measures in the Indian Ocean initially weakened or diverted the slave trade without destroying it, causing serious discontent among slave-holders and traders. This provoked the political crisis that triggered British intervention from 1859–61 without resolving the problem of reconstructing the export economy. The pressure for change gathered momentum following the construction of the Suez Canal and the coming of the steamship. The prospect of opening up the East African coast attracted entrepreneurs and explorers, who saw the potential of improved communication with Europe in addition to closer integration with the Indian Ocean economy.[93] Conversely, the arms trade proliferated as a means of reasserting local authority and resisting the encroachment of competing foreign powers, especially after Britain imposed in 1873 a new treaty outlawing the slave trade, which it enforced by naval and military power.

Arms were then trafficked not only as part of the gun-slave trade nexus, but for their own sake. Soon after the British partitioned the empire of Sultan Seyyid Said in 1861, Anglo-French rivalry led both powers to guarantee by mutual agreement in 1862 the independence of Oman and Zanzibar. By 1891, Britain was challenging a French breach of agreement: French officials in Aden, Madagascar, and Obok were issuing French papers of registration to dhow owners from the Omani port of Sur. In the calculations of the British official mind, these flags and papers could only undermine the Sultan's authority and upset the regional balance of power; worse still, the dhows were still transporting slaves under the protection of the French flag. 'One thing is quite certain,' noted the Senior Naval Officer (SNO) stationed in the Gulf, 'as long as the Native Craft can buy the right to fly French Colours for a mere song, so long the Slave Trade will exist, although no doubt it is decreasing year by year.'[94] As the British authorities and law enforcement agencies intensified their activities, the slave trade gradually diminished. Yet the arms trade steadily replaced it as a means of redressing the internal disequilibrium and meeting the external challenge. The new legacy of violence made it increasingly difficult to determine which was indeed the lesser of two evils:

> In the Middle East, where, besides intensifying anarchy and bloodshed in Central Arabia and in some of the smaller states, it has weakened the authority of the Persian and Turkish Governments and threatens

in the end to produce widespread and incurable disorder, the arms trade is at least as great a public evil as the slave trade; and for this reason it is much to be regretted that joint action by the civilized powers of Europe for its suppression, beyond the zone within which it is already prohibited by the Brussels Act, should have been so long delayed.[95]

It was only a matter of time before the Europeans felt compelled to act again, in efforts to stabilize the crumbling frontier that their 'civilizing' efforts had helped to create.

Most revealing, perhaps, was the paternalistic intervention of the colonial authorities in deciding the legality of the arms trade. Sheikhs, sultans, and shahs were 'persuaded' under European diplomatic pressure to agree, by stages, to arms traffic restrictions or outright prohibition. Native powers would thus become subject to policing by European military or naval personnel throughout their domains and territorial waters: Persia (1881, 1891); Gwadar (1891); Bahrein (1898); Kuwait (1900); Trucial Oman (1902); and Muscat (1898, 1903, and 1912). The British at Muscat believed initially that if supervision were maintained, the effect might be to 'retard the growth of trade'. The notification issued by the British Consul on 24 September 1898 under the Muscat Order-in-Council required British subjects to furnish information regarding arms and ammunition obtained or disposed of by them. In January 1900, the Sultan was induced to adopt a similar policy, supplying the British Consulate with weekly returns of the transactions of the principal Omani arms dealers. Yet the trade of foreigners was excluded; and the registration of British and Omani merchants' transactions soon offered no practical advantage while it actually hampered them in competition with their rivals.[96] Under the arms limitation regulations of 1912, however, the Sultan agreed to establish at Muscat a bonded warehouse, where all imported arms and ammunition would be held until such time as they could be re-exported, and all issues to reliable individuals or their agents would be regulated by special licences prepared by the superintendent, countersigned by the Sultan.[97] In the run-up to 1914, the suppression of the arms trade at Muscat resulted in a marked decline in customs revenue across the board, while the 100,000 rupees paid annually as compensation by the British could not cover the total loss. This made British subsidies under the Canning Award even more important to the Sultan, increasing his dependence upon British backing while circumscribing his policy-making prerogatives.

Local perceptions that these colonial interventions had first ended the slave trade and then turned to its substitute, arms, would have profound repercussions. Obviously, there were fiscal and socio-economic implications, arousing much resentment as the measures systematically deprived indigenous rulers and traders of a major source of revenue while channelling profits into the pockets of European officials and merchants. The British colonial authority's curtailment of a once lucrative trade at Muscat 'threatened Omani economic prosperity, their political autonomy, and their manhood'.[98] It depressed the Omani economy and living standards, since tribesmen at the coast had less money to remit to their villages, subsidies from the Sultan shrank, and purchasing power declined. Some even felt that the arms warehouse was a British device to deny the Omani tribes the use of modern weaponry. But finally, there were military implications as well: in the insurgencies that troubled the Europeans right through the end-game of decolonization, not a few of the old nineteenth-century rifles survived—restored or modified in some form—to be effective sniping weapons against colonial forces in the Gulf.

This chapter has so far shown the extent to which Europe's interactions with the western Indian Ocean periphery created opportunities for cross-border arms trafficking. These arms transfers were conducted on such a scale as precluded all but the most serious attempts at international arms control. The emerging global arms trade phenomenon, with its constantly evolving lines of supply and the climate of insecurity it engendered, offers illuminating insights into the shifting patterns of colonial strategy and policy across the region. Equally, however, it must be acknowledged that the demand for weapons was generated within the context of indigenous crisis and intertribal conflict throughout the Afro-Arab world, Afghanistan, and India's Northwest Frontier. Likewise, the regional arms bazaar of the western Indian Ocean zone was as much an institution of the evolving indigenous military economy, with its own cycles of profit and violence, as it was an interface with global dynamics of military innovation, production, and diffusion.

Indigenous crisis and the arming of the Western zone

The arms transfers that accompanied Europe's quest for trade and dominion cannot be so easily subsumed under a monolithic 'European imperialism'. There is an integral 'analytical space' between broadbrush analyses of Western global empire and the growing capitalist world system, on the one hand, and the history of specific cultures or communities, on the other. Before the European onslaught, large parts of the world were already bound into interregional unities, not only by trade and forms of government, but also by ideas, beliefs, and legends. The Indian Ocean—an arena of production and communication, and a locus of commercial exchange and cultural transmission—had long facilitated the formation of such interregional unities, linking substantial areas of Africa and Asia.[99] These regional zones were no doubt ruptured or redefined by the later policies of empire and the movement of world trade, but it should be remembered that there always was an indigenous dimension with 'inner' dynamics. In the historical interplay between external and internal forces, this indigenous context would prove absolutely critical in shaping both colonial authority and the arms trade.

From the indigenous perspective, the earliest phases of European expansion and arms trading in Africa and Asia had coincided with the period of commercial activity following the establishment of the great land empires in the fifteenth century, and the flowering of both Indian Ocean trade and Chinese trade. The more destructive middle phase, from the seventeenth century, had coincided with the wars of succession and monopoly within the Muslim empires and had aggravated them, not least through the sale of military hardware and services in native bazaars. The later phases of European expansion, which witnessed the onset of European great power rivalry and Britain's ascendancy in the emerging industrial world economy, would overlap with even more volatile periods of socio-economic, political, and military transformation in Africa and Asia.

Throughout the course of the long nineteenth century, developments in indigenous trade and politics would reflect both the atrophy of regional imperial rule and the ascent of new elites: provincial magnates, local chiefs, warrior-traders, and merchant princes. Far from being the static backdrop to European colonial activity—'impotent', 'decadent', or 'stagnant'—the African and Asian lands flanking the Indian Ocean were vulnerable to European colonial authority and arms transfers precisely because they were themselves passing through new patterns of economic development and state formation. C. A. Bayly has described this series of interconnected upheavals as 'a crisis in the relationship between commerce, landed wealth and patrimonial authority comparable with that which convulsed Europe in the first half of the seventeenth century'.[100]

The contours and structures of the Indian Ocean arena were moulded throughout history by the pressure of changing core-periphery relationships. More specifically, the interaction between centralized imperial states, outlying provinces, and coastal regions, on top of the unchanging features of local geography, altogether regulated the political economy of the zones. Ongoing shifts in these core-periphery relationships would mean that the boundaries of the Indian Ocean zones were seldom constant; their porous borders and fluid frontiers responded to fluctuating patterns of human activity on land as well as at sea. Nevertheless, it is historically possible to define two main cultural zones; and it is the history of the western Indian Ocean zone, with its evolving political economy and markets for strategic goods, which concerns us here.

From the earliest antiquity, possibly since the golden age of Solomon (970–930 BCE) and for three centuries after the birth of Islam in the seventh century, the Near East, the Middle East, and parts of the western Indian Ocean had constituted a zone of economic, political, and cultural unification that shared some of the core-periphery relationships of a world system. From the sixteenth century, large political units—the great Muslim empires of the Ottomans, the Safavids, and the Mughals, centred respectively upon Istanbul, Isfahan, and Delhi—dominated the population and wealth of lands flanking the western arc of the Indian Ocean, from North Africa to the borders of Burma. Even if by the late eighteenth century there was increasing fragmentation of political unity with the weakening of these imperial centres and the end of the *Pax Islamica*, there remained nonetheless a striking cultural unity binding their various successor states in East Africa, the Red Sea and Gulf regions, the Northwest Frontier and western India. If anything, the sheer splendour of the empires spread the notion of kingship and new techniques of governance and learning to sections of society which had retained a strong sense of their own identity, expressed in regional solidarity or religious difference. Magnates had beheld the imperial effulgence, the clever combination of military power, money, and claims to legitimacy that the great dynasties had deployed. Now that the cultural mantle of kingship was devolving, they too aspired to be 'world monarchs' in their own regions. In addition, others had benefited from the commercial efflorescence; the emperors had protected and promoted merchant families and scribal communities from non-Muslim ethnic groups—including Jews, Armenians, Parsis, and Hindus—who increasingly formed the basis of an international commercial culture, both on land and at sea.[101]

Local geography also shaped the commercial, political, and military culture of the zone. Typically, it would be lush, fertile lands of the interior that supported the predominantly agrarian economies of the imperial centres; the outlying provinces and coastal areas functioned as intermediate zones of exchange between regions of production and consumption around the ocean. Bounded on the west by a narrow belt of poor, low-rainfall scrub known in Kiswahili as the *nyika* (wilderness), the East African coast could offer no agricultural base to sustain the growth of large polities. Instead, it engendered the Swahili ('Coast') civilization, formed by two cultural streams, one arising from the East Central African interior and one from Arab expansion across the Indian Ocean. Such ethno-linguistic hybridity would facilitate further exchange: the Swahili language gained wide currency in regional commerce, while its mercantile culture engendered a string of coastal city-states. Most of these polities, like Mombasa and Kilwa, were on the continent; a few, like Zanzibar, were confined to offshore islands; all were synchronized with the broader socio-economic structures and rhythms of maritime trade in the Indian Ocean. Arab sheikhs ruled these city-states, although political fragmentation made each mercantile polity susceptible to external attack or internal strife.[102]

Dominating the western Indian Ocean zone to the immediate north of the East African coast were the Gulf states. For centuries before the arrival of the Portuguese, the Gulf had been one of the great arteries of Asiatic trade. Through it passed products from India, China, and the Eastern Archipelago, bound for markets in Persia and the Levant; and back flowed the merchandise of Europe, Arabia, and Persia, to India and the Far East. Mercantile city-states rose and flourished along its shores—Basra, Bahrein, Siraf, Qais, Rishahr, and Hormuz—only to wither and decay when buffeted by waves of political upheaval: the Qamartian revolt, the collapse of the Abbasid Caliphate, the Mongol invasions, the rise of the Safavids, and the Portuguese incursions.[103]

But with the eclipse of Portuguese power, and the general crisis of the Muslim land empires engulfing the Ottomans and the Safavids, the smaller states of the Arabian coast, including Oman, were able to compete for the profits of the mercantilist system and the primacy of the Gulf's waters. Oman's hinterland was more generous than that of the Swahili coast: in the Green Mountains, the *wadis* or dry valleys irrigated by subterranean canals (*aflaj*), and the desert. Indeed, the Omani economy had been initially agricultural and pastoral; the main source of revenue had been the produce tax. During the seventeenth and eighteenth centuries, however, Oman's tribal confederacy and continental theocracy extended Omani control over the coast in order to capitalize on maritime trade. By the mid-eighteenth century, Omani society was clearly mercantile in orientation, with its rulers—the Busaidi Sultans—presiding over the emergence of an expansionist commercial empire that revolved around its twin-centres at Muscat and Zanzibar. Zanzibar was annexed by Oman only in 1744, along with the Omani pacification of the East African coast, but both these port-cities, with their prime locations and deep anchorage, soon provided key access to the most important inland and maritime trading areas of the western Indian Ocean zone.[104]

The history of northwestern India, looking towards Central Asia, was also shaped by unique geopolitics. Inaccessible mountain ranges, which ordinarily barred the passage of military forces, could be breached at the passes—notably the Khyber and the Bolan—that served as conduits through which hostile armies invaded India over the centuries.[105] By the eighteenth century, this region was defined not only by perennially harsh physical conditions, factors of geography and climate, but also intensifying warfare and blood-feuding from the militant frontier tribes of ancient Iranian origin which settled there: the Baluchis and various Pashtun groups, referred to generically as 'Pathans' but including Afridis, Ghilzais, Mahsuds, and Mohmands. The writ of the Ottoman, the Safavid, and the Mughal emperors did not run here. The tribes were even divided against themselves, depending upon the shifting alliances and animosities of their constituent political factions.[106] These fierce tribal communities would be armed by the great merchant caravans, laden with commercial products and strategic goods from the port-cities of the western Indian Ocean. Fearful of being blocked and plundered along the way, the caravans would subsidize and supply the tribes that they encountered as they traversed the mountain passes from Persia through Afghanistan and Central Asia to Mughal India.

By the dawn of the long nineteenth century, the geopolitical context and these core-periphery relationships had altered to such an extent that the societies of the western Indian Ocean zone were drawn into a general crisis evidenced by new patterns of profiteering and power-broking, including higher levels of military production and consumption. The economic nexus of the western Indian Ocean continued to be characterized by its age-old system of monsoons and markets, sailors and ships, entrepreneurs and entrepôts. At sea, the southwesterly monsoon continued to serve traders on voyages from East Africa to the Gulf and western India between April and October, while the northeasterly monsoon facilitated westward and southbound maritime traffic between November and March. Lateen sails remained the standard rigging on maritime craft of the western Indian Ocean, most typical of which was the ubiquitous Arab *dhow* that came in various sizes, depending on the cargo, the crew, and the captain.[107] On land, merchant caravans—with their porters and beasts of burden—still facilitated the exchange of commodities between the Indian Ocean littoral and the interior.

Yet the interposition of regional elites between the old and new sources of global authority—represented by the weakening imperial centres of Africa and Asia, against a strengthening European presence—would create a whole new constellation of mercantile enterprise and state formation. By the 1800s, the zone's trading network had four main terminals: the coast of East Africa; the port-cities of Zanzibar and Muscat; the Makran coast of southern Persia and Baluchistan; and the ports of western India, such as Bombay. A different configuration of power connections thereby crystallized, linking the Arabs of Oman, the Baluchis of Makran, the Hindus of western India, and the Africans of Zanzibar: the Omanis were the political leaders; the Baluchis, the military force; the Indians were merchant capitalists (brokers, bankers, and tax-collectors); and the Africans, slave-traders and slaves.

The 'hollowing-out' of the old empires, as noted previously, occurred in tandem with the growth of successor states that aspired to greatness in their region. But the exercise of their political autonomy and the display of the cultural trappings of kingship could only be sustained by the fruits of economic prosperity, which often entailed the cultivation of slave economies. Among the Europeans, the French had 'gone native' in the footsteps of the Portuguese by employing slave labour on their sugar plantations in the Mascarene islands. This demand certainly stimulated Arab slave-raiding in the African interior during the 1840s–50s.[108] Yet it should be remembered that there was an even wider demand for slaves throughout the regional markets of the western Indian Ocean zone, and the institution of slavery had a far longer lineage in indigenous society.

To satisfy the growing luxury and opulence of regional courts, and typically where labour shortages were most acute, unprecedented numbers of slaves were needed. Slaves were required for pearl-fishing, notably in the coastal waters of the Gulf region; for the cultivation of cloves in East Africa and Zanzibar, dates in Arabia, cotton in Central Asia, and tea and sugar cane in India; and for manual work in the transportation of goods between regions.[109] Slaves, themselves a valuable commodity, were often moved between regions: during the nineteenth century, over a million were shipped in dhows from East African and Red Sea ports; and millions more were shifted around the littoral and interior parts of Africa and Asia. The vast majority of human cargoes brought to the coast originated from tribes around the Lake Nyasa region—the Makua, the Yao, the Ngindo, the Nyanja—with Kilwa as the principal port for slave shipments. Key destinations included Zanzibar, Oman, the Persian Gulf, and north-western India.[110]

Here were the makings of a triangular trade arguably more damaging to indigenous society than that of eighteenth-century West Africa and the Atlantic. Indeed, by suppressing the slave trade on the west coast of Africa during the early nineteenth century, British naval patrols had inadvertently confirmed a new pattern of commerce that flourished further eastwards. In exchange for the slaves from the African interior, the East Africans wanted one product—guns—and in return for these guns, the Europeans wanted ivory. Unfortunately, as the guns became more sophisticated and deadly, ivory supplies became scarcer and sources of slaves dwindled. This resulted in even more aggressive methods of extraction and exploitation, exacerbating the already unstable political climate of the African interior and escalating the crisis of the Afro-Arab world. By embroiling many more peoples in the affairs of Africa's heartlands, the depredations that had begun with the old Atlantic slave trade would increase rather than abate.

Historical studies reiterate that what the East Africans obtained in exchange for slaves in no way equalled the manpower lost to each community. Ivory had minimal intrinsic economic value for East Africans, but what they obtained in exchange were goods that hardly matched the value placed on ivory by the merchant capitalists of India, Europe, and America. What the East Africans obtained in exchange for ivory, slaves, and other valuable raw materials were luxury items, inexpensive consumable goods, and trade-guns typically inferior to the weaponry wielded by the Europeans themselves. Furthermore, the process of integrating the Swahili coast and East Central Africa into the fluid political economy of the western Indian Ocean and the wider world seemed to enrich their Arab and South Asian trading partners—and the European powers—at their expense.[111]

Why then were small arms coveted so highly by these African communities? Politically, such weapons were useful for engrossing prestige and power in the unsettled indigenous context of tribal warfare and state formation. Weapons were needed to fight wars, and war and resistance were a means of securing a new symbiosis between ruler and the ruled, merchants, men, money and markets, of which the method of aggrandizement and resistance was the raid. But this spawned a vicious cycle: guns were used especially in slave-raids, and the enslaved captives from rival tribes were then bartered away for more guns, and so on. Slaves were crucial to local power-broking, as African scholars have stressed, 'the means of getting power, since the human exports were the surest attraction for imported firearms'.[112] Slaves bound for Arab and wider Asian markets were handed over readily to Swahili-Arab traders in exchange for the consignments of European firearms in their caravans. Within this self-perpetuating gun-slave trade cycle, most African chiefs sought to manipulate and monopolize the wider military deployment of small arms: to strengthen the native war machine with a view to state-building.

The use of traditional African weaponry—swords, spears, and bows— was not eclipsed by the new technology, but rather enhanced. The importance of firearms for indigenous military culture was, in part, psychological: the explosive noise of a large, overloaded musket could decide the outcome of raids, skirmishes, and village warfare.[113] Equipped only with muskets, Muslim slave-raiders across the eastern Sudan and Indian Ocean littoral managed to unsettle various non-Muslim tribes of the interior.[114] Small arms could tip balances of power, to either resolve or prolong conflicts. The most powerful recipients quickly grasped the advantages conferred by both old-style muzzle-loaders and modern breech-loaders, which were used to good effect in the hands of sniping, harassing, scouting, and foraging parties, in addition to the batteries and lines of riflemen that so overawed fellow warlords.[115] From a cultural perspective, firearms acquired new socio-religious functions. In some cases, they were regarded as a symbol of power or a badge of office rather than a military asset. Firearms were used widely for ceremonial and festive occasions; for instance, customary gunfire could signal the end of Ramadan and festival prayers, or herald the start of certain administrative and commercial activities. They were sometimes deployed directly against the spirit world, to scare off 'devils'.[116]

But finally, firearms were also instrumental in securing supplies of elephant ivory, the one luxury commodity from the plains and forests of Africa for which European society developed serious cravings. Some scholars have questioned the importance of firearms for ivory procurement, on the basis that there are few references to guns being used in elephant hunts.[117] But the weight of evidence suggests that whereas traditional spears and swords had long been used to hunt elephants, the introduction of firearms transformed the trade: muskets, rifles, and special 'elephant guns' made in Birmingham or Liège did the slaying as well as the guarding after the tusks had been extracted.[118] Aided by the Portuguese and abetted by their own highly skilled blacksmiths, the Chokwe of Angola became one of the earliest societies to adopt muskets for elephant hunts. In a pattern that would be repeated elsewhere, they wiped out all the elephant herds in their own area, before ranging further towards the headwaters of the Congo and Zambezi Rivers.[119] It set the scene for the large-scale slaughter of elephants for their ivory, especially after Western firearms tipped the balance in favour of the hunters. Later technological developments would produce quick-loading, rapid-firing rifles that were effective at very long distances, thereby reducing even further any risk to the hunter.

Ivory had featured as an item in the local political economy long before the arrival of the Europeans. During the early modern period, when Africans had lived in virtual isolation and in harmony with their environment, outlying provinces and smaller principalities dotted with tiny villages had often sent tribute in the form of cattle, slaves, and ivory to the larger, highly structured states, such as Buganda, situated near the great lakes. For centuries, too, Zanzibar had served as 'the chief mart of ivory'—partly from the coast, partly from the interior—to help satisfy the outside world's interest in this commodity.[120]

By the nineteenth century, however, the demand for ivory was proving insatiable, particularly with Europeans in the burgeoning world economy developing a craving for it. In the fast-expanding European market, the best ivory was reserved for making billiard balls; lesser ivory was used in ornaments and handles for cutlery and knives, some of the most exquisite of which were manufactured at Sheffield.[121] Early on, most of the ivory was derived from smaller tusks, belonging to elephants that had inhabited the fringes of forests adjacent to coastal areas. But as ivory stocks were depleted and the herds disappeared, hunters were drawn further inland where larger tusks were found; the biggest, most valuable tusks came from around the vicinity of Lake Tanganyika. Profits from these quickly enabled elephant-hunters to join the ranks of slave-traders and gun-runners as the region's commercial elite. Initially, porterage, not slaves, conveyed the ivory to the coast; the main slave and ivory routes were also fairly distinct. Increasingly, however, as the whole trading system ran amok, slaves were pressed into carrying tusks, while slaves bought in one district might be traded elsewhere for ivory.[122]

The systematic procurement of guns, slaves, and ivory brought Africans and Arabs once remote from one another into regular contact, whether in collaboration or conflict with one another and the wider world. Swahili-Arab penetration of East Africa commenced from the early nineteenth century, when they came as peaceful traders and settled in small compounds near established African chiefs while building up their trade in slaves and ivory. Once these traders had secured sufficient stocks, they would despatch slave and ivory caravans to the coast, where they would complete the transaction and return carrying supplies of new trade goods along with more guns and gunpowder. As the Swahili-Arabs prospered and became better armed, more of their trading posts were turned into military establishments—stockades or forts—in a manner reminiscent of the early European colonial settlements.

By the mid-nineteenth century, the Swahili-Arabs were sufficiently established to overthrow local chiefs and make a bid for political autonomy that owed purely nominal allegiance to Oman and Zanzibar. By the 1870s, the influence of the Arab traders had become so strong that they were involved in serious political disturbances and succession wars within the African interior, a pattern of disruption soon to spread across Central Africa. The new merchant princes quickly gained power and prestige, and eagerly accepted invitations to support one faction or another in internal disputes of forest communities and states, like Kazembe, thus weakening the old order and paving the way for a new configuration. Other East African groups to rise to prominence alongside the Swahili-Arabs would include the Nyamwezi, who by the 1850s were also large-scale caravan operators and power brokers, and the Acholi, who were infamous gun-runners.[123]

Meanwhile, the bigger African polities stockpiled their ivory both for prestige and for the purpose of purchasing more guns, which were then deployed to enlarge their power base, typically through predatory raids upon neighbouring tribes and states. Indeed, it was reported that Buganda, the 'Jewel of Africa', obtained more guns through raiding than trading, its palaces and storehouses filled continually with giant elephant tusks under the ambitious regimes of the Kabakas of Buganda. Yet it was not against hostile foreign powers that the Kabakas—Mutesa I (r.1856–84), then Mwanga II (r.1884–88, 1889–97)—found themselves increasingly hard-pressed to maintain authority, but rather their own Muslim and Christian subjects, polarized through less direct external influences. The arrival of Islamic teachers from the upper reaches of the Nile had enhanced the administration and economy of Buganda through the introduction of writing and record-keeping, but Muslim converts among the Baganda were bound to be pro-Arab by religious affiliation. By contrast, those converted under the ministry of rival Catholic and Protestant missionaries soon became ripe for collaboration with whichever colonial power backed their missionary leaders. The seed of the Christian gospel, sown by European missionaries in a climate of great power rivalry, Islamic revivalism, and indigenous crisis, would yield a harvest more bitter than sweet in Buganda's civil war (1885–92).[124]

The changing temper of trading, raiding, and state formation in the African heartlands was certainly observed by the Europeans. Explorers like David Livingstone, Richard Francis Burton, John Hanning Speke,

H. M. Stanley, and V. L. Cameron had all witnessed the Arab infiltration of the interior, the dynamics of the triangular trade, and the increase in tribal warfare till 'the whole country was in a flame'.[125] The German missionary Ludwig Krapf saw caravans 'from 600 to 1,000 men strong, and mostly armed with muskets' venturing into Masai country, though 'being often nearly all slain'.[126] Handsome profits from the sale of a consignment of fine tusks to the European traders seemed to make such losses worthwhile. Speke, the first European to visit Uganda, in 1862, found to his surprise that his gift of firearms was no novelty to Mutesa, the Kabaka of Buganda.[127] Livingstone and Cameron in the 1870s discovered to the west of Lake Tanganyika parties led by Arabs carrying up to 1,000 flintlock or percussion-type muskets, prompting Cameron to comment: 'When I passed, nearly every village could turn out at least half of its men armed with muskets.'[128]

The European powers, especially the Belgians in the Congo, compounded the problems by delegating a significant portion of their colonial sovereignty to concessionary companies. These commercial organizations were subject to few of the constraints imposed on democratically elected metropolitan governments, and they would escalate the crises of indigenous society and the environment by vigorous participation in the ivory trade. In 1889, it was reported that the last two steamers from the Congo each carried ivory worth £15,000, while the ivory market of Liverpool—hitherto a principal market—was superseded by Antwerp, increasingly the recipient of all ivory extracted from the interior by the Congo Free State.[129] British business interests predicted that this climate of profitability and plunder would lead to 'an ivory famine'. For as surely as the new technology had rendered the old 'Tower' musket obsolete in Europe, it would follow that:

As a marketable commodity [the elephant] will be equally obsolete, whether advancing civilization improves him off the face of the earth, or whether he manages to survive in the savage obscurity of pathless wilds where his foes cannot profitably reach him. So far as he is concerned, therefore, we must be prepared for the possibility that by the time we have abridged the distance between Sheffield and Central Africa, the hunters will be counted by droves and the elephants by pairs. The source will be systematically tapped, but the ivory spring will be practically dry.[130]

It is against such a background of local communities in turmoil, exacerbated by the depredations of a more devastating triangular trade, that we must view the acts of the Berlin and Brussels Conferences, the European partition of Africa, and all subsequent measures to suppress arms trafficking in the western Indian Ocean zone. Having fuelled the demand for ivory, facilitated the slave economy, and flooded the market with guns, the Europeans found themselves striving to establish authority or maintain control in places where the structures and systems of indigenous society were reacting most violently to the pressures of change.

While the interventions of the colonial state would ultimately dislocate the ivory-trading and slave-raiding system, European encroachment into what had become an Arab sphere of influence in East Central Africa did little to end arms trafficking. Such interference actually seemed to provoke wider forms of anti-colonial resistance, including Islamic protest against the incursions of the West. The activities of the Europeans on the Dark Continent would effectively seal the connections between Arab revivalism, the arms trade, and armed revolt. In 1888, an Arab uprising erupted at the northern end of Lake Nyasa; a few months later, there was a resurgence of violence in the Lake Tanganyika region; and, a little later, the Belgians were confronted by large Arab bands in the vicinity of Stanley Falls as well as the Manyema country to the west of the River Semliki. Vast amounts of arms and gunpowder were obtained by these Arabs, both for self-defence as well as in hopes of expanding their trade in slaves and ivory. At Stanley Falls, it was reported that the Arab chief Mwengi Muhara had as many as 10,000 rifles in his possession.[131]

There were other armed manifestations of Islamic revivalism in the year 1888. In Buganda, a Muslim revolt instigated by the Arabs temporarily dethroned the Kabaka, Mwanga, forcing out of his kingdom all the Christian missionaries from Britain and France. Père Lourdel noted that these Arabs were importing huge quantities of gunpowder and many thousands of every type of firearm in the country.[132] In East Africa, the Arab-Swahili rebellion broke out in reaction to prolonged German abuse and rough handling of Muslims—the Arabs called the Germans 'the enemies of God'—with the uprising occurring among Arabs in the coastal district and the Arab chieftains (*walis*) leading the movement. Inspired by the capable Bushiri, and equipped with breech-loaders and a few small cannon, the revolt in its early stages was successful enough to confine the Germans to their two main centres at Dar-es-Salaam and Bagamayo.[133]

Finally, in 1895, antagonism towards the Sultan of Oman and his British connections—connections that had first ended the slave trade and then turned against its substitute, arms—erupted into open raids on the Sultan's domains by tribesmen from the interior. Britain's progressive curtailment of the arms trade at Muscat from 1898 onwards, culminating in the warehouse agreement of 1912, so undermined the Sultan's authority and Omani prosperity that rebellion became almost inevitable. An armed insurgency broke out in 1913, leading to the revival of the Imamate and the growth of British informal authority in Oman.[134] In the eyes of the more devout Omanis, the coercive arms limitation 'agreement' in Muscat came to be seen as 'the final proof of [the Sultan's] total subservience to Christian foreigners and of his heretic tendencies'.[135]

If the Europeans intoned the language of civilizing mission and moral crusade against issues such as slavery and arms trafficking, their zeal was certainly echoed in indigenous society by a long Islamic revivalist discourse, incorporating strands of both anti-colonial protest and *jihad*. Such warfare was waged not only with words and ideas, but also with physical weapons, often sanctioned by religious figures—'saints' and *sufis*—who would become involved in illicit arms transfers.[136] Major Herbert Henry Austin, a soldier-administrator tasked specifically to tackle the Gulf arms trade, examined several local notables but made special mention of an Afghan of 'holy' reputation named Khalifeh Khair Mahomed, 'who had taken a keen interest in gun-running, and emigrated to Persian Baluchistan chiefly for that reason'. Having relocated to a strategic coastal neighbourhood, Mahomed performed a vital arms-brokering function and was 'frequently visited by his countrymen who made the annual pilgrimage to the Gulf for the purchase of arms'.[137] To Arnold Keppel, another British observer who monitored gun-running in the Gulf, it was apparent that 'the fiery exhortations of the mullas' and 'the revived activity of the Hindustani fanatics' were crucial in galvanizing anti-colonial resistance, and the rest of the arms trade with it.[138]

Even as popular sufism and the ultra-fundamentalist Wahhabi movement had embodied the religious dimension of indigenous crisis during the late eighteenth century, so an array of eclectic millenarian and messianic movements would do the same for the late nineteenth century. These later examples of armed insurrection fired by religious extremism may be dated from the rise of the Mahdi and the slaying of General Charles Gordon in the Sudan in 1885, through the jihad of the 'Mad Mullah' of the Dervish State in Somaliland from 1899, to the revival of the Imamate in Oman in 1913.[139] Irrespective of whether the crisis was secular or spiritual in origin, the result was frequently a call to arms: the more modern and potent the weaponry, the better, and supplied by those willing to risk their lives for it.

Interlopers and intermediaries of the Western zone

Anti-colonial resistance and small wars, occurring in the midst of wider political fragmentation and state formation across the Afro-Arab world, altogether paved the way for the activities of various interlopers and intermediaries. All of them profited from these periods of flux and opportunity through their involvement in regional trade networks and arms bazaars. This section examines more closely the interaction between numerous individuals and groups, both external and internal, which made arms transfers in the western Indian Ocean zone such a threat to regional stability over the second half of the nineteenth century.

First, there were the European interlopers: missionaries, explorers, and other free agents who became gun-runners out of necessity or opportunity. Every group had its reasons to value the western Indian Ocean periphery, but all of them came to rely on small arms as indispensable tools of their respective trades. In Buganda, CMS missionaries like the Scotsman Alexander Mackay took great risks to save souls for the Evangelical Protestant faith, just as French priests from the Society of Missionaries of Algiers (the 'White Fathers')—typified by men like Siméon Lourdel—laboured to win native converts to Catholicism. The Kabaka, however, seemed more interested in the missionaries' guns and machines than their competing versions of Christianity.[140] Commercial adventurers—British and German alike—hoped to exploit the riches of the East African hinterland, especially ivory. Emin Pasha, the eminent Austrian botanist and doctor appointed by General Gordon as Governor of the Equatoria Province of the Sudan, both commanded and supplied black Africa's most independent army; he, too, was keen to prise open the African interior, whether on behalf of Britain or Germany. Emin collaborated with the scientist Carl Peters, founder of the German Colonization Society and chief architect of German East Africa, in advancing German designs on Buganda. Backed by a group of Hamburg merchants with large interests in the gin and gun trades, Peters used both products freely to persuade his pliable 'rulers'—amenable petty chiefs—to mark their signatures on blank 'treaty forms' that they could not understand, ceding rights amounting to sovereignty.[141] For all such groups and individuals, firearms were not just instrumental but integral in furtherance of their respective causes, whether used in armed negotiations with native rulers to obtain various concessions, or in armed conflicts to defend a growing colonial interest, which thus entailed the further 'protection' of local communities.

Even more ambivalent and colourful was the type of global free agent embodied by the Irishman Charles Stokes, a lay missionary-turned-rogue trader who came to be regarded as 'the doyen of the gun-runners'.[142] Formerly an employee of the CMS, Stokes had been enticed by the commercial prospects of dealing in arms and ivory. Switching allegiance from God to mammon, he was soon engaging in the shadier aspects of the imperial enterprise. The arms trader's caravans became a strategic lifeline for the monarchs of Buganda, the 'Jewel of Africa' that Europe's colonial powers had begun to covet by the 1880s. Stokes served the British, but also worked for the Germans, collaborated with the French, and defied the Belgians. As the imperial forces advanced, he moved ahead of them into uncharted territory in the African interior, where forest elephants survived and firearms secured handsome profits. Led by Stokes, caravans of slaves and soldiers (*askari*) would snake their way either towards the interior or the coast. In their hands, they carried countless Birmingham muskets and breech-loading rifles (Lee-Enfields, Sniders, Mausers, and Winchesters), ammunition and gunpowder, together with large consignments of ivory and other trade goods. Stokes usually approached Buganda through German territory, where he obtained the bulk of his trade weapons, although he sometimes started at Mombasa; in 1889, he travelled this route with a 2,000-strong caravan.

Stokes brokered deals in guns and pounds of gunpowder by the thousands, charging astronomical prices while arming many groups involved in colonial and tribal conflicts. Although Lugard had offered to 'facilitate [Stokes'] legitimate trade', Lugard confessed he 'could not close with him' in trade; and, conversely, Stokes was 'able to evade [Lugard's] efforts in many ways, as, for instance, by issuing powder to his elephant-hunters at the south of the lake and conniving at their selling it in Uganda'.[143] Nevertheless, Lugard intervened in Buganda's civil war by redeploying weaponry introduced by Stokes, which he reinforced with brand new Maxim guns, while the Kabaka used the munitions Stokes supplied to deal with both domestic and foreign opponents. Stoking the flames of indigenous resistance, the Irishman supplied over 4,000 lbs of gunpowder to Buganda in one year, even as he conducted an extensive trade with Arabs in the Congo. On what was planned as his last trip, the gun-runner finally came to grief in the Congo Free State in the Manyema country. In December 1894, Charles Stokes was arrested by a young Belgian officer named Captain Hubert Lothaire, then summarily executed under the charge of supplying 'powder, cap-guns, caps, and breech-loading guns to [Arab] bands in revolt against the Government of the Independent State of the Congo'.[144]

The 'Stokes Affair' was an international *cause célèbre* that became a scandal in Britain and the rest of Europe.[145] As Wm. Roger Louis has pointed out:

> Stokes' hanging called attention to the crisis through which the Congo State was passing, with the result that the anomalous relationship

between Belgium and King Leopold's African empire was the subject of considerable discussion in the British press. ... Stokes' hanging was an important origin of the Congo reform movement, because it was the first incident to attract widespread attention to the mal-administration of the Congo State.[146]

Yet, with the Fashoda Incident on their hands and the Anglo-Boer War looming on the horizon, the British Government and the British Press allowed the Stokes Affair to be quietly 'forgotten', in Prime Minister Lord Salisbury's own words, at least for the time being.[147] The grim reality was that Africa was still awash with all the munitions needed to arm these conflicts, 'to which Stokes' lot was a mere drop in the ocean'.[148]

Alongside the interlopers from Europe was a conglomeration of even bigger European intermediaries. This wide-ranging 'supply chain' consisted of chartered companies and colonial officials, agency houses, shippers and ship-brokers, who operated in conjunction with the weapons manufacturers and banks, on the one hand, and a variety of native agents, on the other. On the eve of his execution, Charles Stokes had confessed, 'I am a trader and have for sale what has been sold to me and passed through the European Custom-house at Sadaani and bought from officials of the Imperial German Government.'[149] The gun-runner's final statement points to the larger forces at work: even as consignments of small arms, gun-caps, and gunpowder were delivered into his hands 'without any restriction for sale up country', Stokes had been dignified with the rank of assistant *Reichskommissar*, and decorated by the Kaiser with 'the third or fourth class of the Kronen Order' for his services to the German East Africa Company.[150]

Together with the chartered companies and their agents, the European arms-supply chain relied heavily on the cluster of private firms that maintained long-distance trading connections at the periphery. Such arms dealerships were among the early forms of the globalizing enterprise, encompassing both European and indigenous components in wide-ranging multinational circuits. These firms incorporated nationals from countries such as Britain, Germany, France, Belgium, and Russia. All of them established operations at Zanzibar and Muscat, where they could forge commercial partnerships with Afro-Arab mercantile communities and Indian business interests. With their combined resources and reach, they would undoubtedly become the chief locomotives of the arms traffic from Europe.

In the Gulf, the pioneers of the traffic were Messrs. A. & T. J. Malcolm of Bushire, a Persian Armenian firm operating under the aegis of Britain that started importing arms in 1884. They were followed by Fracis, Times & Co., an Anglo-Parsi house that owned agencies at Bushire (1887), Bahrein (1895), and Muscat (1896). Other notable British arms-brokers, shippers, and manufacturers included the London firms of Messrs. G. Bucknall Brothers; E. W. Carling & Co.; F. C. Strick & Co. (in association with the Anglo-Arabian and Persian Steam Navigation Co., owners of the S.S. *Baluchistan*); Jones, Price & Co. (in association with the British and Colonial Steam Navigation Co., owners of the S.S. *Zulu*); Livingstone Muir & Co.; J. C. Hotz & Son; Haji Ali Akbar & Sons; H. M. Assadulla; the Birmingham firms of C. H. Laubenburg & Co.; Isaac Hollis & Sons; T. Bland & Sons; and the Manchester firm of H. C. Dixon & Co. Each of these firms depended upon a shifting network of local agents: Dixon, for instance, had a Persian agent in Bushire; and bills of payment in connection with rifles on the S.S. *Framfield* were drawn on Kal Najap bin Ghalib, another merchant of Bushire, who had been 'for some years in the employ of Messrs. David Sassoon and Company, of Leadenhall Street'.[151] But entering the fray for a share of the profits, and assisted by their own native agents, were German, French, Belgian, and Russian dealers: O'Swald & Co.; A. J. Herz & Sons; Hansing & Co.; M. Goguyer; Dresse, Laloux & Co.; Baijeot & Co., and Keverkoff & Co. Prominent among the export cartridge manufacturers were G. Kynoch & Co., of Birmingham, and Eley Brothers (Ltd.), of London, who were the foremost suppliers of ammunition to the British War Office. But their grip on ammunition markets at the periphery would be contested likewise by the Société Cartoucherie Belge, of Liège. Lloyds Bank, serving as underwriters, afforded solid financial backing for a number of the British shippers, although this would lead eventually to their embarrassing entanglement in another celebrated case, that of the *Baluchistan*.[152]

To gauge the importance of the western Indian Ocean markets for the European arms merchants and their agents, we need only look at the Birmingham manufacturers and the Frenchman M. Goguyer. The rise in the value of arms and ammunition exported from Birmingham to the Gulf was meteoric—from £4,750 (1889) to £37,678 (1894) and £105,000 (1897)—whereas Kynoch stood to lose £1,000 per annum from any stoppage of the trade.[153] Goguyer's personal fortune, amassed over just one decade, was worth £40,000 at the time of his death in 1909, while his store at Muscat contained more than 100,000 small arms (including most patterns of modern magazine rifles) and ten million rounds of ammunition for these weapons.[154] As for shipping arrangements, the bulk of the traffic was already providing ample business for steamers running direct between Europe and the Gulf. For the transhipping stage, Goguyer would give assurances to the leading Indian traders whom he interviewed that 'his operations would be exempt from British interference as he would export arms purchased from him in native vessels flying the French flag'.[155]

Secrecy and subterfuge were the watchwords of the European intermediaries and their agents. Arms were shipped (or transhipped) at the last hour in order to take advantage of confusion or haste. Occasionally, the intention to ship at London was altered because of the vigilance of H.M. Customs, and cases were despatched at Manchester instead. Although the exporters usually received full payment before shipment, the bills of lading even in those instances showed no names of consignees; in such cases, the names were deliberately omitted, as the bills did not pass through a bank and there was no need to adopt the vague generality 'to order'. In one instance, a false name of consignee—Golpalji Walji—was entered, but then repudiated by the latter on arrival. Even the names of exporters were sometimes changed at the last minute in order to divert attention, as when 'Spencer' was substituted for 'Carling & Co.'. Arms were also entered on bills of lading as 'hardware'. When a true copy of the bill was requested on one occasion, the shippers inserted 'merchandise' for 'hardware', which would have been a less inaccurate description, but when the bill was challenged their excuse was that it had been composed 'from memory'. The cases themselves bore marks and not addresses; and even these marks and the destination were altered on board, either at a port-of-call or at sea. The shippers undertook to land not merely at the port originally named, but another port in the region as desired, and the desire of the consignee was again disclosed only at the last moment. Cases of rifles, described in the bills of lading as 'hardware' bound for Bushire, Bahrein, Bandar Abbas, and other Gulf ports, were probably destined for Muscat. As for the importers, they either secured a contract for the collection of the customs, or ensured the goodwill of customs authorities at the periphery through bribes, usually in terms of a special *ad valorem* duty. The receipts from these duties were not then entered as customs receipts, and the greatest care was taken to expunge all marks from the imported weapons. The overall objective was to obfuscate the destination and the names of the recipients, as much as the origin and the contents of the consignments.[156]

Apart from the Europeans, there were numerous indigenous interlopers. Kabarega, the Omukama of Bunyoro, procured firearms and gunpowder for personal use, and acted as a middleman by trading these articles with the Alur in exchange for ivory which he sold to the Arabs at a hefty profit.[157] The Acholi were also adept at gun-running, the means by which firearms were transferred to the Dervishes and the Khalifa in the Sudan.[158] Afro-Arab bands, under the leadership of powerful Arab chiefs like Kilongalonga, Said-bin-Abedi, and Kibonghé, purchased an assortment of muskets, rifles, and shotguns, plus ammunition and gunpowder, to challenge the authority of rival indigenous groups as much as the jurisdiction of the Belgians in the Congo. Indeed, it was specifically for his role in arming these bands and the three chiefs for 'open war' against the Congo Free State that the Belgian authorities had executed Stokes.[159] In a declaration written by hand just prior to his hanging, Stokes had denied selling 'any of these guns, powder or gun-caps ... to the slaves of Kilonga-longa, or the slaves of Kibongi in revolt ... I most certainly did not know of any rupture with the Arabs on the Congo (at the time, December 1892) and the Congo Free State Government'.[160] Nevertheless, apparently incontrovertible evidence turned up in the shape of Stokes' own incriminating correspondence, which confirmed his commercial partnership with Kibonghé.[161]

In the Gulf region, the local arms traders were Omani Arabs, Persians, and members of various Afghan tribes, including Afridis and Ghilzais. According to H. H. Austin,

> [T]hese Afghan gun-runners were men of great enterprise, and richly endowed with daring, cunning, and ingenuity of a high order. They were well provided with money, by the skilful outlay of which they counted on reaping profits of not less than 200 to 300 per cent on the season's operations. Much of the capital collected for the venture it was known was borrowed; so it was pretty certain they would not be balked from their intentions to procure the arms which were in such great demand in their country and on our border.[162]

The Afghan arms traders were masters of subterfuge and disguise. There is evidence that they would assume false identities as Indian banians, Arab horse-dealers, or pious Muslims set on pilgrimage to Arabia, so as to evade the police at ports on the west coast of India. In some cases, however, they were apprehended when thousand-rupee notes were found sewn up between the inner and outer soles of boots and shoes.[163] In the official reports, two figures stand out:

> The leading spirit among the Afghans who went to Bandar Abbas by land bringing their transport with them was at first one Nur-ud-Din

Khan, Lohani; but by 1903 he had been supplanted by a certain
Ghulam Khan, Nasir.[164]

The Ghilzais benefited most from the arms traffic, purchasing large
quantities of rifles and establishing a virtual monopoly of the carrying
trade in arms. They commanded considerable credit among the banians
of Peshawar, Lahore, Delhi, Calcutta and Bombay, and invested huge
sums of money with the principal arms dealers of Muscat.[165]

The local interlopers themselves depended on a thriving network of
indigenous intermediaries: African and Arab caravan traders; Hindu
merchant capitalists; Arab dhow masters; and Baluchi chieftains. The
Nyamwezi, in particular, distinguished themselves from the early nineteenth century as among the most enterprising of African trading
groups, transporting consignments of ivory, beeswax, and some slaves
to the Indian Ocean, to be exchanged for American (*merikani*) cloth
and Venetian glass-beads. They fast acquired a reputation for strength
in carrying heavy loads over great distances. Their long-distance caravans, made up of hundreds if not thousands, would bring the first
manufactured goods of Europe and India from the coast to more
remote societies around the great lakes, such as Buganda, 800 miles
from the Indian Ocean. Their skills and resources as caravan operators
would serve not only themselves but also Europeans, like Charles
Stokes, and other native traders. Conversely, they were seldom able to
deal directly with the mercantile marine of Europe and America in
Zanzibar, because Arab and Indian merchants were already entrenched
on the coast. The obvious strengths and occasional limitations of the
Nyamwezi thus supplied the *raison d'être* of their symbiotic connections with the other trade intermediaries.[166]

It was the renown of the Nyamwezi as inhabitants of a distant land
with many elephants that first drew some of the more daring Arabs to
accompany them on their return journeys. The oral history of the
Nyamwezi describes how their first travellers, reaching the Indian
Ocean at Bagamayo, met 'long-bearded men' (the Arabs) and showed
them their ivory: 'When the Arabs saw this they wanted to go to the
countries where the ivory was obtained.'[167] Rather than enslaving the
Nyamwezi, the Arabs employed them as porters. As the commercial
stakes were raised, Nyamwezi porterage teams were equipped with guns
and hired as armed escorts for the caravans. Once armed, the Nyamwezi gained a certain autonomy: notwithstanding the regular pay for
their services, it became common for parties of Nyamwezi to band
together for mutual benefit, charging standard prices for the food they

supplied to passing caravans, but also levying additional passage tolls (*hongo*) along trade routes. While Burton observed that the Arabs were generally 'too wise to arm a barbarian against themselves', even they could not turn back the tide of the triangular trade; by 1863, the Nyamwezi chief of Unyanyembe had built up an arsenal of around 20,000 guns.[168] In 1871, when Stanley was passing through, he found the Nyamwezi at war with the Sultan of Zanzibar, who was seeking to enforce free passage for his officials. The warlord Mirambo, supreme chief of the Nyamwezi, was adamant about his people's sovereign rights, and his army was sufficiently well-equipped with small arms and spears to conduct a campaign of resistance until he won recognition as an independent ruler, despite attempts by Sultan Seyyid Barghash to impose an arms blockade on him. Mirambo's success stemmed significantly from his acquisition of muskets (*gumeh-gumeh*) and their use by his warriors (*ruga-ruga*). Mirambo's interest in firearms never waned throughout his career, and several of his campaigns were undertaken with the specific objective of replenishing stocks of guns and ammunition.[169]

The Arab-Nyamwezi partnership, motivated by a shared pursuit of profit and power in the midst of turmoil, thereby carried the triangular trade to new heights. Possibly the most renowned of all the Arab merchant princes was Tippu Tib, lord of the Swahili-Arab traders in the interior, who traded nominally under the aegis of the Sultan of Zanzibar while maintaining extensive ties with the Nyamwezi.[170] One of Tippu Tib's mammoth caravans was a procession of 2,000 Nyamwezi porters carrying 2,000 tusks, guarded by 1,000 *askaris*, and hundreds of other followers attracted by its sheer size and prestige. From the Sultan of Zanzibar's monopoly came the trade guns and gunpowder. In 1890 alone, 44,230 lbs of trade powder passed inland through Arab hands, while they controlled large munitions depots within the interior.[171] On the upper reaches of the Congo, Stanley was amazed at the amount of munitions in Arab hands. In the early 1890s, an 800-strong Swahili-Arab colony at Kitangole was responsible for supplying the arms and powder that went through western Uganda to the north.[172] Along with Tippu Tib and his son Bwana Say, who traded munitions to the Congo Free State and the Manyema country beyond the Semliki, there were other notables: Rumaliza, an Arab slave-trader and ivory collector who established at Kasongo in the heart of the Congo an Arab Muslim community; and Sewa Haji, an Indian Muslim merchant capitalist who conducted trade with Uganda from his base at Bagamayo and his inland agency at Bakumbi, just south of Lake Victoria. It was this Arab powder which Lugard blamed for ruining the IBEAC's trade prospects in the region: 'Ivory is all bought up by this powder and it ousts the legitimate trader's goods from the market.'[173]

These decades of local exploitation and collaboration reflected not only the enterprise and fortune of the Arab merchant princes, but also the extensive credit facilities available for backing them at the coast, which frequently crossed ethno-religious lines. Bound by age-old ties, the Swahili and Omani Arabs did business traditionally with indigenous merchant capitalists from British India. An Indian dealer at the coast usually backed the Arab caravan leaders by furnishing trade goods and supplies, receiving in exchange a note bearing a high rate of interest. The banians, Hindu merchant capitalists originally from western India, played a crucial role in financing the arms trade by providing the capital on credit. As dealers, they in turn obtained their supplies from British and German firms in Zanzibar, or British and French firms in Muscat, though there were exceptions like the Muscat firm of Gopalji Walji, which did its own importing.[174] The growth of the banian population at Zanzibar attests to their rising importance in the local economy: they numbered around 1,000 in 1844, but were possibly as numerous as 6,000 by 1860.[175] By the time of Tippu Tib's third and longest venture, Sultan Seyyid Majid of Zanzibar was poised to intervene so as to obtain the $50,000 worth of trade goods he required out of the financial backing of the expedition. Even if this figure was exceptional, it provides some indication of the amount of Indian capital that might be locked up in the interior at any moment.[176]

Once the consignments of ivory and slaves reached the coast, they were delivered into the hands of respective buyers, sold off and shipped away. In the case of ivory, the Indian merchants would go to Bagamoyo to meet the caravan-conductors and settle business: the cost of the caravan plus 15 per cent per annum was charged to the Arab, while the Indian took the ivory to Zanzibar, sold it on behalf of the Arab, and paid the latter the balance. Arriving in Zanzibar typically between July and August, the ivory was either despatched by the Indian merchants direct to Bombay or London, or sold to Hamburg or American merchants on the spot.[177] Slaves, on the other hand, were shipped by Arab dhow masters (*nakhodas*), from ports like Kilwa to destinations such as Zanzibar and Muscat (for Omani markets), Sur and Bushire (for Persian markets), and Basra and Mohammera (for Baghdad and other centres of the Ottoman Empire). The Arab traders from the Gulf would then utilize the northeast monsoon to carry dates—in a very literal sense, the fruits of a slave economy—from Sur, Bussorah, and other ports on the Arabian coast, to Aden and the north Somali ports, where they would be discharged.[178] The nakhodas would progress to Jibuti, where arms were purchased with the proceeds from these sales and from incidental piracy along the way. The arms were then taken down and sold at the Benadir ports en route to Zanzibar, where they normally went for a return freight for the southwest monsoon, which gave the dhows a fair wind back to the Gulf. To the Admiralty, there was little doubt that the nakhodas were 'legitimate traders, but intended to add profits from the sales of rifles and ammunition to their legitimate cargo of dates'.[179] A few dhows were caught jettisoning their consignments of rifles, but dozens more evaded British naval patrols; their Persian and Afghan customers at the Makran coast continued to depend on them, laden with trade weapons from Zanzibar and Muscat, as late as the monsoon season of 1909–10.[180]

These dhows were highly manoeuvrable sailing craft whose traditional design had evolved specifically to harness trade winds and provide regular, swift transportation in the western Indian Ocean. As H. H. Austin noted, 'little more than a happy fluke, such as a dhow being becalmed within the beat of a patrolling ship, was in the least likely to lead to an important capture'.[181] A dhow had every advantage over a man-of-war in escaping detection: the masts or smoke of a cruiser were visible to the dhow long before the cruiser either sighted the dhow or was within range; the dhow could run arms ashore and conceal them under the pretext of calling for water at any of the countless small ports and watering-places along the coast; and there it could lie quiet until nightfall, when it could watch the approaching man-of-war and keep a lookout for its navigation lights. Moreover, dhows could be easily adapted for arms concealment in false bottoms, below cargoes of dates, cloth, and timber, or even within: rifles in sacks of flour and grain, or bales of cloth; and Mauser pistols in baskets of dates, or full tins of kerosene. Alternatively, rifles could be tied up with rope in bundles of ten or a dozen, the other end of which was secured to the outside of the keel of the dhow: whenever there was a danger of being caught by a pursuing ship, the bundles would go overboard until the boarding party were safely out of sight, when the rifles could be hauled aboard again. Finally, bundles of rifles could be bound with large balks of wood; and, when dumped, could be retrieved afterwards once the dhow was released, when nothing incriminating was discovered either above or below the surface.[182]

Awaiting these dhows on the opposite Indian Ocean coastline were multiple caravans of camels, led there from Kabul or Herat during the cold season by Afghan clientele. Other than the aforementioned gun-runner Khalifeh Khair Mahomed, many of these Afghan arms purchasers would at first proceed by rail through northern India to Karachi or Bombay. With borrowed money from the banians cunningly concealed about their persons, they were ready to board the steamer to Muscat, where weapons could be bought openly from various wholesale arms merchants. The consignments would be left with these merchants until the Afghans had finalized arrangements for them to be shipped by dhows to the opposite coast. These voyages took place typically during the hours of darkness, to a secluded cove somewhere along the 350 miles of coastline between Minab (in the west) and Gwadar (in the east) that had been deemed most likely to elude the vigilance of naval patrols. But rarely were there precise reports detailing the arrival of consignments from Muscat on the Makran coast, let alone providing the actual number of arms imported, as during the summer of 1907.[183]

Arms Imported from Muscat into the Makran Coast, 1907

Date of Landing	Place of Landing	Number of Rifles
1 May	Near Humdan	76
2 May	Rapch	100
12 May	Darak	30
23 May	20 miles west of Galag	150
25 May	Kunarak (near Chahbar)	100
25 May	Tank	200
26 May	Darak	100
29 May	Darak	100
30 May	Tank	10
30 May	Galag	200
23 June	Kunarak	30
9 July	Sadaich	50
21 July	Gurdim	50
28 July	Kunarak	50
18 August	Gurdim	100
22 August	Kunarak	1,000

Source: IO, L/P&S/18, Memorandum D 182, Appendix T, 'The Arms and Ammunition Traffic in the Gulfs of Persia and Oman'.

Most of the time, the surveillance systems were artfully dodged. Once the arms were unloaded, they could be removed inland before daylight and stored securely under the charge of friendly Baluchi chieftains. The Baluchi *sirdars* in occupation of Persian Baluchistan were paid a commission on every rifle and pistol landed within their territory by the gun-runners. Occasionally, they provided armed escort for the carriage of merchandise northwards. When the various parties of helpers arrived with their camels, they took possession of the merchandise, concealed it within the framework of camel-saddles, combined into a single large caravan, and returned via the safest routes through Persian (rather than British) territory towards Herat or Kandahar. The mysterious disappearance of the camels was sometimes the only trace left by these transactions; at other times, decoy camels were deliberately 'parked' at one spot to mislead the authorities while the munitions were carried off from another point. Once within their own borders, the parties again split up before delivering the arms and ammunition to the most promising home markets.[184]

Since the stakes were so high, smuggling tactics as well as travel routes and timing had to be varied in order to give caravans and dhows the best possible chances. Forward planning, which decided the arrangement of purchases, transport, routes, and rendezvous-points for the following season, became increasingly important as the *cordon sanitaire* imposed by the Europeans tightened. For the players participating in this game of risk, the prospective winnings were apparently well worth the veil of secrecy and recourse to methods of high ingenuity. For the European officers who were ever at their wits' ends to stay ahead of the game and avoid getting outmanoeuvred, the growing sophistication of such activities was duly acknowledged. H. H. Austin would thus conclude: 'In the preceding pages an attempt has been made to give some idea of the complexity of the situation, to disclose the network of intrigue, and to prove the complicity of practically every chieftain and tribesman on the Arabian and Persian coastline in this trade. Its ramifications were widely distributed, and covered a huge tract of country.'[185] It was a business that could prove just as tricky for those whose support was enlisted in order to apprehend the gun-runners. As Austin pointed out, 'This demanded the employment of secret agents in India, Afghanistan, Arabia, and Persia, in order to unravel its intricacies, and to lay bare the machinations of those who utilized every form of device to hoodwink the opposition to the traffic, initiated by our Government. And these agents ... ever carried their lives in their hands; for one false move, which might give them away, would probably result in their having their throats cut.'[186]

Whatever the attendant dangers, there could be no doubt that the triangular trade was a highly lucrative business, radiating wealth from the port-cities on which it was based. It constituted a formidable vested interest for the African warlords, the Arab merchant princes, the banians, the nakhodas, and the sirdars involved in its profits and plunder. It rewarded the raiders, the smugglers, the financiers, and the dealers; and it more than covered the commercial cost of caravans, dhows, food provisions, customs duties, the death of many slaves, and the confiscation of many guns en route to the bazaars. At average market prices in the late 1880s, £15 could buy up to three adult slaves, one 36 lb tusk, or between 20 and 30 Sniders.[187]

The *Baluchistan* affair

For the numerous local communities who earned their livelihood from the commerce of the western Indian Ocean zone, smuggling was partly the continuation of older indigenous patterns of exchange now forced to become clandestine because of state regulation. For the colonial authorities, contraband was the sale by non-state agents of commodities legally subjected to state monopolies or restrictions, a definition that would encompass the import and export of munitions across the western Indian Ocean. In that historical milieu, perhaps no single case of arms smuggling and contraband seizure turned out to be as revealing as the affair of the steamship rather aptly named *Baluchistan*. In the *Baluchistan* incident of 1898, the arms trade dilemma unfolded in such a manner that collaboration and crisis in indigenous society were matched by controversy in the imperial public sphere. This last section will demonstrate how indigenous, colonial, and metropolitan circles each developed different perceptions of the arms trade, all tied up with conflicts of interest which were never easy to resolve.

Shortly after the outbreak of the Second Anglo-Afghan War in 1878, the men on the spot had started writing reports about the large consignments of precision weaponry leaving European shores to be traded to bazaars in the Gulf region, which were then transferred to Afghan forces at Herat and elsewhere.[188] The Government of India had introduced arms prohibition as far back as 1880, in consideration of this heightened security threat. The Persian Government issued its own prohibition in 1881. Yet these prohibitions were apparently flouted for nearly two decades, with the knowledge of both governments. Why this was so has remained something of a mystery.

What is beyond doubt is that societies flanking the shores of the western Indian Ocean had been transformed by over a century of crisis, effectively undermining old land empires like Persia and creating new opportunities for maritime states like Oman. By the late nineteenth century, the weakness of Safavid central authority *vis-à-vis* the outer provinces was evident in the administration of the Persian Gulf, where the Shah had delegated his authority to regional governors who bought their offices and ruled as free agents in all but name. It rapidly produced a dichotomy between theory and practice in the administration of Persian law. In theory, the law was the will of the Shah: in his prohibition of 1881, the Shah established that the right to purchase arms and ammunition was a state monopoly, and he appointed officials to ensure that his decree was carried out.[189] But it was also customary for the officials to pay for the positions they held, to enforce an exorbitant tariff, and then to pass the prohibited articles through the customs house. In 1890, the governors paid the Shah £12,000 for their posts; by 1897, the going rate was £60,000. After the trade had been allowed to continue, and the customs-farmers had reimbursed themselves from its proceeds, a sizeable proportion of merchandise upon which duty had been paid was then sequestered as contraband, benefiting both the Shah and his officials.[190] The appointment of a Persian preventive officer in January 1896, under the title *Amin-i-Aslihah*, made no difference to the traffic; it was soon apparent that his real function was to tap the stream of bribes paid by arms dealers for the benefit of high Persian officials and the Persian Government itself. Bushire had by this time developed into a major arms emporium; it was estimated that at least 30,000 rifles were landed there in 1897, on which the local governor recovered duty at the rate of 8 to 10 per cent *ad valorem*. Shiraz had become an important distributing centre dependent on Bushire; and Bahrein, an emporium of secondary importance.[191] Not to be outdone in neighbouring Muscat, its Sultan also sought to capitalize on the flourishing arms trade by raising his own duty from 5 to 7.5 per cent.[192]

It was such 'laxity and venality', to use Curzon's words, that so drew the ire of the British public. When a delegation of Birmingham gunmakers approached Curzon at the Foreign Office, one of their representatives declared: 'It was a trick of the wily Oriental, and seemed more like a chapter out of the *Arabian Nights* than a piece of modern statesmanship.'[193] In British colonial circles, this product of retrograde 'oriental despotism' was perceived as nothing more than 'a bogus prohibition'. All the consular reports issued during 1881–91 refused to accept that any legal restriction really existed, and the report for 1895 framed the situation more starkly yet: 'Theoretically, this trade is prohibited by the Persian Government, but, like all similar prohibitions in Persia, this practically only substitutes an arbitrary impost for a fixed duty.'[194] The British Government was also in an awkward position *vis-à-vis* Muscat, where the Sultan had agreed in 1891 to restrict importation of arms into his Makran enclave of Gwadar, but had refused to do so the following year in relation to Oman itself.[195]

This growing demand for arms via the coastal gateways was, of course, a piece of the larger jigsaw of indigenous crisis and conflict. The weakening of centralized authority had encouraged tribal autonomy in the interior, often manifested in terms of local unrest or warfare. Observers in the British colonial sphere had little doubt that the relaxed and rapacious attitude of the Shah, the Sultan, the sheikhs, and their subordinates towards arms trafficking had led directly to the arming of militant tribes in southern Persia and Oman:

> In the neighbourhood, especially of Bushehr, no grown man any longer enjoyed social consideration or could obtain a wife who did not possess a rifle. The Bushehr-Shiraz road was frequently closed by the heavily armed tribes of Tangistan, who, if they had been united among themselves, might even have been able at this time to capture the town of Bushehr. In fact, as aptly remarked by a Persian, 'Martini Khan' was now Shah of the south of Persia.[196]

Empowered by firearms, the tribes of Makran made several assaults on the telegraph system of the Indo-European Telegraph Company, which culminated in the murder of an English telegraph agent in December 1897. The Persian Government and the local chiefs, bound to give protection under agreements with the Government of India, soon found themselves impotent against these armed 'robbers' of southern Persia. Uprisings in Dhofar and Mattra against the authority of the Sultan of Oman proved likewise that the tribesmen were armed and dangerous. Finally, the recrudescence of indigenous 'piratical attacks' on British Indian shipping in the Shatt-el-Arab and the Gulf, emboldened through arms transfers, was really the maritime equivalent of tribal breakout.[197]

By the end of 1897, the scenario was poised for a sea-change, and it took the eruption of just one more armed conflict near the core of British colonial rule to expose all the scandalous twists and turns of international arms trafficking. If Africa was largely unknown to the Europeans until the late nineteenth century, and Arabia was unstable, the untamed frontiers of Afghanistan proved just as needy for weapons. The tribes of Afghanistan and the Northwest Frontier were characteristically factious and would not yield to central authority. Britain's ploy of installing a puppet regime had failed twice, leading to anticolonial resistance and outright war in 1839–42 and 1878–80. Between the British occupation of the Punjab from 1849 and the outbreak of the First World War in 1914, some 52 punitive military expeditions were mounted into tribal no-man's-land, in addition to ongoing small-scale engagements along the shifting frontier line between Afghanistan and British India. According to T. R. Moreman: 'Such campaigns were undertaken as the independent Pathan tribes posed a constant military threat to the security of India, given their proclivity to raid trans-Indus districts, disrupt the tribal areas, and cause trouble with both the government and tribes of Afghanistan.'[198] From the early 1880s, the Northwest Frontier became a matter of grave concern to both military and civil authorities in India. In 1893, the negotiation of the Durand Line set a boundary for Indian and Afghan influence over the tribes of the Northwest Frontier. Yet the instability persisted: it was a volatile combination of deep mistrust of the British; perceived threat from Russia; fears of Ottoman or Persian aggression; Islamic revivalism and growing militancy among the frontier tribes; plus a further manifestation of forward policy. By the 1890s, it was apparent to the men on the spot that significant numbers of 'arms of precision' were finding their way into the hands of the tribesmen, thus offsetting the British superiority in firepower. After the Afridi revolt of 1897, it was reported that the flow of arms traffic veered round to the Northwest Frontier; consequently, large quantities of modern rifles began to replace the traditional long rifled-muskets (*jezails*) with which the Afridis had been armed hitherto.[199]

It was against such a background that the *Baluchistan* affair occurred. The S.S. *Baluchistan* was a freighter owned by Messrs. F. C. Strick & Co. of London, otherwise known as the Anglo-Arabian and Persian Steam Navigation Company. Her cargo in late 1897 consisted of 306 cases of cartridges and 280 cases of rifles. These arms consignments had been manufactured by several well-known Birmingham gunmakers, including Messrs. Isaac Hollis & Sons. They had been underwritten by the Sea Insurance Co. Ltd. and Lloyds Bank; and had been procured by Messrs. Fracis, Times & Co., an Anglo-Parsi house with agencies established at Bushire, Bahrein, and Muscat. In the ship's manifest and bills of lading, the cargo was described as 'hardware' and mostly marked 'Bahrein via Bushire—optional Muscat'.[200]

With everything apparently in order, the voyage of the *Baluchistan* was ready to begin. On 10 December 1897, she sailed from the newly-constructed Manchester Ship Canal with her large consignment of arms and ammunition. But while travelling en route to Marseilles, the next port-of-call, her owners received word that the Persian authorities had suddenly revived the strictest terms of their prohibition. On 7 December 1897, operations had commenced with the seizure in Bushire of 150 rifles and 149,000 cartridges belonging to Fracis, Times & Co. By 15 December, the Shah had authorized British cruisers to search all ships trading in the Gulf under the Persian flag, any contraband found on board being handed over to his officials. All shippers now became anxious to divert their consignments from the Gulf. Indeed, the owners of the *Baluchistan* took the additional step of writing to the Foreign Office in reference to the orders given by the Persian Government. They appealed to Lord Salisbury 'to give such instructions as will enable our steamer to have protection in this matter'; and they admitted that the Resident, Colonel Meade, had informed their agent that 'the arms are liable to be seized and confiscated in Persian waters'.[201] Meanwhile, at Marseilles, the freighter picked up an additional 272 cases of Russian-made rifles for Ethiopia via Jibuti, resuming her journey to the Gulf on 27 December 1897 via Port Said and the Suez Canal. At Port Said, the captain of the *Baluchistan* received telegraphic instructions to amend the ship's manifest and the port-marks of all merchandise bound for Persian destinations: all port-marks were erased; if 'Muscat' already appeared, it was left; and if not, it was added.

Unknown to them, however, in pursuance of the Anglo-Persian understanding, the Sultan of Oman had been approached by Colonel Meade, and was persuaded to issue a similar prohibition throughout his domains and territorial waters (dated 13 January 1898). On 15 January 1898, 250 cases of magazine rifles were successfully transhipped at Jibuti to the S.S. *Dwina* (under the French flag), but on 24 January, the *Baluchistan* was intercepted two miles off the coast of Muscat by H.M.S. *Lapwing*, and from her were seized over 400 cases of 'contraband'—7,856 rifles and 700,000 rounds of ammunition. On 15 April 1898, a court convened by the Sultan of Oman ruled that they had been justly confiscated, and they were held in the British Consulate until the spring of 1900, when the key legal actions arising out of the affair had been decided and the contraband was made over to the Sultan.[202]

News of the Government's action at the periphery provoked a domestic backlash that began with incomprehension and ended in litigation. The seizures at Bushire and Muscat were both sensationalized and criticized by the firms and traders whose interests were affected. The Foreign Office in London was flooded with letters and besieged with representations from the manufacturers, the exporters, the ship-owners, and the underwriters whose pockets had been hurt in various ways. Charles Playfair, Chairman of the Guardians of the Birmingham Proof House, warned that the British Government's assistance in the seizure had been 'arbitrary and without justification', while the consequent stoppage of the trade would not only seriously reduce the revenue of the proof house, but also damage a major branch of the industry.[203] Even more critical was the eminent Birmingham industrialist who now presided over an ammunition firm with a major interest in the Gulf trade:

Mr. Arthur Chamberlain in his speech at the Kynoch Annual Meeting comments most unfavourably upon the Government's way of telling British merchants and manufacturers to be enterprising and to push their commerce abroad when they accentuate their policy by seizing their goods and ruining a flourishing industry.[204]

The fact that he was brother of the Secretary of State for the Colonies, Joseph Chamberlain, added an ironic twist to the tensions between private, state, and imperial interests. Lloyds, on the other hand, was prepared to go more quietly: as underwriters claiming they had insured the arms in ignorance of their destination, all they requested was that they 'would not be allowed to suffer loss'.[205] Yet all the aggrieved parties agreed that the British Government's latest action actually harmed the national interest, for the Government could furnish no substantive evidence in 1898 to prove that Birmingham's exports were going as far as the frontier tribes to the detriment of Britain's colonial interests.[206] 'For that matter,' as one trade journal commented, 'nobody knows to this day why it was right to send arms even to Persia in November, 1897, and wrong in December.'[207]

Legal proceedings commenced against the government officers involved. The aggrieved parties were depicted as participants in 'an honest trade', carried on for nearly two decades. That was now threatened with extinction by the 'rapacity and greed' of an autocratic foreign regime whose legal prohibition had been 'a dead letter' for years, not to mention the caprice of a British Government which had actually 'assisted in this wanton act of spoliation and oppression'.[208] Fracis, Times & Co. filed a suit against Commander Carr of H.M.S. *Lapwing*, for the value of their arms seized on the *Baluchistan*. They filed another suit against Colonel

Meade, in relation to the confiscations at Bushire as well as those at Bahrein in January 1898, where the Sheikh had followed the example of Persia and Oman in seizing the stock owned by the plaintiffs. In both suits, defence was undertaken by the Treasury; the final result in each case was favourable to the defendants and the Government. In the suit against Carr, the final ruling of the House of Lords on 8 July 1901 established that no British tribunal could overturn the decision of the Muscat court which held that the munitions seized were intended for Persia; that nothing had been done contrary to the law of Muscat; and that the seizure was legal. In Meade's case, judgement was given for the defendant on 22 May 1901, with high credit for having 'acted not only with great discretion and ability, but also with the strictest regard to the requirements of the law'. Conversely, these proceedings would reduce Fracis, Times & Co. to bankruptcy: in the *Baluchistan* affair alone, their losses amounted to between £30,000 and £40,000. Part of the losses fell upon the underwriters in London, whose efforts to obtain indemnification did not end until 1906, when their claims were finally rejected by the Government.[209]

In retrospect, what seemed most peculiar about the whole case was the fact that the *Baluchistan* had been allowed to journey onward to Muscat, even though the nature of her cargo was known before she sailed. There was no doubt in the British official mind that the arms trade had become, by the late 1890s, a 'nefarious business'.[210] It was Curzon's view, as Under-Secretary for Foreign Affairs, that the 'system of altering port marks proved that they well knew they were carrying on a deceptive and illicit trade'.[211] Yet full details of the contents of all cases actually had been declared beforehand to H.M. Customs, the cases had been opened for inspection prior to shipment, and British government officials had cleared the goods for export.[212] Even if the consignees listed in the bills of lading subsequently denied all knowledge of the transaction, who would not, given the circumstances of the seizure? More importantly, what could possibly account for the British Government's puzzling behaviour in letting the goods slip through the metropolitan net in the first place?

The evidence suggests that the British Government found themselves on the horns of a dilemma, caught between the burden of proof and the need to apportion blame so as not to tarnish its own reputation. In response to the question of proof, Curzon had explained in the House of Commons on 21 February 1898 that there was some reason to believe that British-made weapons had reached the frontier tribes against which British troops were engaged in conflict; 'such was the opinion of high officials on the spot', though Curzon himself had 'not yet seen the evidence on which the opinion was based'.[213] The Birmingham gunmakers were scandalized at the 'excuses' of the Government's 'logic-chopping Under Secretary'. They maintained that the Government was the credulous victim of press propaganda, 'the newspaper campaign which was started in India, and then propagated in England, to the effect that the Birmingham gunmakers were arming the hostile tribes on the Indian frontier'; or, alternatively, it was an official conspiracy such that 'the whole thing was arranged from Whitehall in the supposed interests of the British Government'.[214]

Back in December 1897, the gunmakers had been more generous in their admission:

> That the Afridis have some British manufactured arms in their possession is more than probable. It is known that they have rifles and carbines of very many systems, and the wonder would be if, in such a collection, Birmingham productions were not represented. Since arms do find their way to the Northwest frontier, we may be sure that some of them originally came from Birmingham. It is known for a fact that Martini Government rifles, and possibly Lee-Metfords also, are included in the munitions of the tribesmen, but we have not yet heard the British Government accused of supplying them. The truth is that the arms have been obtained surreptitiously, and in almost all cases the original makers have had nothing whatever to do with the transaction.[215]

As the debate grew more heated, however, W. H. Hughes, Chairman of the Birmingham and Provincial Gunmakers' Union, spoke plainly:

> We have no reason to doubt but that the arms shipped to Persia and to Muscat are for the use of the numerous tribes on either side of the Persian Gulf. We believe that the statements which have been freely circulated to the effect that the rifles have been supplied to the Afridis are without foundation. We have been anxiously looking for the appearance of some explanation of these erroneous statements but none has been forthcoming, and indeed, the only evidence available (which includes a special report which the Government Arsenal at Rawal Pindi has been called upon to make) does not in the least degree support such a theory. Until some convincing proof is forthcoming, the Birmingham gun trade refuses to believe that these rifles have reached the frontier, or that they were purchased with such intention. ...

We have no hesitation in saying that the Birmingham makers inter-ested in the Persian gun trade have been perfectly innocent of any intention to supply arms to native tribes who are at warfare with this country.[216]

In further support of their arguments, the gunmakers cited an article in *The Times of India* which held that,

The Afridis and other clans know nothing of the Martinis which, not so long ago, were bought and sold so freely in Southern Persia. They rely upon the professional rifle-thieves, or upon their skill in rushing sentries, for the supply of breech-loaders. Since the war came to an end there was a brisk trade in rifles captured from our troops in action, Lee-Metfords and Martinis, over 300 in all, and these have become widely distributed. ... We may take it, there-fore, that the Birmingham and Belgian gunmakers have never, even indirectly, supplied the Pathan clans with breech-loaders.[217]

One trade journal lampooned the authorities in an amusing rhyme of 'What Happened', when, having obtained government permission, the 'Babu journalist'

Hurree Chunder Mookerjee sought the gunsmith and
Bought the tubes of Lancaster, Ballard, Dean and Bland,
Bought a shiny bowie knife, bought a town-made sword,
Jingled like a carriage horse when he went abroad ...[218]

The upshot was that the whole outfit was stolen from him by natives and sold 'across the border'. The gun trade concluded that the Govern-ment 'got bitten with the craze that the Afridis were getting arms and ammunition through Persia'.[219] This assertion was not far wrong.

Since the Second Anglo-Afghan War of 1878–80, circumstantial evid-ence furnished by men on the spot had linked weapons in Afghanistan and the Northwest Frontier to the Gulf trade, but scepticism pervaded official circles well into the 1890s.[220] By 1896–97, however, the Govern-ment of India and the Government at home had effectively decided, without waiting for definite proofs that the Northwest Frontier was affected, to take measures against an 'illicit trade between England and Persia—an evil in itself of no slight magnitude'.[221] In late October 1897, the India Office directed the British Consul for Kerman, Captain Sykes, to clarify the connections. Although Sykes' findings still lacked conclusive evidence, some ammunition recovered from Afridi territory was apparently manufactured by Kynoch and the Société Cartoucherie Belge, possibly imported through the Gulf.[222] For Curzon, this was sufficient; the real problem now was that there was 'no provision in English law under which the sailing of the *Beluchistan* with arms intended for illicit import into Persia could have been prevented'.[223] Therefore, out of intuition rather than certainty, the British Government moved swiftly to protect imperial interests over domestic concerns: the *modus operandi* would be to involve indigenous rulers in the enforcement of indigenous laws, with the blame falling squarely upon them; the hard evidence, if any, would have to come later.

When the Shah decided to revive his prohibition in November 1897, the British Government seized the chance they were waiting for. The official line was that Britain 'acquiesced and assisted in that proceeding. ... As a matter of international comity, her Majesty's Government could not refuse that invitation' from Persia.[224] The Birmingham gunmakers, on the other hand, regarded it as a case of late Victorian gunboat diplomacy: 'So far as the *Beluchistan* is concerned, the Persian Government has nothing to do with the seizure, which was effected by a British gun-boat under orders from home.'[225] Moreover, following on from the Anglo-Persian understanding, the Sultan of Oman had been pressed by Colonel Meade, on the British Government's behalf, to issue an edict outlawing the arms trade at Muscat.[226] Defending their actions before the aggrieved parties, Curzon could now rightfully claim that 'it was not the habit of the Government to give notice that they were going to enforce the law'.[227] Whereas British officialdom had not been always above reproach in its conduct *vis-à-vis* the Gulf arms trade in the 1890s, a more consistent, hard-line policy was definitely being formulated by the close of the decade, confirmed in the selection of Curzon as the next Viceroy upon Elgin's retirement.[228]

Whether the end justified the means might remain a matter of debate. What is certain is that the various authorities became involved in the affair for their own ends. The Shah was re-arming his army—and the seizures were manoeuvres apparently designed to supply his troops at no expense—while his former governor of Bushire, the *Malik-ut-Tujjar*, was delighted to perform the operations on condition that he could keep as a reward one-third of the arms confiscated. The Sultan of Oman had little choice but to accept the Anglo-Persian overtures, if he valued continuing British military aid against tribal revolts from the interior. Following the seizures from the *Baluchistan*, he was reluctant to appoint a court to try the case, 'being at first anxious to throw the entire responsibility on the

British Government'.[229] The British Government and law-courts, for their part, sought to exonerate British officers, while shifting ultimate responsibility to the sovereign, if arbitrary, policies of Tehran and Muscat.[230]

Nonetheless, hard evidence did start to turn up. In February 1898, it was proven that Afghan emissaries were awaiting the arrival of the *Baluchistan* at Muscat; in one of their homes at Pasni were found 20 packets of Martini-Henry ammunition, with paper showing it was made in Belgium. Liège manufacturers had obtained the exact measurements of the British Martini-Henry rifle only the previous year, such that any ammunition made by them could be used by tribes with access to those rifles. If the exporters had not already removed all marks from the rifles and cartridge cases, it was still official knowledge that caravans from Makran, bound for Afghanistan and the Northwest Frontier, surrendered only 'worthless, stolen, or specially-made arms, and no rifles have been captured in the field, because rifles as well as the bodies of the slain are removed by the tribesmen'.[231]

Then, in late November 1898, a Martini-Henry rifle marked 'Fracis, Times & Co., 27 Leadenhall Street, London' was actually purchased from a local Ghilzai trader by Captain George Roos-Keppel, the British Political Officer stationed in the Kurram Valley. Roos-Keppel observed that many similar weapons were available in the vicinity. In the summer of 1901, cartridges dropped by a Mahsud raiding party and used by other frontier tribes were found to bear the initials 'F.T.C.' and a double-headed eagle, the emblem of Fracis, Times & Co. The British Political Officer of South Waziristan reported the ease with which he was able to buy such cartridges locally, at the rate of 5,000 a week. Revolvers and Lee-Metford rifles, some of which were identified similarly as supplied by Fracis, Times & Co., were also found in the possession of local chiefs and their followers.[232] By 1902, arms were being captured in the Northwest Frontier that carried, beyond all doubt, the stamp of Muscat exporters.[233]

Just as alarming, too, was the fact that a 'semi-official' avenue for arms trafficking had opened up across the Afghan border. Initially, British cash subsidies and arms supplies were intended to centralize and strengthen the authority of a friendly native ruler in Afghanistan. But the matter of 'open and unrestrained' arms transfers from British India to the Amir of Afghanistan grew increasingly problematic, especially with Indian troops re-arming with modern weaponry by the turn of the century. As Afghan loyalties became questionable, threat perceptions were shaped by the 'leakage of rifles from Cabul to the hostile tribes of the frontier'.[234] Persisting instability at the frontier seemed to vindicate, in part, the prohibitions enforced by the various governments that had interests in the western Indian Ocean zone. Regardless of the shadier aspects of arms supply and distribution, the authorities faced a regional crisis that was very real; cross-border conflicts were indeed being escalated by weaponry supplied through an arms transfer system of increasing spread and sophistication, in which the *Baluchistan* itself was merely a pawn in the hands of key players.

Looking back in retrospect, the *Baluchistan* affair opens a window on long-term patterns of trade and politics, as well as more fundamental shifts in the political economy and societies of the western Indian Ocean. In this present age of rapid global communication and nebulous asymmetric conflict, we are perhaps no longer accustomed to viewing steamships as gun-running vessels. But the dhow trade continues to fascinate, associated as it has been with subterfuge and the smuggling of contraband for hundreds of years. Dhows played an important role in facilitating the arms trade of the colonial period, and they still prove their adaptability through ongoing participation in the murkier international commerce—if not piracy—of the modern world.[235] Despite the fall in elephant numbers, the proliferation of modern firearms would guarantee a virtually uninterrupted supply of ivory through darkened underworlds, even as poaching survives in many parts of Africa.[236] The combined legacy of colonial activity, indigenous crisis, and arms trafficking is also evident in the political instability gripping societies from East Central Africa to Afghanistan, perpetuated by warlord armies high on drugs, alcohol, and gun-related violence.[237] Even if the triangular trade of the western Indian Ocean no longer operates as once it did, it has indeed cast long shadows; the echoes of gunfire are still being heard in the tribalism and terror of today.

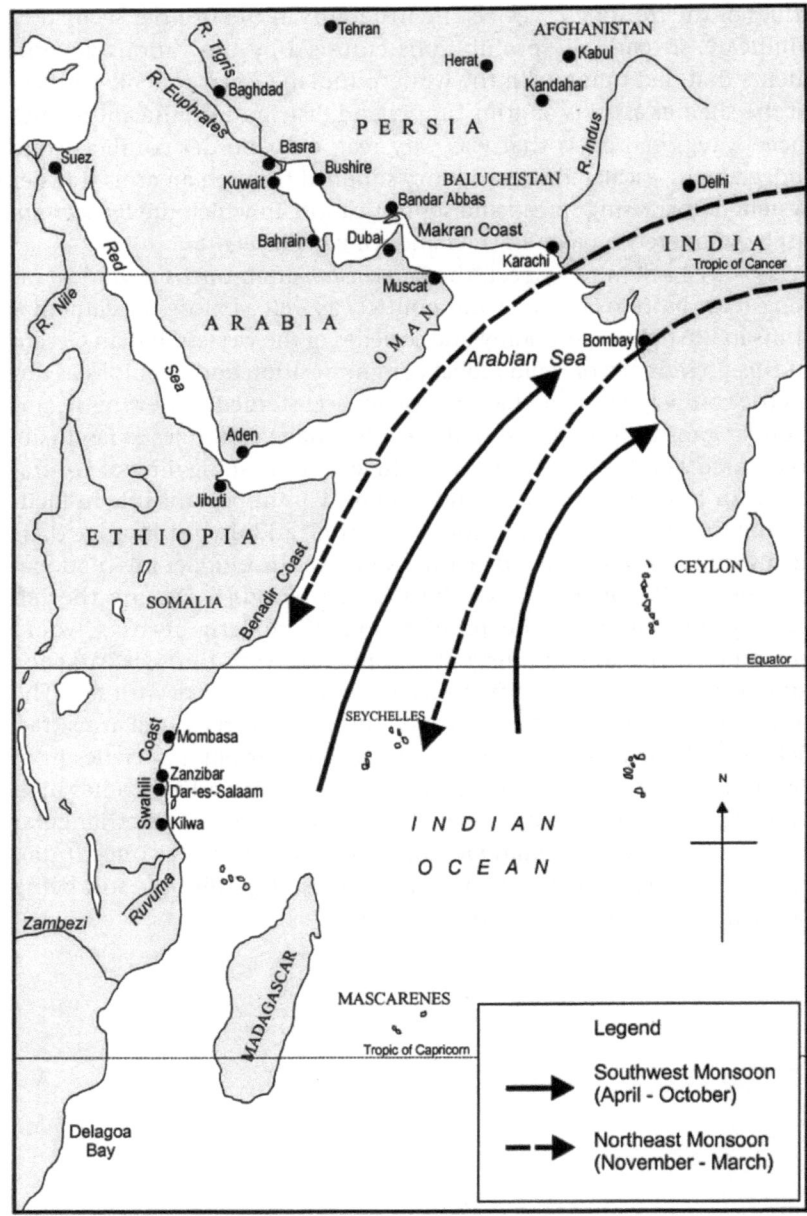

The Western Indian Ocean Zone

4
The Arms Trade in the Eastern Indian Ocean

Singapore is now viewed as the general Place of supply of Gunpowder and Firearms to all Pirates and disorderly Persons infesting the Straits of Malacca.

Robert Fullerton, 1830[1]

Decades before the gun-running scenarios of the western Indian Ocean had even materialized, a substantial trade in 'warlike stores' was already being conducted across the frontiers of the eastern Indian Ocean. British trading outposts at Penang, Malacca, and Singapore, established between 1786 and 1824, were soon cast as commercial entrepôts in the bygone tradition of the great maritime emporia. Reconstituted as the British Straits Settlements in 1826, these bastions of free trade were generally allowed to develop extensive business connections and facilities for arms distribution. Historically, port-cities like these would prove instrumental in sustaining the supply of munitions and other contraband to societies throughout Southeast Asia. Incidentally, as transhipment nodes of a wider transoceanic network, their reach could also extend westward as far as the Gulf region, or further eastward over the South China Sea to the Chinese mainland.

The island-colony of Penang was thus pivotal in the distribution of firearms and explosives, via the Straits of Malacca, to parts of Sumatra including the Sultanate of Aceh. The old port-city of Malacca, formerly one of the region's greatest emporia, continued to serve as a conduit for arms transfers albeit on a lesser scale.[2] But it was cosmopolitan Singapore, the most commercial and strategic of the Straits Settlements, which became the principal node of an arms trade network enmeshing the entire maritime zone then labelled the 'Eastern Archipelago' or 'Eastern Seas'.[3] Two other transhipment nodes that thrived on a symbiotic relationship with

Singapore were Labuan (off the coast of North Borneo) and Jolo (in the Sulu Archipelago); both supplied exotic local products to Singaporean markets, in return for the munitions that would be sold in their own native bazaars. The European chartered companies were themselves major arms dealers; the English East India Company sealed a lion's share of the deals under licence from its Court of Directors in London, alongside myriad other intermediaries. Various Europeans, Americans, Chinese, Malays, Indians, Arabs, Bugis, Tausug, and other indigenous groups participated in the multinational circuits and intercontinental transit of this globalizing network. Their cargoes of muskets, rifles, ammunition, and explosives were offered in warehouses and bazaars, whether behind closed doors or in open stalls. Their customers, drawn from nearly every ethnic category, were scattered across the entire region.

But episodic regional instability or conflict required the periodic application of trade restrictions. Governor Fullerton's remarks at the beginning of this chapter are a sharp reminder of how the rapid growth of arms traffic at a strategically positioned British free port came to be regarded historically as a security menace. The arms trade at early colonial Singapore flourished to such a degree that visiting warrior-traders and sea-robbers, raiders and rebels, could all stock up on munitions with relative ease. This would present a major asymmetric challenge to the authority of the colonial state in the region. Barely a decade after the settlement had been founded upon free trade principles, the British colonial administration at Singapore saw fit to introduce arms prohibition and implement surveillance on smuggling activities on the waterfront. In the weeks leading up to Fullerton's statement in March 1830, reports were filed concerning the surreptitious movement and storage of munitions in warehouses along the Singapore River, mostly intended for transhipment elsewhere.[4]

Again, the introduction of arms control achieved redirection in the distribution pattern rather than reduction in the overall magnitude of arms traffic, with the flows of contraband veering to areas outside the zone of prohibition or beyond the orbit of the European colonial spheres. Places further eastward in the Indonesian Archipelago, such as Portuguese Timor and New Guinea, or further northward from Siam to southern China, yielded fresh outlets for arms trafficking. Ports along the coast of British Burma, or French Vietnam and Cambodia, afforded alternative transhipment facilities until colonial legislation attempted to close those channels as well.[5] Even then, munitions could be somehow re-exported from distribution centres in the Eastern Seas to destinations thousands of miles away in the Arabian Peninsula, via the Persian Gulf.[6] Such fluidity and flexibility underscored both the interconnected nature and extraordinary subtlety of the evolving arms transfer system; it was developing a capacity to link arms-producing centres in metropolitan Europe to an intermediate transhipping nexus in the eastern Indian Ocean, before consignments finally reached—and armed—clientele in the western Indian Ocean.

Arms prohibition in the Eastern Seas was bedevilled by two main problems: in the geographical context of fluid maritime boundaries and porous border areas, the arms control regime could never be truly watertight; and, in the economic context of profit-maximizing colonial free trade policies, it could hardly be sustained at any consistent level. Only the imminent threat of regional instability or conflict seemed to warrant any conspicuous state intervention against the free passage of munitions through the Straits of Malacca and Singapore. Under such circumstances, it was impolitic to allow strategic goods to fall into the wrong hands, since this might empower piratical attacks, anti-colonial revolts, or other forms of indigenous warfare. Restrictions would become more urgent as the military technological innovations of the 1850s–80s engendered shiploads of increasingly sophisticated, lethal weaponry. It was then that the region witnessed significant movement from largely competitive policy frameworks to cooperative regimes: the shift from limited arms traffic regulation, shaped by constant vacillation between *laissez-faire* diplomatic and commercial initiatives (emphasizing revenue and profit), to stricter international arms control, based on multilateral cooperation (emphasizing security).

Arms bans were re-imposed throughout the later nineteenth century, primarily to safeguard British and Dutch colonial interests at times of regional crisis or cross-cultural collision, such as the Taiping Rebellion in China (1851–64) or the Aceh War in northern Sumatra (1873–1904). These European powers found themselves labouring to dismantle the combined legacies of largely unregulated arms traffic under decades-old colonial free trade policies, and centuries of open trading activity under pre-colonial regimes. Arms prohibition was an uneven process, with metropolitan, colonial, local, and personal factors all affecting the speed and degree of enforcement. Only when vastly expanded supplies began to arm insurgents in the Philippine Revolution (1896–98), and threaten wider anti-colonial resistance at the height of the new imperialism, did colonial regimes finally overcome entrenched economic interests to attempt international arms control in the broader interests of state and regional security. This need for cooperation was made explicit in the Jolo Protocol of 1897, introduced by the Spanish colonial authorities with a view to pacifying the Sulu Archipelago, but involving both the well-established British authorities and German newcomers looking toward the Pacific for their 'place in the sun':

> A considerable trade in arms and munitions has for a long time been carried on, encouraging internal strife among the inhabitants of the Archipelago and furnishing to the Philippine rebels the means of combating our sovereignty.

> By the explanatory protocol now signed, a serious evil will be removed ... the Spanish Government will accomplish in Joló what was agreed on in the Conferences of Berlin and in the Brussels Act with regard to the importation of alcohol and Arms in Africa and in Ocean dependencies.[7]

Just as rampant gun-running in the western Indian Ocean zone demanded international arms regulation in the Brussels Conference Act of 1890, calling for the implementation of naval blockades and other law enforcement measures, a multilateral approach would be required to tackle a similar problem in the eastern Indian Ocean.

This chapter considers the extent to which the arms trade in the Eastern Seas followed its own developmental path, yet also operated as an integral component of a wider arms transfer system across the Indian Ocean arena. The pattern and volume of arms traffic, from the fringes of the Bay of Bengal to archipelagic waters in Southeast Asia that opened out onto the South China Sea, were determined as much by peculiar fixed features of regional geography as by more variable processes of cross-cultural interaction. The porous nature of this uniquely fluid, fragmented geographic region—encompassing two of the world's largest island chains, further enhanced by the allure of exotic products—not simply encouraged but actually hastened the global interpenetration of colonialism, racism, and capitalism. All in all, underlying factors such as these would account for the particular timing and nature of the European colonial impact; the critical stresses and ruptures in indigenous societies; the specific composition of maritime trade and production; and the various intermediaries who participated in age-old monsoonal cycles of seaborne commerce.

With the intercontinental spice and drug trades attracting early European interest in the eastern zone's smaller-scale and more vulnerable societies, the chronological sequence of indigenous crisis and response to European incursions (and associated arms transfers) was bound to differ from scenarios in East Africa, the Gulf, and the Northwest Frontier. Crucially though, whatever the apparent differences of scale and scope, arms trading activities in both halves of the Indian Ocean were linked systemically in ways that proved mutually reinforcing. Cross-border flows of people, commodities, money, information, and techniques in this expanding transoceanic milieu would constantly open up new worlds of illegality and new underworlds of criminality; the emerging problem of global arms trafficking would call for international responses, demonstrating fresh purpose and resolve in bringing about collective security.

European expansion into the Eastern zone

As Europeans arrived in the East Indies in pursuit of commercial profit, colonial ambition, and Christian mission, the early phases of their imperial venture also saw the first transfers of Western firearms to the eastern Indian Ocean. From the sixteenth century, the Portuguese *Estado da India* exercised its disruptive influence through force of arms and the establishment of many factories. Portuguese fortresses sprang up at Malacca, Bantam, and Timor, controlling the main trade routes in the Eastern Archipelago; at Amboina and Ternate, reaping notable profits from clove and nutmeg harvests channelled out of the Indonesian 'spice islands'; and across the South China Sea, at Macau, a gateway to commerce with China and Japan.[8] Following their capture of Malacca in 1511, the Portuguese became major importers of firearms in maritime Southeast Asia. Portuguese soldier-administrators and merchants were fairly reluctant arms dealers, constrained by the monopolistic character of a mercantilist system that sought to preserve the colonial regime's advantage in deploying firearms while denying potential opponents access to those very weapons. Yet Portuguese mercenaries armed with their own muskets and cannon did not hesitate to enter the services of Southeast Asian kingdoms, assisting the spread of the knowledge of firearms and being partly responsible for aggravating political rivalries around the region.[9]

From the seventeenth century, the commercial and military presence of the Europeans grew increasingly restrictive and destructive as the Dutch and the English battled to monopolize and redirect trade. As the Dutch gained the initiative by seizing the reins of authority from the Portuguese, the factories of the Dutch East India Company (VOC) were soon operating in coastal Ceylon, Malabar, Malacca, Makassar, and Batavia, dominating much of Java.[10] During the 1620s, the Dutch set up an important gunpowder-manufacturing centre at Pulicat (on India's Coromandel Coast), which stocked Dutch factories in the East Indies until gunpowder production was decentralized and localized in the 1660s. This process was to continue well into the nineteenth century, as Stamford Raffles later observed, with the Dutch supervising the local population at Batavia in the manufacture of gunpowder.[11] Like the Portuguese, the Dutch sought to adhere to broad mercantilist principles as far as arms transfers were concerned. But owing to their ambitions in the region and their rivalries with other Europeans, they were often pushed into arms deals with local communities, eventually introducing some of the newer models of Western firearms to Southeast Asia. The VOC began replacing matchlocks with flintlocks from the 1680s, and these firearms would appear in Javanese arsenals from the 1700s. The Dutch started supplying flintlock muskets to the rulers of Surabaya during the 1710s and 1720s, at a time when the standardization of the flintlock by Britain and France was both accelerating and making more affordable the production and provision of such weapons to other states in the region.[12]

By the dawn of the long nineteenth century, the global projection of more competitive, aggressive forms of European imperialism was set to alter the entire complexion of international arms transfers. Legitimated by notions of transcendent law and sovereignty, and evidenced by growing numbers of soldiers, administrators, settlers, and merchants, European global expansion would ensure the arming of both colonial and indigenous forces through a worldwide web of industrial enterprise and commercial intrigue. The rising tempo of European great power competition, plus the increasing scope of cross-cultural collaboration and collision, engendered a volatile international climate that would, time and again, create new markets for military merchandise. The conclusion of the French Revolutionary and Napoleonic Wars in Europe would lead to a renewal of Anglo-Dutch rivalry in the Eastern Seas, as chartered companies and colonial forces vied for commercial and political mastery.

Critically, the timing and extent of such colonial penetration would give almost immediate impetus to the arms trade in the eastern Indian Ocean zone, ahead of arms transfer scenarios in the western Indian Ocean. Admittedly, it took the better part of the nineteenth century for the British and the Dutch to consolidate their hold over the Eastern Archipelago. Yet, from the outset, the impact of European influence was felt more keenly throughout the littoral and maritime domains of Southeast Asia than it would be along the coastal fringes of East Africa and the Gulf, or across the provinces of the great Islamic land empires in Southwest and South Asia. Whereas European colonial activity in the western zone was limited mostly to sporadic settlements, plantations, and commercial treaties with native powers until the mid-nineteenth century, and the protectorates and consulates in East Africa and the Gulf did not materialize until the 1890s, the process of economic exploitation and the politics of administrative control were more advanced in Southeast Asia.

The Europeans found it comparatively less difficult to dominate the smaller, fragmented polities of the Eastern Archipelago. The early European coastal presence was allowed to establish footholds further inland, and European colonial authority was able to exert a stronger, disruptive influence on regional trade and politics. Back in the days of the spice trade, Dutch monopolistic policies had been enforced systematically by the draconian *hongi-tochten* (law-enforcement ships).[13] With British incursions into the Archipelago enabling their territorial acquisition of Penang (1786), Malacca (1795), Java (1811) and Singapore (1819), the frontiers of the colonial state began rolling forward again. The Anglo-Dutch Treaty of 1824 would effectively reduce the region to two colonial spheres of influence, partitioning the Eastern Archipelago through the Straits of Malacca and Singapore, with the British taking the territories north of the agreed dividing line and the Dutch taking territories to the south.[14] The Straits themselves, a major artery of maritime commerce since pre-colonial times, were steadily transformed into an international waterway of global importance amid imperial rivalries and indigenous crises. By contrast, it was only in 1886 that East Africa was partitioned into British and German colonial spheres; it was only after the opening of the Suez Canal, amid diplomatic flux in the era of new imperialism between 1870 and 1914, that the waterways of the Gulf region assumed global significance.

By that time, the agents of the European chartered companies and the officials of the European colonial states had gone much further in consolidating their interest in the East. The British fought three wars with Burma (1824–26, 1852, 1885), resulting in the pacification of the entire Burmese kingdom. They defeated China in two Opium Wars (1839–42, 1856–60), thereby imposing unequal treaties that secured Hong Kong, various 'treaty ports', and virtually unrestricted commercial access to the Chinese interior. They also presided over forward movements into the states of the Malay Peninsula (between 1874 and 1914), underpinned by the progressive introduction of a Residential system. British sub-imperialism reared its head, too, in the colourful career of 'Rajah' James Brooke: he created his own kingdom from a tangled portion of rainforest and mangrove in Sarawak, which the Sultan of Brunei had awarded him in 1841 for military assistance rendered in the suppression of a Dayak uprising. The British annexed Labuan in 1846, and placed all of North Borneo under the protection of a British chartered company from the 1880s.[15] Meanwhile, the Dutch started extending administrative control to the outer islands of the Indonesian Archipelago, setting up a series of 'border residencies' from the 1840s to the 1870s.[16] The French consolidated their own *Union Indochinoise* after establishing protectorates over Cochin China (1858) and Cambodia (1863), followed by Annam and Tonkin (1884).[17] The Spanish, long-entrenched in the Philippines, decided to launch military expeditions against the Sulu islands in 1845 and 1848; the pretext was the annihilation of Sulu's 'pirate nests', but these campaigns were really intended to thwart Dutch and British ambitions in an area that Spain located within its sphere of influence.[18] Spain imposed its own unequal treaty on the Sulu Sultanate in 1851, deliberately excluding the commerce of other European powers while monopolizing 'the purchase and use of all kinds of firearms'.[19]

The advanced nature and adverse competitive dynamics of European expansion in the eastern Indian Ocean zone would reveal, at an earlier stage, both the potential and the peril of arms transfers. In the search for concessions and other competitive advantages, and in the transition from mercantilist monopolies to free trade, the European chartered companies had no qualms about brokering firearms and other munitions. But in the contest for spheres of influence and the establishment of colonial authority, the European colonial regimes soon appreciated the limits of free trade where transfers of strategic goods were concerned, particularly when facing resistance and warfare from armed indigenous groups, plus diplomatic pressure from other Europeans whose interests were affected. In the consolidation of colonial rule, the colonial regimes would have to strike a balance between the commercial and strategic value of arms transfers, on the one hand, and the quest for stable revenues and settled frontiers, on the other. Ultimately, they observed a growing need for arms control with greater emphasis on cooperation, whether enforced by police authorities or naval patrols.

From the mid-eighteenth century, European expansion into the zone began to culminate in overlapping (if not clashing) commercial and political interests. In contending for mastery of the Indian subcontinent, the East India Companies of Britain and France quickly assumed a major role as intermediaries in its military economy. In searching for commercial privileges and port facilities along the trade routes between India and China, the British also found themselves promoting arms transfers while proclaiming the virtues of free trade. Official agents of the English East India Company, and growing numbers of private 'country' traders, engaged more freely in arms trafficking from the shores of Burma to the waters of the Sulu Sea. Firearms were presented as gifts to the Burmese kings at Ava, partly with the intention of countering French influence in neighbouring Siam and Cochin China. Muskets, in addition to heavy ordnance and gunpowder, comprised the key articles of exchange in the early munitions traffic between the English Company and the Sulu Sultanate, partly to offset Dutch influence in the East Indies and Spanish influence in the Philippines.[20]

Alexander Dalrymple, for example, was a principal factor of the English East India Company who explored the Burmese and Vietnamese coasts before acquiring Balambangan in the Sulu Archipelago on behalf of his masters. He felt that arms transfers coordinated from Balambangan afforded the best means of advancing the Company's interests against rival European competitors in the area. In Dalrymple's assessment, the prospect of high profits and political partnerships with well-armed Sulu allies would strengthen the fledgling settlement against possible attacks from neighbouring Dutch or Spanish colonies: 'If the Company should make Balambangan a mart for arms and ammunition the demand would be very great and nothing would bring strangers thither sooner.'[21] The idea was to manipulate military supplies to indigenous powers, as was the practice in India, thus obtaining local goodwill and cooperation in matters commercial and strategic, and thereby acquiring concessions and arming friendly powers. Unfortunately, such practices tended to generate more instability than security in the long run. The Balambangan settlement ended up as a short-lived venture, whether backed by arms transfers or not, but Company agents were actively exploring other avenues.

The British supplied weaponry to Aceh in northern Sumatra, while searching for suitable locations to develop a viable commercial entrepôt and naval base to facilitate the growing China trade. In efforts to impress the local rulers and strengthen them against potential Dutch or French interference, munitions had been shipped to the Acehnese coast back in the 1760s.[22] At the height of the Napoleonic Wars, Lord Minto, the Governor-General of India, personally instructed Major Archibald Campbell to convey a gift of 100 flintlock muskets—procured from the Company's workshops at Birmingham and London—to the Sultan of Aceh, Jauhar al-Alam.[23] Meanwhile, the free enterprise of

country traders at nearby Penang had led to far bigger arms acquisitions by the Sultan, alongside unofficial transfers to the 'pepper rajas'
who were his nominal vassals and determined to maintain their autonomy. Although the British colonial authorities at Penang started
banning the export of arms to Aceh in 1811, early prohibition did little
to actually stem the tide.[24] In March 1813, the Sultan's attempts to exercise sovereign prerogatives over the rebel chiefs (and maritime traffic
thought to be circumventing port regulations) resulted in a major
incident, in which a British-registered vessel—the *Anapoorny*—was seized
by mistake.[25]

William Petrie, the Governor of Penang, was livid. In a letter to
Sultan Jauhar al-Alam, he underscored the military-strategic assistance
Britain had rendered in the shape of arms supplies to the Sultan but
arms restrictions against the Sultan's opponents:

> Your majesty has received on various Occasions from the British
> Govt. the most unequivocal testimonies of its anxiety and desire to
> promote your interests and to assist you as far as is consistent with
> the security of our possessions and your majesty has received on
> more than one Occasion actual and undeniable Testimony of the
> Sincerity of this Desire in the presents which have been sent to you
> of arms & ammunition from the Supreme Govt. & from the Govt. at
> this Presidency and every obstacle had been placed in the Way of
> Supply to your Enemies of the means of acting offensively against
> you by prohibitory Decrees in force at this island against the expor-
> tation of Gunpowder and warlike Stores except for your majesty's
> service.[26]

Petrie expressed consternation over what he regarded as unfair reprisals
against free and fair commerce:

> With these Sentiments & these principles which have been invari-
> ably acted upon by the British Govt. on the faith of an expected rec-
> iprocal Return from your majesty it is a matter of regret and concern
> that I feel myself called upon to notice the Reports which have been
> made to me of the line of Conduct pursued by your majesty & your
> ministers agt. the Ships & Properties of British Subjects trading to
> the territories of Acheen with friendly & commercial views.[27]

The Penang authorities reintroduced arms prohibition, condemning
the Sultan's seizure of the *Anapoorny*—together with its cargo of pepper

and benzoin resin—as an unlawful act of piracy.[28] In August 1813, the British man-of-war H.M.S. *Africaine* was despatched from Penang to recover the *Anapoorny* and bring the Sultan to his senses. But in a dramatic turn of events, amid diplomatic flux and domestic unrest in Aceh, Jauhar al-Alam was overthrown following the revolt of three powerful chiefs (*Panglima Sagis*) and their armed retainers in October 1814.[29]

The Penang authorities relented, offering temporary refuge to the deposed ruler while permitting further shipments of munitions through the merchants with whom he had regular business dealings. The English East India Company reiterated its policy of arms prohibition 'to any place upon the Continent of Asia between the River Indus and the Town of Malacca or the Peninsula of Malacca inclusive, or in any Island under the Government of the said Company, situate to the North of the Equator, or to the said Company's Factory of Bencoolen in the Island of Sumatra or its dependencies; save only the said United Company or such as shall obtain their special leave and licence in writing, or ... in writing under their authority for that purpose'.[30] Again, such official pronouncements did not prevent Acehnese merchants from acquiring the munitions surreptitiously by sailing to Penang, 'where firearms and ammunition are always in great demand'.[31] Jauhar al-Alam managed to regain his throne in April 1819, but not without the aid of fresh weaponry procured from Penang and the arbitration of the Lieutenant-Governor of Bencoolen, Sir Stamford Raffles, in Aceh's civil war.[32]

The involvement of Stamford Raffles in Acehnese affairs stemmed from his ambition to counter resurgent Dutch influence in maritime Southeast Asia following the conclusion of the Napoleonic Wars in Europe. An early proponent of the 'imperialism of free trade' and one of the first empire-builders to promote a regenerative programme using Benthamite ideas, Raffles seemed especially perturbed about the monopolistic tendencies of his Dutch counterparts.[33] He advocated British intervention to end the civil war in Aceh, signing a treaty with whoever was the accepted ruler, in order to secure a base opposite Penang that might give Britain strategic command of the northern approaches to the Straits of Malacca. He further proposed the establishment of another base at the southern end of the Straits. The successful implementation of both proposals would virtually guarantee British control of that strategic waterway and free passage for shipping.[34] In February 1819, while en route to Aceh, Raffles took the opportunity to conclude a separate agreement with two Malay princes, Tengku Long (recognized thereafter as Sultan Hussein Shah) and Temenggong Abdul Rahman of the Riau-Johor kingdom, thus obtaining exclusive rights to establish a British factory in Singapore. In fulfilment of its founder's expansive liberal-utilitarianism and aggressive reformism, colonial Singapore was to develop rapidly as a free trade entrepôt of rising importance to international commerce.[35] Equally, however, it possessed the perfect location and environment to function as the leading arms transhipment centre of the eastern Indian Ocean.

Whereas territories like Aceh would be drawn into the orbit of a typically monopolistic Dutch colonial sphere, an alternative British controlled 'free trade zone' was to emerge in the Eastern Archipelago under the terms of the 1824 Anglo-Dutch Treaty. Singapore was amalgamated with Penang and Malacca to form the Straits Settlements in 1826, and the free-wheeling capitalist dynamic of the Straits Settlements was further augmented by Britain's annexation of Labuan in 1846. Positioned at the crossroads of Asia, astride waterways connecting the vast expanses of the Indian Ocean, the South China Sea, and the Pacific Ocean, these British free trade colonies would do much to facilitate arms transfers across the eastern Indian Ocean. They developed extensive business connections and infrastructure for arms distribution, with the merchandise offered in bazaars and warehouses along the waterfront. Penang continued as the major distribution centre for markets in Sumatra, while Labuan performed the same function for markets in Borneo and the Sulu Sea. Malacca channelled munitions on a smaller scale than either Penang or Labuan, whereas Singapore re-exported arms consignments on a larger scale and over greater distances than any of the others.

Of course, munitions were no ordinary merchandise. Occasionally, the perceived dangers from unchecked arms traffic would elevate them from being mere commercial objects to articles of exceptional strategic significance. Despite the status of the settlements as British 'free ports'—the first to be designated as such throughout the Indian Ocean periphery—one conspicuous area of official intervention was the periodic restriction on the sale and transhipment of weaponry, especially at times of pronounced instability or conflict in the region.[36] More often than not, however, the principled rhetoric of free trade tended to mask the more sinister whisperings of commercial exchanges in the depths of a fast-growing colonial underworld. Eric Tagliacozzo has analysed the critical factors that made Singapore ideal for 'under-trading':

> The city was a crucial haven for smugglers because of its size and chaotic complexity as well. Here the vision and reach of the state,

supposedly at its strongest at the seat of regional imperial power, were actually diminished in its own backyard. There were never enough coast guard cutters or police; there were always too many sampans or dark alleys. This amaurosis was very dangerous from the perspective of area regimes. When the coercive power of government was brought down upon local smugglers to stop their illegal activities, these groups resisted domination in various ways. False shipping papers, false destinations, hidden cargo spaces, and small, fast sailing craft were all utilized alongside other means of avoidance, including loopholes in the law and the use of fall guys by well-organized syndicates.[37]

For many decades after its founding in 1819, the port of Singapore was a place where the 'manic chaos of free trade commerce' prevailed. Even though it was the capital of the Straits Settlements and the centre of British authority in the Eastern Seas, the island-colony remained a relative safe haven for all who wanted to dodge the arms control mechanisms of the colonial state on either side of the Straits.[38]

This intermediate phase of the evolving arms transfer system begins to highlight several glaring paradoxes of empire, which surfaced well before the imperial effulgence of the late nineteenth century. It is immediately indicative of the creative tension between ideology, profitability, and security: officials were already debating the pros and cons of free trade or monopoly; arms consignments were part and parcel of an imperial game bent on profit and power-broking; and consignments remained unchecked so long as they would increase profit margins and not imperil the colonial project. Again, it illustrates the complex interplay (as well as conflicts of interest) between 'metropolis' and 'periphery', involving metropolitan governments and their colonial regimes in areas of policy formulation and implementation, on the one hand, and on the other, a private sector whose interests were expressed in the public sphere and whose enterprise was evident in the constantly adapting arms-supply chain that linked Europe to the East.

In line with the earlier pattern in the Indian subcontinent, a chartered organization such as the English East India Company was given virtually free rein to operate as an arms provider from its base in the Straits Settlements. The only proviso was that the increasingly widespread trade in munitions had to be regulated on paper by the gentlemanly capitalists of Leadenhall Street, the Company's Court of Directors in London. The Company was licensed to supply munitions in 'deserving' cases, including indigenous rulers who needed such aid to control 'piracy', but it did so with the aim of securing broader commercial and political advantages. Robert Fullerton, appointed by the Company to serve as its first Straits Governor, was personally supportive of what he deemed to be a judicious use of cash subsidies and arms transfers for advancing the rule of law and British interests in the Malay Peninsula:

> Upon the conclusion of Captain Low's negotiations at Perak, the Raja solicited a small loan of money and supply of arms to enable him to support his authority more effectually and to cooperate in the expulsion of the Pirates which infested the several rivers within his Territory. Being desirous of securing his cordial cooperation, without which the measure could not be readily effected, we advanced to his agents the sum of 3500 Dollars and some muskets for which an acknowledgement was taken. ... A perusal of the Rajah's letters subsequent to the receipt of this assistance will prove how beneficial it has been.[39]

Conversely, it was unwise to allow strategic material to fall into the hands of the common inhabitants of the Eastern Archipelago, since indiscriminate transfers and unregulated traffic could increase disorder in the shape of piratical attacks, tribal uprisings, anti-colonial skirmishes, and even more protracted small wars.[40]

British colonial policy on arms trading in the Eastern Seas thus vacillated over the next few decades, depending on the severity of regional conflicts as much as the strength of diplomatic pressure from the Dutch colonial authorities and opposition from private mercantile interests. It was a balancing act that the authorities at home and abroad did not always get right; the system was riddled with inconsistencies, even contradictions, inside British policy-making circles. Although the import of arms and ammunition into Singapore was outlawed on 8 August 1828, reflecting the phenomenal expansion of such traffic from the early days of the settlement, arms transhipment went unchecked until an increase in piratical attacks throughout the region persuaded the Straits Government in 1829 to impose a licence system on their export.[41] The colonial authorities also discovered that large consignments were entering Singapore as 'Sporting Powder', a procedure apparently sanctioned by the Court of Directors in London.[42] The Court was pressured to adopt a stricter stance on the issue of licences, though there remained many ways to circumvent regulations. In April 1830, for instance, the Court refused to grant a licence to a Singapore merchant for the import of 1,000 muskets with bayonets, on top of 300 barrels of gunpowder.[43]

The merchant immediately petitioned the Board of Commissioners, which decided to permit an unrestricted trade in munitions at Singapore on the premise that any such restriction was useless since the denizens of the Archipelago could still obtain their supplies from American and French traders.[44]

The freedom to buy and sell weaponry was restored until the late 1850s, when Singapore's lucrative arms trade was threatened by arms control legislation arising out of regional instability in both India and China. When the authority of the English East India Company was dissolved after the great Indian Mutiny-Rebellion of 1857, the Straits Settlements came under the control of the India Office in London. The Indian Legislative Council, which had jurisdiction over the affairs of the Straits Settlements, acted to tighten security by passing a law to control the importation, manufacture, and sale of arms and ammunition in British colonial dependencies across the eastern Indian Ocean.[45] The Straits Governor, Edmund Blundell, approved of the act in principle but accepted that it would be resisted strongly in the Straits, where 'there is scarce a mercantile firm in the place, English or foreign, that does not import largely guns, small arms, military stores and ammunition'.[46] Blundell also warned Calcutta on several occasions that large consignments of arms were being transhipped to China over the course of the Taiping Rebellion (1851–64) and the Second Opium War (1856–60). Yet the 'iniquitous transactions' with the Chinese rebels went largely unchecked, and little attempt was made in practice to halt the arms traffic until after the end of the cross-cultural conflict, when the British finally began to assist Chinese imperial troops in suppressing domestic unrest. As it turned out, 'old Arms disposed of at the Ordnance sales' by the War Office, and 'at Auction, or by Tender, at the Tower of London', were re-sold by British merchants to a massive Chinese market and many of the weapons ended up arming regional forces hostile to Britain. From the second half of 1856 through the first half of 1857, over 26,000 muskets, carbines, and rifles were shipped from London Docks to Shanghai via Singapore, while numerous other shipments went unrecorded.[47]

As the political situation in China deteriorated, Blundell wrote from Singapore 'to call the attention of the Government of India to the necessity of checking the present immoderate trade in munitions of war':

It must be recollected that this is and always has been a perfectly fine port that its traders are unaccustomed to be interfered with in any way in their transactions and that the trade in munitions of war is one which realizes large profits and which those engaged in, will not easily forego. ... Very large quantities of munitions of war are sold in this place and conveyed to China by the Junks which annually visit us. ... I have reason to believe that much of it will find its way into the hands of our enemies. When a war is regularly proclaimed, perhaps this subject will be taken into consideration and provision be made for prohibiting all supplies of ammunition to the Chinese from this place.[48]

It was only in 1863 that Blundell's superiors issued the order that no munitions could be exported from Singapore except for the use of the Chinese Imperial Government.[49] The restriction was calculated to prevent military supplies from reaching not only the Taiping rebels on the Chinese mainland but also the 'pirates' infesting the South China Sea, who availed themselves of the opportunity to obtain large quantities of arms from Singapore.[50]

Even then, an alliance of key commercial interests would systematically undermine the case for official prohibition. Since the wording of the proclamation had explicitly prohibited the export of arms and ammunition 'to all places' except China, it drew swift criticism from the free trade lobby in both the metropolis and the periphery. *The Ironmonger* reported in 1863 the great 'inconvenience' caused to the Birmingham gun trade by the new controls, articulated in complaints from the Birmingham Chamber of Commerce to Sir Charles Wood, the Secretary of State for India.[51] The Singapore Chamber of Commerce echoed those views in the colonial sphere, such that the Straits Government relented, permitting once more the export of these articles to the Malay Archipelago and to Siam under licence.[52] The Singapore Chamber of Commerce also protested against the regulation of the arms trade at Singapore itself, for fear that it might injure the overall commercial standing of the port, but the Straits authorities would not withdraw the interdiction there. A trade-off now seemed inevitable: the overall effect of the restriction was to encourage a contraband trade that armed the Chinese insurgents.[53] Furthermore, munitions initially obtained at Singapore were re-exported from other British settlements in the region, such as Labuan.[54]

At Labuan, the sale of arms and ammunition had been unrestricted since the founding of that island-colony in 1846. The early governors of Labuan observed that warrior-traders from the Sulu Archipelago made regular purchases of munitions from the settlement's market.[55]

Although their piratical activities off the coast of northwestern Borneo would be curbed by the mid-1860s, the marauders continued to procure supplies indirectly, alongside other indigenous groups, through a network of legitimate traders who called at Singapore and Labuan.[56] Governor Callaghan drew the attention of the Colonial Office to the fact that 'a large trade—for such a small place—was carried on here in these articles'.[57] Among the munitions listed as 'imports into Labuan' were muskets and gunpowder: in 1862, the reports set the figure at nine cases of muskets and 13,733 lbs of gunpowder; in 1863, 30 cases of muskets and 4,800 lbs of gunpowder; and in 1864, 24 cases of muskets and 14,933 lbs of gunpowder. The actual figures would again have been higher, considering the quantities of munitions that evaded official detection. Port officials conceded as much in acknowledging that there was no attempt to register exports of these articles in native vessels, 'owing to the difficulty there would be in obtaining reliable returns of them'.[58]

There was little incentive to implement more thoroughgoing restrictions at Labuan. Local shopkeepers felt that legislation would have no practical effect as long as there were superior facilities for the arms trade in Singapore. When the matter of Labuan's arms traffic was again brought before the British Government in 1865, a law was passed to control 'the manufacture, importation, transport and sale of arms and ammunition'.[59] But such legislation served merely to allay the fears of government critics at home; it was not enforced at the periphery since it was scarcely in the interest of the local economy to ban the export of munitions. Small arms obtained from Singapore continued to appreciate in value, becoming Labuan's most valuable export in the light increasing Sulu resistance against Spanish colonial authority in the 1870s–80s: Enfield and Spencer rifles, in particular, along with a whole assortment of pistols.[60] If the market price of the articles was high, as an official report of the 1880s would suggest, then little could be done to 'prevent the running into the Company's territories of any quantity of them by native vessels from Sarawak ... and from Spanish possessions in Sulu, and from Dutch Borneo. ... As moreover, two steamers, one of which is under the Spanish flag, are at present running from Singapore via Labuan and Sandakan to Sulu, [there was] nothing to prevent the importation of such stores into the islands of the Sulu Archipelago and neighbouring countries direct from Singapore, unless the Government of the Straits Settlements also prohibited their export to the Company's territories'.[61]

Over the course of the long nineteenth century, the arms trade in the eastern Indian Ocean was conducted almost entirely across Anglo-Dutch colonial frontiers in the Eastern Archipelago. From an administrative angle, it would seem that the most logical solution to the region's arms trade conundrum lay in bilateral cooperation, for both colonial powers to devise a comprehensive system of laws to regulate and restrict the flow of munitions. At least on paper, British policy on arms trafficking via the Straits Settlements was shaped by pressure from the Dutch authorities in the Netherlands East Indies. Between the Anglo-Dutch partition of the Archipelago in 1824 and the conclusion of the Aceh War in 1904, the British promulgated a whole series of bans on arms exports to the Dutch colonial sphere. But even if 'an imposing juridical edifice was constructed by both Singapore and Batavia to deal with the transit of unregistered weapons across the border,' as Eric Tagliacozzo has observed, 'the degree of energy with which these stipulations were enforced was not equal in both Dutch and English spheres'.[62]

By the second half of the nineteenth century, arms control legislation in the British sphere had begun to provide what Batavia desired in terms of monitoring and checking the illicit export of weaponry. Although legal loopholes still enabled private individuals to import weapons in small quantities for their own use, India Act No. 31 of 1860 stipulated several years' hard labour and massive fines for the sale or manufacture of firearms without a licence. A licence was required for the storage of anything in excess of 5 lbs of gunpowder, and dealers themselves were not permitted to retain more than 50 lbs of gunpowder at any one time. When the Straits Settlements were transferred from the jurisdiction of the India Office to the Colonial Office, becoming a Crown Colony administered by a Governor in Singapore, the prospects for direct arms regulation were further improved. Act No. 13 of 1867 empowered the Governor of the Straits Settlements, with the backing of the new Straits Legislative Council, to ban all manner of arms exports. From 1868, the masters of incoming vessels were also required by law to declare their munitions cargoes upon entry, with additional measures instituted to regulate the storage of gunpowder within city limits.[63] Nevertheless, each ban had to be renewed by the Straits Governor on a yearly basis, which was not assured unless there was some threat or escalation of conflict perceived as inimical to British interests.

The outbreak of the Aceh War in 1873 would certainly encourage arms control initiatives in the Eastern Archipelago, with the Dutch colonial authorities prohibiting arms imports into northern Sumatra and pressing their British counterparts in the Straits Settlements to do likewise. The Straits Governor, Sir Harry Ord, was happy to oblige because these restrictions were consonant with his own hawkish convictions. The British

Government warned the Governor against being overly sympathetic to the Dutch cause, but officials at Whitehall were nonetheless willing to justify the adoption of tougher measures at this time:

[Since] the trade of the Colony and the interests of British subjects engaged in it were exposed to serious injury when wars or disturbances break out in the Native States adjacent to the Settlements, more especially as under such circumstances persons resident in the Colony are in the habit of supplying arms and ammunition to the Belligerents and of thus prolonging the struggle with its injurious effects; that consequently when such recurrences come to the notice of the Colonial Government, it is customary to issue a Proclamation forbidding altogether this export to the scene of the disturbances.[64]

The British colonial ban on arms exports to Aceh, proclaimed in 1873, remained legally binding until the turn of the century.[65] The scope of prohibition was broadened in 1879 to encompass arms shipments to the rest of the Netherlands East Indies, followed up by a ban on arms exports to Labuan in 1880.[66] Ordinance No. 18 of 1887 empowered the authorities to search for firearms aboard vessels in Straits harbours, while new regulations were introduced in 1899 to address security threats emanating from cross-border transfers of explosives.[67]

Even so, it was evident that metropolitan, colonial, local, and personal interests would collectively determine the timing and degree of enforcement. 'As impressive as all of this jurisprudence was on the books,' observes Eric Tagliacozzo, 'in reality the vast compendium of British arms legislation was rarely exercised to the full extent of the law, as was pointed out by both British and Dutch observers of the time.'[68] The British Consul at Uleelheue, Aceh, informed his superiors in Singapore that no copies of the various arms control acts and ordinances were available to him on a visit to Penang, whether from the local magistrate, police, or government store. These arms regulations had not apparently been translated into local languages such as Telegu, Malay, or Chinese, and were also conspicuously absent from the shops of local arms dealers.[69] George Lavino, the Dutch Consul in the Straits Settlements, was even more critical of the inadequate enforcement and routine evasion of Straits arms control measures. He noted the manner in which customary practices—such as allowing arms smugglers to go free if they could not pay the fines imposed for their smuggling—made a mockery of British interdiction efforts. Lavino despatched copies of Straits arms control legislation to the British Colonial Secretary, indicating where certain provisions were being overlooked and highlighting clauses that seemed to obstruct the judicial process: for instance, stipulations that confiscated arms shipments had to be directly proven to belong to a specific individual on a ship before the vessel could be held accountable.[70]

Notwithstanding the official prohibition of arms exports to the Dutch colonial sphere, it was obvious that the British authorities continued to permit arms transfers within their own sphere in order to achieve various policy objectives, as they had previously. Considering the state of flux in colonial geopolitics, these intermittent flows of weaponry could still lead to consequences beyond the original scope of the colonial state. When the British prepared for formal intervention in the Malay states in 1874, Ord's gubernatorial successor Sir Andrew Clarke did not hesitate to support the Datu Klana in Sungei Ujong as 'a willing instrument of British policy': first by recognizing him as the legitimate ruler, and then releasing arms and ammunition for the purpose of keeping the river open and free of illegal toll-stations.[71] But there would be no actual check once the authorities released the weapons. Although the overarching aim of the colonial state was to keep firearms beyond the reach of local communities, the evolution of separate administrative entities in the British sphere—the Straits Settlements, the North Borneo Company, the 'White Rajahs' of Sarawak, and the Federated Malay States—altogether compounded problems of coordination and consistency through the application of different rules and regulations.[72] Conversely, sibling rivalries and shifting allegiances within the fractured Malay polity ensured that the small arms would continue to change hands, escalating the small wars that were fought subsequently in Sungei Ujong and elsewhere.

The fragmented nature of British colonial authority would actually hinder efforts to regulate arms transfers. Ironically, there was by the 1880s a greater commitment within policy-making circles to prevent munitions from leaking into the Dutch sphere, in response to external diplomatic pressure, than to stop arms traffic across internal British divisions, given the host of internal rivalries that marred cooperation in the British sphere. There were those at the Colonial Office in London who had long expressed apprehension over the sizeable arms purchases from Singapore and Labuan, but there were many other officials who saw no direct benefit to the Straits Settlements in enforcing the arms bans *vis-à-vis* the wider Archipelago. Some lobbied for less stringent controls and lower fines for apprehended arms traffickers, while others even turned a blind eye towards gun-running activities.[73] Despite appeals by the North

Borneo Company for arms control legislation to be enacted through-out Britain's Crown Colonies in the region, the Colonial Office would only approve a more limited ban on arms exports from Labuan in 1884. The Colonial Office sought to use this issue as leverage against the Company's Directors when negotiating issues of greater importance to the Crown. Apart from internal politics, there was also an underlying desire to skim the profits of free trade while they lasted, with minimal concern for the problems of arms trafficking when the violence over-flowed into Company territory.[74]

The political will to sustain the British arms control measures was further eroded by powerful forces of the British free trade lobby, both at home and in the Straits, ever ready to pursue the commercial oppor-tunities and financial rewards afforded by the new imperialism. After all, the immediate fallout from the prohibition of arms exports to northern Sumatra was financial in nature. Export statistics again suggest how lucra-tive the trade had become by this stage, in addition to how much was being lost: the value of Singapore's arms exports to the Netherlands East Indies fell from $13,211 in 1873 to less than $4,000 in 1874, while ammunition revenues fell from $14,331 to $6,202 in the same period.[75] Since the real figures would have been very likely higher than the official estimates, it was no surprise that the commercial lobby vociferously opposed the arms bans of the 1870s–80s, just as they had resisted earlier arms control legislation in the Straits during the 1850s–60s. When an order was issued in 1873 stipulating that commercial vessels could no longer carry armaments, even for self-protection, this was quickly rescinded when Straits merchants nearly rioted on the grounds that it would leave them vulnerable to piratical attacks.[76] The Straits mercantile community also expressed outrage over the treatment of British trading vessels entering Dutch colonial waters, arguing that Batavia was using arms checks as a pretext to expose and spoil British cargoes, thus advanc-ing Dutch commercial interests.[77]

While British colonial policies and commercial adventurism were key determinants in the dynamics of regional arms transfers, it is instruc-tive to make comparisons with parallel developments in the Dutch sphere. In the decades following the Anglo-Dutch Treaty of 1824, shifts in Dutch colonial policies would also contribute to small arms prolifer-ation. In 1837, Staatsblad No. 2/43 largely outlawed arms importation into the Netherlands Indies, permitting only a limited trade in hunting guns, finely-worked pistols, and other collectors' pieces; the latter were subject to heavy taxation, normally at the rate of 30 guilders per piece, plus an additional 15 guilders to obtain a certificate of ownership. But changes in the tariff system from the mid-1860s would make firearms instantly more affordable and accessible. From 1866, the 100-guilder minimum-value requirement on imported weapons was dropped and a flat 6 per cent tariff was levied regardless of the value of the weapons; special stipulations about packaging were also discarded in an attempt to generate more revenue for the colonial state. In effect, this created a vast new market that arms dealers were all too eager to exploit: carbines, rifles, pistols, and revolvers began to inundate the Netherlands Indies, either transhipped from facilities in the Straits Settlements or shipped directly from factories in metropolitan Europe. The value of arms imports into Java, which averaged 45,000 guilders per annum over the years 1864–68, rose dramatically above 116,000 guilders in 1869, the year that also marked the opening of the Suez Canal. In Semarang, a major port-city on the north coast of Java, the value of arms imports spiralled from 18,108 guilders in 1868 to 82,790 guilders in 1869.[78]

While the new tariff system in the Dutch sphere proved to be an instant money-spinner, it also threw on the market larger quantities of weapons, only some of which Batavia was able to monitor. It was not long before criminal violence and anti-colonial protest were linked to the emergence of a 'gun culture' in the Javanese countryside. Even the Dutch Government was alarmed; the Minister for the Colonies soon reprimanded the Governor-General at Batavia for the appalling lack of information and interest on the ground. In September 1870, Batavia finally responded by issuing an order to all residencies in Java to monitor arms trafficking at the local level. Conversely, the overall limitations of Dutch resources dedicated to surveillance and interdiction, set against the outstretched and porous character of Dutch colonial territory, would hinder the enforcement of arms control legislation. Arms continued to flow from coastal areas—like Semarang—to other parts of the Archipelago, or to pockets of the interior still formally under indigenous rule, from Aceh in northern Sumatra to the fringes of northern Borneo.[79] In Tagliacozzo's assessment, 'this combination of British ambivalence and Dutch inability allowed arms to pour over the frontier for several decades at the end of the nineteenth century'.[80] Indeed, any partial restriction to the arms traffic would intensify black-market conditions, with spot-shortages inflating prices. Gunpowder, which retailed at $2 per keg before the Aceh War erupted in 1873, went for $10 per keg in 1876 and nearly $20 per keg in the early 1880s, depending on how far it was conveyed into the interior.[81]

Of equal concern was the sheer improvement in the quality of the weapons. Shiploads of the new weaponry, flowing from the 'Breech-loader Revolution' in the Western hemisphere, were shortly being documented upon arrival at (or from) Singapore. Enfield, Snider, Martini-Henry, and Mauser rifles, and even Winchester repeating rifles, made their appearance alongside the 'Brown Bess' muskets that were mostly British War Office surplus, decades old. Consignments of up to 600 weapons were complemented by shipments of fine machine-made powder and other propellants, plus the matching percussion caps and cartridges.[82] As early as 1858, the Straits Governor expressed reservations over the fact that the Temenggong of Johor was arming his local militia with the new Enfield rifles.[83] Dutch consular reports soon confirmed that local populations in Sumatra and Borneo, along with maritime groups such as the Bugis, were acquiring Enfield, Mauser, Winchester, and other breech-loading rifles from ports on both sides of the Straits. Even Beaumont rifles, issued to the crew of Dutch naval vessels enforcing the blockade around Aceh in the 1870s, seemed to find their way into the hands of indigenous opponents, the barrels modified and adapted to suit local requirements. Ammunition was frequently supplied in boxes measuring one cubic foot apiece, each holding up to 250 small tins of 250 percussion caps (or 62,500 caps in total). One such tin, retrieved from the body of a dead Acehnese warrior, contained percussion caps manufactured by the Birmingham firm of Kynoch; with Chinese characters imprinted on the obverse side, some of which also represented Malay words, it furnished irrefutable proof that Western arms manufacturers were producing items with an understanding of who their customers were.[84] As the global small arms production rate fluctuated to meet demand, the patronage of such clientele proved critical to the long-term viability of arms manufacturers situated half a world away in metropolitan centres like Birmingham or Liège.

For the Western powers at the periphery, the damage potential of the new weaponry could either work for or against them. The fears of its repercussions on colonial authority generated alarm and urgency, but the conflicting priorities of colonial governments and the characteristic porosity of this largely archipelagic region once more implied that no arms control regime could be rendered watertight. New patterns of arms trading would evolve as arms traffic in maritime Southeast Asia was more frequently diverted to other areas than it was actually suppressed. Against the loopholes and limitations of piecemeal arms control legislation, numerous consignments would be transhipped from Singapore to regional arms trading centres like Labuan or Jolo, for re-export to areas situated outside the zone of prohibition or beyond the orbit of the European colonial spheres: Portuguese Timor and New Guinea; Hainan island, off the southern coast of China; Siam and its outlying provinces, Patani and Trengganu; and British Burma, or French Vietnam and Cambodia.[85]

To the east of the British and Dutch colonial spheres, attempts by the Spanish authorities to curb gun-running in the Sulu Archipelago by introducing a customs house and naval blockades during the 1870s would likewise have a wider redistributive effect. Here, as elsewhere, the regulatory channels and instruments of the colonial state would create lucrative black-market conditions that were readily exploited by German and local Chinese gun-runners based at Singapore, Labuan, and Sulu itself.[86] This was not helped by the ambiguity of the Sulu Protocol of 1885, signed by Britain, Germany, and Spain, which virtually guaranteed the commercial passage of all goods in the zone—irrespective of their nature—in yet another policy flip-flop predicated upon the need to uphold free trade principles.[87] The Spanish authorities continued their naval patrols in contravention of the agreement, but many arms consignments still evaded their interdiction efforts. In 1888, the Dutch Vice-Consul in Singapore submitted a report to the Dutch Minister for Foreign Affairs at The Hague, describing the spread of arms trafficking from Singapore to other ports in the Netherlands Indies: Jambi, Siak, and Bengkalis (around Sumatra); Batavia, Surabaya, Cirebon, and Rembang (around Java); Kaili (in Sulawesi); Pontianak (in Borneo); and Bali (just east of Java).[88] In 1894, the Straits Governor Sir Charles Mitchell likewise informed the Colonial Office about 'a clandestine export of arms and ammunition from Singapore to Lombok' (just east of Bali).[89] Munitions were also understood to be 'leaking' westwards, when cases of rifles and ammunition were shipped from Singapore to the Arabian Peninsula, in an atmosphere of heightened regional instability and colonial insecurity. Spanning colonial spheres across an entire ocean, the dangers linked to an increasingly interconnected arms trade and the direct threat to British interests in the Gulf region in the mid-1890s would finally persuade the British Prime Minister, Lord Salisbury, and his Colonial Secretary, Joseph Chamberlain, to propose a more comprehensive ban.[90] In 1897, a legal framework for international arms prohibition was drafted by three signatory powers—again, the Spanish, the British, and the Germans—and duly ratified as the Jolo Protocol.[91]

Yet the Jolo Protocol was never properly instituted. The eruption of the Spanish-American War in 1898 soon put paid to any notion of multilateral arms control, for gun-running became hugely profitable once more amid the political and economic turmoil of the Sulu

Archipelago. Local chiefs and warrior-traders treated the colonial contest as another opportunity to strengthen their hand and expand their power bases in indigenous society. The British and the Dutch issued fresh arms bans in their respective colonial spheres, but the Spanish, in extinguishing coastal beacons across the Philippines as a defensive tactic against the American fleet, unwittingly enhanced the silent passage of arms traffic in the dark. Despite resolute efforts from the turn of the century to convert European colonial frontiers into tighter borders and boundaries, Singapore and Batavia only managed to slow down—but never stop—the transit of contraband weapons through Southeast Asia. The outbreak of the First World War in 1914 would shatter all remaining hope, escalating flows of the new weaponry and new technology over frontiers old and new, while funnelling small arms to altogether bigger conflicts in the global arena.[92]

Indigenous crisis and the arming of the Eastern zone

Even as the arms trade in the eastern Indian Ocean was driven by the mechanics and momentum of European expansion, it bears repeating that European colonial activity and the arms-supply chain were both shaped by the chronology and character of indigenous crisis and transformation at the periphery. The course of European expansion and arms transfers were, at every point, moulded by the indigenous dynamic of changing core-periphery relationships—more specifically, the interaction between centralized dynastic states, outlying provinces, and coastal regions—against the unchanging features of local geography.

From the earliest antiquity, the Indian Ocean had been an arena of 'archaic globalization'.[93] It coalesced into two related but distinct zones of economic, political, and cultural unification, each sharing some of the core-periphery relationships of a world system. Just as the regions extending across East Africa, the Middle East, and India's Northwest Frontier comprised one zone, so for over two millennia, large parts of South Asia, Southeast Asia, and East Asia were interconnected in ways that would constitute an eastern Indian Ocean zone. By the sixth century BCE, Southeast Asia had already evolved advanced maritime traditions and technologies, in addition to its own complex cultural and economic system based on interactions among hunter-gatherers, shifting cultivators, and rice farmers. Over many centuries, the cycles of monsoon winds and long-distance voyages further integrated the trade and politics of this region with those of the Indian subcontinent and southern China.[94] The commercial networks of the eastern Indian

Ocean were thus increasingly animated by a system of monsoons and markets, sailors and ships, entrepreneurs and emporia, perhaps best exemplified by the romance and sophistication of the ancient maritime Silk Road. Even as *dhow* was used as a generic term for all indigenous sailing craft in the western Indian Ocean, so the word *prahu* would typify the traditional shipping of the Eastern Seas.[95] Every year, and with astonishing regularity even after the advent of steam navigation centuries later, the southwesterly monsoon continued to serve seaborne traders travelling from India to China between April and October; the northeasterly monsoon facilitated southbound and westward voyages between November and March.

And yet, compared to large swathes of the western Indian Ocean zone, the societies further eastward tended to be smaller in scale and more vulnerable, with the cycle of crises and ruptures occurring earlier. Differences in regional geopolitics would help to explain the timing and degree of interpenetration by the agents of colonial expansion and the arms transfers from Europe. The western Indian Ocean was dominated between the sixteenth and eighteenth centuries by the wealth and sophistication of large political units—the empires of the Ottomans, the Safavids, and the Mughals—but polities of comparable size and power never dominated the eastern Indian Ocean. China's 'Celestial Empire' radiated its mostly benign influence from afar, while mainland Southeast Asia remained a patchwork of fiercely independent, centralized dynastic states—the Burmese, Thai, Cambodian, Laotian, and Vietnamese kingdoms—that became proficient in the use of firearms. In the *Suma Oriental* of Tomé Pires, offering careful analysis of the Indian Ocean political economy around 1515, the Portuguese apothecary-turned-ambassador highlighted Vietnam's 'countless musketeers' who were likely armed with Chinese fire-lances introduced a century earlier. In sixteenth-century Burma, it is estimated that one-third to one-half of the soldiers serving under the third Toungoo king Bayinnaung (r.1551–81) were equipped with muskets.[96]

In maritime Southeast Asia, the topography of the region contributed significantly to the difficulty of achieving stability. The Eastern Archipelago was an island world comprising numerous coastal trading states, only a few of which had access to rice-rich hinterlands, like the interior of Java. The political structures of these states were relatively fragile; some states were ephemeral, with their diminutive size and exposed position rendering them especially vulnerable to attack from the sea. The primacy of the Malacca Sultanate from the fifteenth century, which marked the golden age of the Malay entrepôt state, imparted political and economic cohesion to this highly fluid, fragmented geographic region. But the fall of Malacca to the Portuguese in 1511, and finally the murder of the last direct descendant of the Malacca dynastic line in 1699, led to the eventual termination of the Malay imperium over an extended period of dislocation and readjustment.

Between the sixteenth and eighteenth centuries, the pattern of alternating fragmentation and integration created opportunities for profit and plunder, which were readily exploited by regional elites and successor states. Malacca's immediate heir, the Riau-Johor kingdom, never quite managed to reconstitute the Malay imperium, particularly after the regicide of 1699 plunged its legitimacy into crisis. Without a recognized Malay overlord, the more ambitious states in the Archipelago (Brunei, Sulu, and Aceh) aspired to inherit the mantle of authority bequeathed by Malacca, while the smaller kingdoms (Perak and the other peninsular Malay states) preoccupied themselves with more basic issues of political autonomy and economic viability. Brunei, which had initially seized the opportunity to become a dominant entrepôt after the fall of Malacca, was by the eighteenth century eclipsed by Sulu, whose growing authority was seen in the allegiance it could command from numerous itinerant maritime groups in the region, such as the *orang laut* ('strand and sea peoples') from the western half of the Archipelago. Two migrant groups also exploited the ruptures within the Malay world: the Minangkabau of Sumatra, and the Bugis of Sulawesi, formidable coalitions of warrior-traders whose connections and influence extended across the Archipelago. The Bugis especially had a reputation for martial prowess, reinforced by war chants and dances, chain mail armour, and assorted weaponry that included the deployment of firearms. The backing of Bugis mercenaries was often decisive in determining the outcome of battle and balance of power among Malay princes.[97] Yet the collective trauma of regicide and a Malay-Bugis power struggle paved the way for the final disintegration of old sources of indigenous authority, the sack of Riau in 1784, and the emergence of new Malay states thereafter.

The hollowing-out effect of these crises not only engendered a new pattern of indigenous small state formation in the eastern Indian Ocean zone, but also enabled earlier European colonial penetration that in turn established a longer tradition of indigenous resistance. Portuguese, Dutch, and British colonial wars, fought between the sixteenth and eighteenth centuries, vividly reflect indigenous practices of opposition, evasion, and bargaining with a superior power and disruptive forces from the outside world. If wide geographical dispersal, weakening central institutions, a factious yet powerful indigenous aristocracy, and disputed successions all made the Eastern Archipelago more susceptible to internal disequilibrium, these same geopolitical factors also made the region more vulnerable to external intervention. The fall of Riau in 1784 was initially a Malay capitulation to Bugis power that subsequently enabled a 'rescue' by Dutch military force; the Riau-Johor kingdom, the cultural centre of the Malay world, was thus converted into a colonial state of the VOC. Political fragmentation and sibling rivalries throughout the Malay world further set the scene for the founding of British settlements in Penang (1786) and Singapore (1819), thereby foreshadowing the Anglo-Dutch Treaty of 1824 that set aside centuries of indigenous history by partitioning the Eastern Archipelago through the Malacca Straits.[98] John Crawfurd, the second British Resident of Singapore, concluded exclusive arrangements in 1824 with the Malay rulers as well as the Dutch. To his superiors he communicated both the import and the impact of this diplomatic coup, which 'virtually amounts to a dismemberment of the principality of Johore, and must thus be productive of some embarrassment and confusion'.[99] In fact, the Riau-Johor kingdom was forever dismantled, while the cultural unity of east coast Sumatra and the Malay Peninsula was left arbitrarily disrupted.

Such ruptures in the geopolitical firmament were exacerbated by large-scale changes in the patterns of commercial production and consumption, which brought about fundamental shifts in the global balance of trade and a reorientation of regional economic circuits. The infrastructure of European commerce in monsoon Asia was mostly built around a succession of 'drug' trades that ranged from spices (in the sixteenth century) to tea and opium (in the eighteenth and nineteenth centuries). Most of this commerce had been characterized by the one-way flow of commodities—compellingly exotic, compulsively addictive—that could either be monopolized by brute force or purchased with strategic goods and bullion.[100] Opium, however, gained the dubious honour of being the first 'indigenous' drug to be marketed primarily in indigenous societies. This single commodity, above all, enabled the balance to be tipped in Europe's favour, as Carl Trocki would suggest:

> By the eighteenth century, Europeans dominated the flow of opium from India to China and had created in China something entirely new under the Eastern sun, a veritable 'drug plague'. Also, for the first time in history, Western traders had something besides silver to offer for the riches of Asia. This shift in circumstances, in itself, made all the difference.[101]

Through the Commutation Act of 1784, the British Prime Minister William Pitt bailed out the East India Company from its American losses by reducing the duties on Chinese tea.[102] The tea trade generated, in turn, a remunerative commerce in opium that was extracted from the Bengal poppy. Opium started redressing the trade imbalance from the dawn of the long nineteenth century, financing a substantial part of colonial administrative costs, and facilitating the growth of both European and Asian capitalism.

As opium increased in value, it became part and parcel of an eastern Indian Ocean triangular trade revolving around drugs, guns, and slaves. Once more, there is a key parallel to be drawn with the triangular trade of eighteenth-century West Africa and the Atlantic. In the Atlantic trade, the muskets and rum brought by Western traders encouraged the 'Barracoon' states of West Africa to provide them with slaves, who were then traded for sugar and tobacco from the West Indies and thereby became the means of reproducing the supply of sugar and tobacco. In eastern Indian Ocean commerce, drugs, guns, and slavery also went hand-in-hand: the opium and munitions brought by Western traders encouraged Southeast Asian states to supply the traders with exotic products accepted in the China trade, and secure simultaneously the slaves who became the means of procuring those exotic products. By the 1790s, nearly one-third of Bengal's opium output was going to Southeast Asia alone, where opium and arms were together exchanged for tin, pepper, gold, silver, ivory, pearls, mother-of-pearl, sea-slugs (*tripang*), edible bird's-nests, and all manner of 'Straits produce'.[103] At Sulu, the profit reaped from opium traffic would likely exceed 100 per cent, as would the profits from the sale of muskets, flints, shot, gunpowder, and saltpetre.[104] At Singapore, the opium trade figured prominently in Raffles' rationale for the founding of a British settlement; the supplementary income from opium would support the workings of free trade there, and help to finance all of the colonial regimes in Southeast Asia.[105] The arms trade was never far behind, remaining a vital component of the equation. These drugs and guns supplied by the British and other Western agents would infuse greater volatility into the trade and politics of the East, leading to further militarization and to conflicts typified by 'piracy' and 'opium wars'.

Marauding, in particular, became the great scourge of the Eastern Archipelago during the nineteenth century. Far from being the simple product of indigenous societies in stages of decay, as portrayed in so many historical studies on 'Malay piracy' since the time of Raffles and Brooke, the reality was more complex. It was the cumulative result of expanding world trade, growing European involvement, and disintegrating indigenous authority in the littoral areas and on the seas.[106]

J. L. Anderson has noted that nineteenth-century 'piracy' in the eastern Indian Ocean zone, extending from the waters of the Malay Archipelago to the South China Sea, could be classified under at least three headings: parasitic, episodic, and intrinsic. The character of any particular armed foray against seaborne commerce or coastal dwellers could contain elements of all three forms, but each type had its different cause and effect.[107]

First, there was the 'parasitic piracy' that fed directly or indirectly on the growing volume of commercial shipping transiting the sea-lanes of the region. Initially, the expansion of maritime commerce between East and West had meant simply that there were more vessels and goods to plunder and therefore more profit in marauding. But as regional economic circuits were integrated into the world economy, local polities had to reorient their political structures and forms of production. The late eighteenth-century expansion of European commerce in the Eastern Seas—especially in tea—in turn generated demand for the maritime and jungle produce of Southeast Asia that were desirable for the China trade.[108] In order to secure those commodities, and more crucially the labour needed for their collection, the Sulu Sultanate became a centre for piratical activity and slave-raiding on an escalating scale. The Iranun and Balangingi, ranging in all directions from their mangrove-enclosed bases in the Sulu Archipelago, would become by the mid-nineteenth century the most powerful and feared seaborne raiders of the Malay world.[109]

Such marauding was, however, suppressed by European naval power when further increases in trade convinced colonial regimes of the need to safeguard commercial shipping. For the first time in Indian Ocean history, there would be actual legal frameworks for prohibition and protection, backed up with patrols to hunt 'pirates' and expeditions to destroy 'pirate nests'. Once the 'sea-robbers', their perceived 'banditry', and indigenous institutions of 'slavery' had been demonized through the liberal-humanitarian lenses of Western colonial stereotyping, it was only a matter of time before the outlawing of various indigenous maritime groups and the interdiction of contraband would serve to further advance Western conceptions of law and sovereignty in the region.[110]

Ironically, the expanding European presence actually encouraged a type of 'episodic piracy' by disrupting older patterns of trade and systematically monopolizing new ones. In diverting to Malacca, Batavia, and other settlements the commerce that had for centuries gone direct to China, the Europeans had steadily exacerbated the problem of marauding in the region. Indigenous maritime groups, driven from their traditional means of livelihood by the European commercial monopolies, were forced to survive by preying upon weaker parties:

> They were a proud people, accustomed to freedom; they resented most bitterly the injustice shown them and the restraints imposed by those whom they regarded as white barbarians. They were accustomed to the sea, and under the leadership of their princes they turned their ways to piracy and plunder, until in course of time this guerrilla warfare by sea developed for many into an habitual mode of life, more lucrative and certainly more exciting than their former ways of peace.

> Piracy became looked upon as an honourable occupation, so that any chief who wished to improve his fortune could collect about him a handful of restless followers and settle with them upon some secluded island in the Archipelago; thence he could sail out to attack ships and villages. If he were successful he would gain fresh adherents soon enough; his settlement would become a little town, strongly fortified and stockaded, while his fleet would become large enough to be divided into several squadrons.[111]

Bugis piracy was an example of this, and it waned with the passing of the economic hardship or political circumstances that created it.[112]

Equally, it is important not to overlook the fact that a form of 'intrinsic piracy' had evolved over the long term as an adjunct to indigenous state formation. With the fall of the Malay imperium, the disintegration of centralized authority, and the ascent of new regional elites in the maritime domain, the marauding that had been endemic to the politics of the Malay world entered an epidemic phase during the seventeenth and eighteenth centuries. This was neither a marginal nor occasional activity, but an indispensable component of public finance:

> Many of the characteristically small political units of much of the Malay Peninsula and Indonesian Archipelago were located at or near river mouths. Lacking extensive, revenue producing lands, the rulers of those states had to rely for their finances on tribute exacted from dwellers in the remoter areas, levies on water-borne goods, and on piracy.[113]

Once more, geopolitics and topography favoured such activities: the rulers and their retainers presided over states or districts that were small, insular, or littoral, lacking the resource base from which regular taxes could be raised, and therefore relying (at best) on tribute levied on passers-by and (at worst) on outright piracy. Retainers, such as the *orang laut*, spent much of their time cruising local waters to protect native traders and to harass other shipping: in the fifteenth and seventeenth centuries, this had proved instrumental in the ascent of Malacca and then Riau-Johor. As long as the political system of the region was operative, the activities of the sea peoples had been violent but perfectly legitimate pursuits. Some groups of *orang laut* did not, however, come under recognized rulers. These were *perompak*, wanderers and renegades that included hereditary outlaw bands of no fixed abode. There were also *perompak* who were temporary bands of outlaws under less reputable chiefs and foreign adventurers.[114] These opportunists had no qualms about sponsoring or engaging directly in a broad spectrum of piratical activities, which extended frequently to slave-raiding and trafficking.

By the eighteenth century, the nomads of the Eastern Seas had been transformed into 'sea-robbers', while 'piracy' was increasingly perpetrated by the aristocratic scions of the Malay royal houses (*anak raja*).[115] Such recourse to piratical activities, and to the warfare of which they were often an expression, formed a rational economic strategy for most polities in the Eastern Archipelago, especially during periods of dislocation and readjustment. Raffles, when questioning the legitimacy of this apparently predatory behaviour as a means of livelihood, was informed somewhat ironically by the prince whom he helped to install as Sultan of Johor that 'piracy is our birthright and so brings no disgrace'.[116] Rear-Admiral Sir Edward Owen, as Commander-in-Chief, East Indies Station, commented in 1830 that such maritime pursuits were 'as important a buttress of state or tribal exchequers as are certain national lotteries of our own day'.[117] In the labour-scarce economies of maritime Southeast Asia, where men and women in various degrees of dependency were valued as contributions to the prestige, power, and productivity of a small state, and where slaves were highly marketable commodities, state-sponsored marauding was conducted on a regular basis for the sake of economic viability and political survival.

Such variegated, volatile interactions between imperial expansion and indigenous crisis fuelled the demand for small arms in the eastern Indian Ocean zone. The arms trade was bound up with the ruptures caused by colonial intervention, local resistance, and regional conflict, in which 'slavery', 'banditry', and 'piracy' were as much symptomatic of the general crisis of the Indo-Islamic world as products of Western colonial stereotyping.[118] Like other parts of the Indian Ocean arena during the nineteenth century, firearms were procured in maritime Southeast Asia as a means of engrossing real political and economic power, where the method of aggrandizement and resistance was the raid. Arms transfers were driven along by the 'pirate wind': the easterly wind during the southwest monsoon that carried the Iranun and the Balangingi over the waters and thus sustained their marauding activities.[119] When piratical pursuits had become by the late eighteenth century a common sideline of Bugis and Malay chiefs, firearms, gunpowder, and ammunition were likewise sought to equip local militia and piratical fleets, to plunder and to enslave. This same weaponry could be used to fend off the hostile advances of Western or indigenous opponents.

Surprisingly, when surveying the historiography of the period, it is not immediately obvious that firearms played a major role in the military-strategic or marauding operations of the zone. Nineteenth-century European writers were still fascinated by 'the Orient', tending to emphasize the wider usage of traditional weaponry in Southeast Asian piracy, an essentially exotic array of swords, knives, daggers, spears, and blowpipes.[120] The dreaded Iranun were noted for their adept use of the traditional throwing-spear (*lembing*), the traditional dagger (*kris*), and the traditional hand-and-a-half-sword (*kampilan*).[121] Contemporary scholars have also commented on the cultural symbolism and ceremonial usage of firearms, which might enhance the aura and authority of rulers. Some local communities viewed firearms—big and small—in terms of the phallic representation of male fertility (*lingam*). One indigenous community in western Borneo venerated an old VOC cannon as a totem. Guns were valued not only for their firepower but also their potential in amassing spiritual power; gunfire marked religious festivals as well as political events, including the accession of a new ruler or the arrival of honoured visitors. Of course, firearms could be deployed for more mundane tasks. In northern Sumatra, local chiefs equipped their retainers with guns to kill wild animals that raided their pepper gardens by night.[122]

Closer scrutiny of documentary and physical evidence will, however, reveal that firearms filled an important niche in the indigenous arsenal. Apart from the heavier brass guns—anything from six to 24 pounders—mounted in the embrasures of some larger prahus, even more commonly seen were the brass swivel-guns (*lela*) of varying calibre, light and manoeuvrable, and often carried on solid upright supports along the sides and upper sections of native craft. In the tangled realities of an island world teeming with rivers, lagoons, and mangrove swamps, which favoured the traditional skirmishing tactics of indigenous warfare, these locally-made guns of intermediate size were admirably suited for use at sea, as well as in stockades: easily transported, deployed, supplied with ammunition (typically grapeshot), and repaired. They were not superseded by Western artillery pieces, but always prized as part of the spoils of war or plunder, and even reckoned as a standard of economic value.[123]

Smaller firearms were also deployed in local warfare and marauding. In an age of small state formation in the Eastern Archipelago, the time-honoured craft of kris-making furnished techniques that were transferable to the manufacture of small arms. By the early seventeenth century, the Acehnese had achieved renown for the production and use of arquebuses, along with swivel-guns and cannon, through their association with Ottoman Turkey. The Bugis, the Minangkabau, and the Malays soon followed suit, owing to the frequency of wars between them. By the eighteenth century, the Bugis and the Minangkabau were making muskets of a quality much admired by European observers. While Bugis warriors continued to use the pike, the sword, and the bow, their elite fighters came to rely increasingly on musketry capable of breaching chainmail armour. Malay raiding parties carried the musket (*senapang*) together with the traditional dagger, sword, and throwing-spear.[124] When crisis in Brunei from the mid-eighteenth century gave rise to marauding and inter-tribal conflicts among the Bugis, the Malays, and the Dayaks in the nineteenth century, James Brooke encountered piratical fleets armed with muskets and rifles, in addition to swivel-guns, spears, and swords.[125] Meanwhile, the Spanish colonial authorities noted that Iranun and Balangingi raiders were also deploying 'cannon and muskets' on their vessels.[126]

By the nineteenth century, firearms had become integral to the new patterns of raiding and resistance in maritime Southeast Asia. Within the first decade of the founding of colonial Singapore, vessels were assaulted in full view of its waterfront by marauders who traded openly in arms and loot in the town. What was needed to arm their own fleets and stockades was retained, while anything surplus to requirements was re-exported from Singapore and sold at a substantial profit.[127] Through dealings with these local gun-runners, British and American merchants, and corrupt Dutch officials, Islamic insurgents under the charismatic Prince Diponegoro were able to procure hundreds of muskets, together with bayonets, ball cartridges, and gunpowder. Combining the use of imported firearms with traditional staves and slings, they mounted effective guerrilla offensives against Dutch colonial forces in the Java War (1825–30).[128] Several decades later, at Labuan, Governor Thomas Callaghan would have ample reason to comment on the progressive arming of 'pirates' and other local communities across the Archipelago:

[T]he Sulu Pirates are generally supplied with arms and ammunition by the traders of their country who visit Singapore and Labuan, and by native merchants from Borneo who trade to the Ports on the Eastern Coast of the Island. ... A large portion of the arms and ammunition sold here is bought by traders for purposes of protection on their voyages along the Coast.

The rural population are also purchasers to some extent as the natives of Borneo generally are gradually becoming more habituated to the use of fire-arms and getting fonder of keeping them. ... A large quantity of the powder sold is used in firing salutes, of which the natives are inordinately fond.[129]

Munitions were transferred from Labuan to the coast of northeast Borneo, and via 'several independent Rivers ... to the lawless inhabitants of the Tawi Tawi and other Islands of the Sulu Archipelago'. The authority of the Sultans of Brunei and Sulu was mediated through headmen of the rivers, who 'having nothing to gain by enforcing the order of their suzerains, but little to fear and much to gain by disobeying them, would act according to their interests and allow firearms and gunpowder received from Brunei and Sulu to be sold into the adjoining territory of the British North Borneo Company'.[130] By 1897, it was clear even to the policy-makers in Whitehall that 'the weapons and munitions so imported eventually reach the hands of natives who keep up a border warfare against the people in the North Borneo territory'.[131]

On the Malay Peninsula, a similar pattern could be discerned. Even during the five decades of supposed 'non-intervention' between the 1820s and 1874, colonial and indigenous conflicts were escalated by the cross-border proliferation of European small arms. Just prior to the outbreak of the Naning War in 1831, 'the news that hostilities were intended had leaked out in the bazaar, and a Portuguese dealer in fire-arms managed to quietly part with a good deal of his stock in trade, most of which found its way into the interior'.[132] Such transfers of weaponry emboldened Malay resistance during the conflict, empowering them to repel two battalions of British colonial forces before eventually capitulating.[133] Even then, there began a series of small wars between neighbouring states in Negri Sembilan, which kept the area in ferment over the next few decades. In 1865, the British intimated to the Penghulu of Johol that they would not interfere in the trade between Malacca and Gemenchi, although the rampant traffic in munitions increasingly compelled them to act.[134] Fuelled by the availability of firearms, civil disorder across Negri Sembilan continued intermittently until the British intervened officially in 1877, roughly contemporaneous with other forward movements into the Malay Peninsula. British intervention commenced in Perak in 1874, and there also arms transfers would escalate the Perak and Sungei Ujong crises of 1874–76. The Penang authorities seemed powerless to stem the flow of firearms to combatants in the Larut Wars. Notwithstanding the prohibitions on arms exports to the west coast of the Malay Peninsula in 1873, armed rebellion in Perak drove the Straits Government to sterner measures 'consequent upon the murder of Mr Birch [the first British Resident at Perak, in November 1875] and the employment by Government of troops in Perak'.[135]

In maritime Southeast Asia, much of the marauding and warfare that characterized state formation at this time was tied to the fortunes of the triangular trade in drugs, guns, and slaves. The expansion of world trade from the late eighteenth century had stimulated the procurement of Straits produce at unprecedented levels. By the early nineteenth century, the regional political economy was largely dependent on slave labour to procure the exotic commodities sought by Europeans to balance the China trade. That, in turn, entailed the acquisition of firearms to equip the marauders who would engage in slave-raiding on their behalf. The general crisis of the Malay world and the new configuration of trade in the eastern Indian Ocean zone were mutually reinforcing; the availability of opium and firearms, and the procurement of slaves and other scarce resources, would prove especially addictive to those who built their local political economies on them, but destructive to all who lost out in the scramble. The survival of the Sulu Sultanate, for instance, not only depended upon the reorientation and expansion of its slave economy, but also its ability to perpetuate that economy through a localized monopoly of munitions supplies and marauding. The economic role of Sulu's chiefs (*datus*) as redistributors of vital resources—arms, ammunition, and gunpowder—was as crucial to the maintenance of their hegemony in the region as it was to the livelihood of other maritime groups. In order to meet the growing demand for slave labour in the Sultanate, the *datus* equipped Balangingi vessels and extended credit to the Iranun, in terms of advances in munitions, opium, rice, boats, and additional crew. Everything would be repaid in captured slaves—Malays, Chinese, and Javanese—from all over the Eastern Archipelago.[136]

Such reconfiguration of political economy was everywhere apparent in the Archipelago during this period. The Iranun and Balangingi sometimes obtained their supplies from, and sold their services to, Malay rulers. The biggest slave market in peninsular Malaya was Endau, where the Iranun offered captives from Borneo and Java for sale as labourers in Pahang's gold and tin mines.[137] For their part, many Malay rulers—some as distinguished as the Temenggong of Johor—actively sponsored marauding and slave-raiding by outfitting and arming piratical fleets. Payment for slaves was frequently made in terms of firearms.[138] This was also true of the Minangkabau, in general, and the Bugis, in particular, who operated as 'pirates', gun-runners, and slave-traders from their bases at Kampong Glam (in Singapore) as well as Borneo, often supplying Western traders with Straits produce and local chiefs with slaves, all in return for opium and firearms.[139] Even if the opium and slave traffic had largely abated by the 1890s, the trade in munitions, which now included breech-loading rifles and cartridges, remained very much alive:

[T]he Sultan of Brunei is attempting to obtain permission to import breech loading Guns and Ammunition from Singapore through a Mr. Cox. ... The Sultan of Brunei has no soldiers and no Police, and the weapons cannot be required in any way to defend himself or his Kingdom. ... The weapons referred to, if allowed to enter Brunei, would inevitably reach the Natives of this State, either as barter for slaves or as weapons to be used in their capture.[140]

In a related consular report, the British colonial authorities observed the correlation between transfers of modern weaponry from the West and traditional modes of warfare or power-broking in the region:

Whether it is Lawas or Brunei itself the fact still remains that arms and ammunition are freely sold to the wild tribes in His Highness the Sultan's territory, and so long as this continues so long will head hunting and slavery flourish in the upper Padas.[141]

It is impossible to comprehend the persistence of arms trading in the Eastern Archipelago without appreciating the centrality of that wider triangular trade in indigenous state formation and economic reorientation during the nineteenth century.[142]

The dynamics of arms transfers on the mainland were slightly different, contingent upon regional variations in the impact of European colonial authority as much as the character of indigenous crisis in largely continental polities. Britain's colonial policy towards centralized dynastic states like Burma and Siam was shaped by the proximity of those states to British interests in India and Malaya, French influence in Indo-China, and alternative trade routes to China. For their part, the kingdoms of Burma, Siam, and Indo-China were riddled with political disputes, ethnic unrest, and open warfare that seemed to be contained only by patrimonial authority asserted through diplomacy and judicious application of the new weaponry.[143] The expansion of the colonial state and the growing commercial presence of the Europeans would undermine even those traditional premises for native governance; arms supplied directly or transhipped from regional port-cities, such as Singapore, would further destabilize indigenous societies, if not weaken terminally the rule of native dynasts.[144]

Like so many of the African and Asian lands flanking the Indian Ocean, Burma underwent its own crisis in the relationship between commerce, landed wealth, and patrimonial authority. Throughout the upheavals of the seventeenth and eighteenth centuries, Burma had deployed imported small arms and heavier armaments against traditional foes, which included provincial rebels as well as restive vassals.[145] By the opening decades of the nineteenth century, Burmese monarchs required native and foreign traders to make payment in small arms for licences to use their vessels on the Irrawaddy.[146] But a combination of external pressures—the inroads made by European commerce, the failure of diplomacy, and Burma's defeat by Britain in two wars by the mid-nineteenth century—plus internal instability, altogether increased the demand for firearms. The proportion of Burmese soldiers using firearms would rise accordingly: one-half of a 60,000-strong army engaged in the First Anglo-Burmese War (1824–26) carried muskets into battle—indeed the Burmese used British muskets against the British in 1824—while 80 per cent of the 5,000-strong infantry encountered during the unsuccessful palace rebellion (1866) were armed with muskets.[147] Nonetheless, in the wake of the Second Anglo-Burmese War (1852), the Burmese monarchy was forced to cede the lower basin of the Irrawaddy and all remaining coastal areas to the British, which made it increasingly difficult for the Konbaung kings to maintain their monopoly over the procurement and deployment of firearms.

The reform-minded King Mindon Min (r.1853–78) initiated a phase of military modernization that entailed adaptation of the new technology and further development of the local arms industry. The Burmese had gunpowder-making capability, but factories under the charge of Mindon's brother, the Kanaung Prince, started manufacturing rifles and ammunition to replace any outmoded musketry still in use.[148] Mindon possessed vast quantities of muskets, and sought to produce if not procure the newer breech-loading rifles. By 1862, the British Commissioner Arthur Phayre actually advocated a free movement of arms and ammunition across native frontiers, with the full knowledge of British officialdom, rather than face the massive clandestine operations of gun-runners who hoped to profit from such flux and opportunity.[149] From 1866, Mindon actively sought to acquire more sophisticated weaponry in attempts to end episodic banditry in Upper Burma and restore order following armed uprisings. The latter included the Shan rebellion and the palace rebellion mounted by two disaffected princes who assassinated the Kanaung Prince (for which they received support from the British, now increasingly alarmed by Kanaung's military modernization programme). Ironically, Mindon purchased 10,000 Enfield rifles from Britain in 1867, and agreed to revise the existing commercial treaty in Britain's favour.

The rump of the Burmese kingdom, treated progressively as a client state of the British Raj, eventually faced a complete restriction of arms imports via Rangoon in the 1870s. This ultimately drove Burma's Konbaung rulers into the military-strategic embrace of Britain's French rivals, who were exploring ways to expand their colonial state in Indo-China.[150] Percussion caps smuggled into Upper Burma were soon retailing in the Mandalay bazaar at 35 rupees per thousand, while gun-flints retailed at 15 rupees per thousand. The British Prime Minister Lord Salisbury, Viceroy Lord Dufferin, and Secretary of State for India Lord Randolf Churchill were especially alarmed by 'the transport through the province of Tonquin to Burma of arms of various kinds, ammunition and military stores generally'.[151] In the hands of Mindon's young and inexperienced successor Thibaw, the new weaponry inspired a false sense of security that engendered extreme risk-taking; it prompted Konbaung Burma's disastrous confrontation with the Raj in a Third Anglo-Burmese War (1885), which thereafter precipitated the dissolution of the Burmese monarchy.[152]

In the case of Siam, the arms trade would be also linked to the changing complexion of indigenous politics and European incursions in the region. A key incentive to the Thais in the large-scale production and procurement of firearms had been the persistent threat from the Burmese between the sixteenth and early nineteenth centuries.[153] The regional balance of power altered significantly with the coming of the Europeans. Siam found itself in need of weapons not only for the preservation of sovereignty against native opponents, but also against the advances of various Western nations. Most of the time, the provision of these weapons was a role that Britain was happy to fulfil. For the security of Britain's expanding interests in India, it was imperative that Siam be maintained as an independent buffer state between 'British' Burma and 'French' Indo-China. On the other hand, Anglo-Siamese relations were occasionally thrown into turmoil over the disputed political status of the northern Malay states—Kedah, Kelantan, Perlis, and Trengganu—which were traditional vassal states of Siam, but where Britain was rapidly acquiring a vested interest in consequence of British expansion in Malaya.

After establishing their settlement at Penang in 1786, the British had found it expedient to cultivate friendly relations with the Raja of Kedah. The Raja of Kedah, who presided over much of Penang's hinterland, took the opportunity to obtain muskets and gunpowder from the British. These munitions were mostly intended to buttress the Raja's authority and enhance domestic security arrangements, although some weaponry was transhipped to Burma for use against Siam.[154] The signing of the 1826 Burney Treaty between Britain and Siam marked a turning-point, however; it opened official channels through which Siam could apply diplomatic pressure on Britain to halt arms transfers to Kedah, thereby compelling the Malays to import munitions from elsewhere in order to assert their autonomy from Siam.[155] King Mongkut (r.1851–68) further received a gift of Minié rifles, shortly after the Siamese had conceded special trading rights to Britain in the Bowring Treaty of 1855.[156] Whether guided by strategic considerations or the invisible hand of commerce, Siam gained leverage over subsequent arms shipments from British territories and was emboldened to facilitate the diffusion of munitions from Siamese shores. Siamese ports and coastal principalities, such as Patani and Trengganu, would become alternative arms transhipment centres, especially in the case of consignments bound for the Netherlands Indies. Although Siam was pressured by several European powers to issue its own arms prohibition in 1885, the use of Siamese ships and other vessels under the Siamese flag for gun-running activities persisted well into the next century.[157]

It was Britain's insistence on supplying munitions to the northern Malay states, in addition to Siam, that complicated Anglo-Siamese relations and shaped the evolution of British Malaya. Writing from Singapore in 1885, the Straits Governor Cecil Clementi-Smith used the matter of arms transfers to draw the attention of the Colonial Office to old and festering disputes, as much as new issues of sovereignty:

[T]he Consul for Siam at this Port has pressed this Government not to grant permits to export Gunpowder to Tringanu without his counter-signature. ... The question of having permits for the export of Gun-powder countersigned by the Siamese Consul is seemingly a small one, but it is the first attempt, I believe, on the part of Siam to interfere in the trade between the Colony and any of the Malay States on the East Coast.

The peace, prosperity and progress of the States under our protection is attracting more and more attention ... the neighbouring Rajahs and peoples, who will seek our aid to throw off all connection with Siam. It would be of manifest interest to them and to British interests that such connection should cease, and hence I would again advise that nothing be done to recognize the claims of the Siamese to interfere with the trade and administration of Tringanu or any other State similarly situated.[158]

Changes of administration at Whitehall (1885–86) did not disrupt continuity in colonial policy. The Conservative Colonial Secretary Lord Derby and his Liberal successor Lord Granville both concurred with Clementi-Smith: Siam could not be allowed to interfere in the affairs of native states seeking British protection.[159] The crisis was only resolved when Siam at last relinquished its suzerain rights over the northern Malay states in 1909, thus enabling them to be brought under British colonial administration. Over a century later, these politics of arms transfers supply retrospective insight into the capacity of nineteenth-century Siam to bargain with superior powers and 'bend with the wind', thereby withstanding disruptive forces from the outside world. Whereas Burma, Indo-China, and the Malay states each succumbed to a combination of Western arms, armed rebels, and diplomatic pressure, the Thai kingdom retained its sovereignty and continued to reinvent itself for survival in the modern world.

Interlopers and intermediaries of the Eastern zone

Again, it is apparent how various agents of the European chartered companies became major players in an expanding web of commercial activity and intrigue. Yet they operated alongside other interlopers, a multiplicity of arms trade intermediaries equally wanting to capitalize on fresh opportunities presented throughout this period of flux. In the arms transfer system of the eastern Indian Ocean, Europeans, Americans, Chinese, Malays, Indians, Arabs, Bugis, Tausug, and other indigenous groups were all involved at some stage, sometimes intermingling, whether as arms brokers, financiers, shippers, or smugglers. Their consignments of muskets, rifles, ammunition, and explosives were distributed from warehouses and bazaars alike. Their clientele were drawn from the entire region and from nearly every ethnic category. Combined Chinese and Indian networks of trade and finance would support the crucial intermediate 'bazaar nexus' that linked European capital to such diverse indigenous communities. This section examines, in greater detail, the interaction between various 'outsiders' and 'insiders' who turned the arms trade into a growing menace to the security of the region.

In the earlier, more tentative phases of European expansion, traders from the West had already begun to appreciate that the intra-Asiatic trade around the Indian Ocean could be a substitute for American silver within the political economy of the East. Whether they were royal agents, company servants, or private traders, Europeans trading beyond the quasi-official capacity of the chartered companies soon found that the cross-cultural commerce of the Indian Ocean afforded myriad opportunities to turn over their capital. The markets of the Red Sea and Persian Gulf ports were fair game, as were the nodes and networks along the coasts of India, Southeast Asia, and China. Among the primary commodities to be sold in that so-called 'country trade' were Indian textiles and opium, in addition to European arms and ammunition. Following their takeover of Bengal in the 1760s, the British pushed opium to the forefront as the primary exchange commodity in Asia, although guns and gunpowder were never far behind in the equation.

The British 'country traders' became the unofficial vehicle through which the servants of the English East India Company managed their personal fortunes. They were autonomous English—and more often Scottish—merchants who had established private firms in India and continued to ply their trade with the Company's permission. They operated just beyond the fringe of what was formally acceptable, entering into partnerships with Hindu, Muslim, and Parsi merchants, doing business with the Portuguese and the Dutch, and offering a variety of commercial products (including strategic goods) to local communities. Until the middle of the nineteenth century, such traders performed an important function for the British colonial authorities: they were able to deal in goods that were inconvenient for the Company, or else illegal. They were thus allowed to play a crucial intermediary role in the sale of munitions and other commodities that could be classified eventually as 'contraband'. In the eastern Indian Ocean zone, their aim was to turn over a significant portion of the cargo in transit through Southeast Asia, exchanging textiles, opium, arms and ammunition for tin, silver, spices, and any Straits produce that would be in demand in China. In search of profits, the country traders called at numerous ports around the Eastern Archipelago, including Penang, Malacca, Singapore, Labuan, Kedah, Trengganu, Patani, Aceh, Riau, Batavia, Bali, Brunei, Pontianak, Banjarmasin, Makassar, Maluku, and Jolo. These ports-of-call constituted a formidable circuit that ran through colonial spheres as well as autonomous spaces.[160]

The country trade proved to be effective and highly remunerative, blazing a trail for other intermediaries in the increasingly sophisticated commercial operations of the eastern Indian Ocean zone. By the mid-nineteenth century, agency houses and shipping firms were no longer confined to India but had been established elsewhere, linking arms transfers from manufacturing centres in Europe to markets in the East.[161] There were European firms in Singapore with general import-export businesses, which subsequently developed more specialized functions including the transhipment of munitions: William Macdonald & Co.; William Spottiswood & Co.; John Little & Co.; George Armstrong & Co.; Syme & Co. (founded 1823); Maclane, Fraser, & Co. (founded 1827); and Jose D'Almeida & Sons.[162] In the late nineteenth century, when arms prohibition, regional conflict, and anti-colonial resistance altogether heightened the demand for arms, firms as reputable as John Little & Co. would openly advertise fresh supplies of gunpowder, ammunition, and breech-loading rifles—Sniders, Martini-Henrys, and Winchester Expresses—in local newspapers such as *The Straits Times* in addition to the Malay-language press.[163]

Besides engaging the attention of British country traders and European agency houses, the burgeoning commerce of the Eastern Seas had attracted interlopers from even further westward. As early as 1811, Raffles had bemoaned the activities of commercial competitors from a fledgling United States. These private traders had no qualms about exploiting the demand for munitions in indigenous markets, seeming as indifferent to security concerns as they were to ethical issues in their pursuit of profit:

> The Americans, wherever they go, as they have no object but commercial adventure, are by no means scrupulous how they acquire their profits, and as firearms are in the highest request, especially among the more Eastern Isles, these would be considered as the

most profitable articles. They have already filled the different clusters of islands in the South Seas with firearms, and they would not
fail to do the same in the different Eastern Islands. These considerations seem obviously to point to a line of policy, respecting the trade
of the Eastern Islands, which in some respects coincides closely with
that adopted by the Dutch, while in others it differs from it entirely
in ultimate principles.[164]

Desperately wanting to secure the exotic products sought by China,
island-hopping traders from Salem, New England, began supplying the
denizens of Sulu with small arms, gunpowder, and ammunition, which
thus became their stock-in-trade by the 1820s:

> Almost all the vessels leaving the ports of the U.S. for China convey
> muskets, pistols, swords and gunpowder, which they sell on their
> outward passage to the natives and other inhabitants of Palawan
> and Magindano. ... [T]here are Americans living on both these
> islands who make the necessary arrangements, and who have several
> small native vessels, constantly sailing about the usual tracks of
> ships to China in the proper season, for the sole purpose of meeting
> American vessels and taking from them their cargoes of warlike
> stores.[165]

It was fairly common to find adventurous American and British mariners
in the employ of European agency houses, criss-crossing the Archipelago
in native craft as well as small brigs and schooners. By using a Spanish
brig from Manila—the *Leonidas*—one American mariner, Gamaliel Ward,
successfully engaged in gun-running under the Spanish flag.[166] A British
mariner, William Wyndham, captained his own schooner—the *Velocipede*
—that facilitated the exchange of arms and opium from Singapore for
tortoise shell and mother-of-pearl from Sulu.[167] Alongside the country
traders from Bengal, with their cargoes of munitions, opium, and cotton
textiles, it was the Salem skippers and free agents like Ward and Wyndham who took the lead in the triangular trade of the Eastern Seas until
the mid-nineteenth century.[168] While the English East India Company
drifted into terminal decline after the Indian Mutiny-Rebellion of 1857,
the pioneering role of those traders who flourished under its wings, the
risks they were prepared to undertake, and the prospect of even greater
rewards would embolden others to follow in their wake.

Among the latter was Captain John Dill Ross, an American merchant
who from the 1860s dominated the carrying trade between Singapore

and Labuan. Ross' yearly average of seven voyages between Singapore and the Borneo coast not only transported mail, Sulu and Chinese passengers, and their cargoes, but also freights of guns and opium, the proceeds of which were reinvested in wider commerce.[169] Simultaneously, there were other Western interlopers who vied for a share in the carrying trade. One of Ross' commercial rivals was William Frederick Schuck, a former member of the German consular service, who associated himself with the Singapore-German firm of Schomburg and acted as an intermediary between the German Government and the Sulu Sultanate. While Schuck shipped consignments of arms, opium, textiles, and tobacco on behalf of the trading house of Schomburg, he also forwarded correspondence between Bismarck and the Sultan that enabled Schomburg to establish warehouses and residences in Sulu's Sandakan Bay, with two steamers under the German flag. By promoting German commercial interests, in particular the transhipment of arms and opium from Singapore and the procurement of exotic produce from Sulu, the new trading depot posed a serious challenge to Labuan.[170] When further competition was presented by Portuguese merchants who facilitated the export of munitions from East Timor in the late 1880s, much to the irritation of British, Dutch, and German interests, there was a sense in which the arms trade in the Eastern Archipelago had come full circle.[171]

Respectable local entrepreneurs also played a supporting role in the arms trade, tapping into its profits to varying degrees. Prominent among them were the Straits Chinese (*Peranakan*) merchants, a commercial elite from an acculturated diasporic Chinese community, who thus interposed themselves as middlemen between European capitalists and indigenous communities around the region. Several of them expanded commercial operations from Malacca to Singapore, including Choon Bock & Co., Tan Kim Ching, and Tan Kim Seng. By the 1860s, these *Peranakan* merchants had proven their credentials as agents of early globalization, extending their business empires beyond the Straits Settlements to dominate the carrying trade across Anglo-Dutch colonial frontiers into the twentieth century.[172]

This section has so far surveyed the arms trade intermediaries who found ways to flog their wares in the open market, and especially those who managed to do so through various establishment connections. Yet there were numerous others who resorted to 'under-trading' in order to circumvent the restrictions imposed by arms control legislation. The documented cases of gun-running in the eastern Indian Ocean point to a tangled underworld of shady dealings, skulduggery, and subterfuge, in which gun-runners devised cunning methods of trafficking that were often ahead of the colonial state's designs to enhance surveillance and interdiction operations.

As with cases of smuggling in the western Indian Ocean, many European and American sea-captains who plied the archipelagic waters of the East simply refused to furnish detailed freight lists to their consular representatives at the various ports-of-call. The U.S. Consul at Manila highlighted this conspicuous lack of cooperation in a letter to John Quincy Adams at the State Department in 1823:

> I find it extremely difficult to ascertain the exact amount of the importation and exportation from this island. The commanders and supercargoes of American vessels are very unwilling to give me a list of their cargoes and I have no instructions to compel them to give me the exact amount.[173]

Moreover, export manifests rarely revealed the true contents of a vessel's cargo. Listings such as 'sundries', 'ironmongery', 'ballast' or 'hardware' were used regularly to mask arms shipments. Munitions were concealed under the false bottoms of ships, camouflaged by Straits produce and other commercial products, or contained within an ingenious assortment of casings, which ranged from kerosene tins to tobacco sacks, rice bales, hollowed-out table legs, and piano cases:

> Pianos were perfect commodities in which to hide this illegal passage; as big-ticket items, their import duties were paid in Europe, meaning they were sealed into cases in ports like Rotterdam and went uninspected on the long journey out to the Indies. Dutch smugglers took advantage of this fact and hid sizable numbers of guns in the large shipping crates that stored these objects, bargaining that they would not be opened until they had reached their eventual buyers. This ruse seems to have worked for some time until the authorities became aware of it. Eventually, however, letters went out from the highest levels of Dutch government to the ship captains, dock loaders, and agents of certain steamers, informing them that such commerce was strictly illegal.[174]

Tagliacozzo has further illustrated the extent to which unloading operations were smoothly organized across ethnic lines:

> Syndicates used fall men, usually indigent coolies, to take the blame if an operation went sour; these men would go to prison sometimes,

but their families miraculously would wind up much richer by the time they got out of jail. ...

One of the most lyrical descriptions of an attempt to smuggle arms into Singapore can be found around 1915, when a Straits policeman noticed a coolie disembarking from a docking steamer. Something about the lilt or cadence of the coolie under his load was wrong; the object seemed too light for the way the man was straining. The detective asked the man to stop, but he dropped his burden and sprinted off into the crowd and was lost. When the policeman disassembled the wooden folding table the man had been carrying, a dozen pistols and several hundred rounds of ammunition spilled out of the hollowed-out legs of the table, falling onto the ground.[175]

Whereas colonial regimes found it challenging to coordinate arms control efforts between them, even when confronted by crisis or conflict, there is every indication that arms traffickers could operate cross-culturally over their own ethnic business preferences, wherever the anticipated reward exceeded calculated risk.[176]

In mainland Southeast Asia, it was not uncommon for arms to be smuggled across the Burmese border from British India, the very seat of British colonial authority in Asia. Thus, in May 1877, arriving at Mandalay Post Office by parcel post from Calcutta were nearly 500,000 percussion caps of all sizes, 2,000 gun-flints, and 2,000 gun-nipples. A Parsi firm had despatched these munitions to several fictitious addresses in Mandalay, and a local agent went up to Mandalay to take delivery. But the parcels produced suspicious rattling noises when shaken, which led to the discovery of their contents by the Postmaster. They were not finally delivered, being returned on the orders of the Resident, Colonel Duncan, to the Postmaster-General's office at Calcutta for disposal. Upon further investigation, it was suspected that a local postman (originally from Chittagong) had been bribed to deliver the packages once they arrived, for none in Mandalay came forward to claim the parcels after the discovery. Although preventive officers on the Mandalay line made subsequent smuggling 'rather hazardous' for the Mandalay traders, supplies continued to 'arm the native' in Upper Burma.[177]

To gather precise information about indigenous gun-runners of diverse ethnic backgrounds was a tricky affair, since they were even less likely than their Western counterparts to comply with appointed trading channels and regulations. The Dutch Consul George Lavino was informed by the Harbour Master at Penang that 'little reliance can be placed in the accuracy of returns given by Chinese and Natives' who were especially instrumental in the arms trade with northern Sumatra, since they 'frequently have motives for not stating the actualities of goods they ship or receive'.[178] Subterfuge was routine. At least until the early 1860s, the Sulu marauders would regularly divest their vessels of military equipment and leave most of their crew on nearby islands when they visited Labuan 'in the guise of peaceful traders' to purchase firearms, gunpowder, and ammunition.[179] Local fishing boats and pepper-trading prahus operating off the coast of Borneo or Sumatra enjoyed a similar degree of flexibility in running weapons cargoes across the Straits. Whatever the country of origin, all vessels were actually permitted to carry limited stores of gunpowder for self-defence, even when it became apparent that such supplies were sold for profit on countless voyages and then restocked at the next port-of-call.[180] The rewards from arms trafficking were obviously substantial enough to encourage gun-runners—Western and indigenous—to take great risks in regional waters.

But finally, embedded within the business histories and banking records of Indian Ocean societies is evidence that drug money was used to finance the business of gun-running. Many of the commercial voyages in the Straits were underwritten by Chettiar money-lending houses, and Indo-Islamic mercantile communities were frequently involved behind the scenes in raising capital for such ventures.[181] But a closer inspection of the activities and accounts of the bigger merchant capitalists would suggest that profits from opium and the wider triangular trade of the Eastern Seas provided critical financial support for the arms transfer system across the entire Indian Ocean arena. It was through revenue-farming and money-lending that a number of key Indian merchants acquired the capital and the influence to dominate the opium trade from the large regional centres of the subcontinent. These wealthy entrepreneurs often reinvested the proceeds from their investments in the East by supplying capital to various mercantile groups located in western India. Not only did this reinvestment lay the foundation for a more dynamic economy in western Indian cities, such as Bombay, or forge commercial links with Parsi, Jewish, and European merchant firms, it also backed the traditional credit facilities provided by the merchant capitalists of western India, Muscat, Zanzibar, and East Africa.[182] This capital would be used later by those entrepreneurs to purchase munitions and slaves in the western Indian Ocean zone, as well as fund the gun-running exploits of the Arab dhow masters and caravan merchants examined in the previous chapter.

One such example was the firm of Jairam Sewji, the leading pioneer among the merchant capitalists of Zanzibar during the nineteenth century. In 1835, it gained control of the influential and lucrative post of customs collectors to the Sultan of Zanzibar, which enabled it to develop an extensive business network—and its own networks of indebtedness—throughout the region. By 1873, the firm had accumulated a capital of nearly £500,000 invested in loans and mortgages in East Africa, of which an estimated £60,000 had been advanced to the Sultan and his family, £200,000 to Arab traders in the interior and Zanzibar, £140,000 to Europeans and Americans, and £100,000 to other Indian merchants. Moreover, these were only the African assets of the firm, and excluded the capital and stock-in-trade of the parent company in Mandvi and Bombay, which far exceeded the African investments.[183] Yet it is inconceivable that Jairam Sewji could have expanded business operations to Zanzibar in the early days without the initial capital outlay advanced on credit by the wealthier mercantile establishments that had prospered through India's opium trade to the East.[184] Nor, for that matter, could two other prominent merchant capitalists: Sewa Haji, who served as banker and supplier to both Tippu Tib and Charles Stokes; and Allidina Visram, who handled commercial affairs (including the procurement of firearms) in Zanzibar for the Kabaka of Buganda, and went on to establish the first regular commercial bank in British Uganda.[185]

This potent combination of gun-runners, slave-traders, warlord armies, piratical fleets, and opium and ivory revenues would thereby become the bane of the Indian Ocean world in the nineteenth century. It was clear that any political order or commercial organization capable of controlling the shipment or marketing of drugs, firearms, and manpower could then rely on such product lines as political and economic resources. At this level, there seems little difference between the chartered company, the colonial regime, the mercantile house, and the marauders, on the one hand, and on the other, by extension, the drug cartels, the Golden Triangle, and the South China Sea pirates of today.[186] Around the Indian Ocean arena, drugs, guns, and slavery paved the way: by destroying the old order and financing the new, they laid the foundations for the growth of capitalism and a new militancy in Africa and Asia.

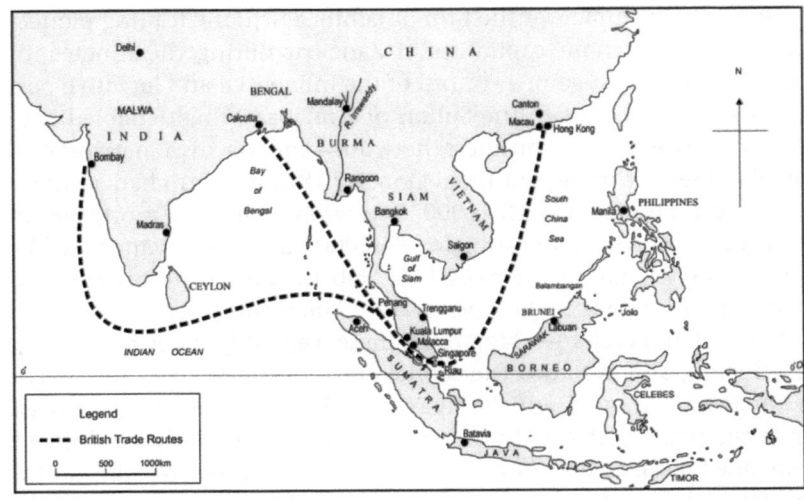

The Eastern Indian Ocean Zone

5
The Arms Trade and War in the Indian Ocean

So she followed her red-coats, whatever they did,
From the heights of Quebec to the plains of Assaye,
From Gibraltar to Acre, Cape Town and Madrid,
And nothing about her was changed on the way...
(But most of the Empire which we now possess
Was won through those years by old-fashioned Brown Bess.)

From *Brown Bess* by Rudyard Kipling

With breathtaking poetic licence, Kipling overstates the case for the old standard firearm of the British Army. It was true that since the eighteenth century, British redcoats had successfully wielded 'Brown Bess'-type muskets on the battlefields of Europe, America, Asia, and Africa. It was also true that until the mid-nineteenth century—when a great deal of the Empire had been acquired—very little technological innovation had altered their basic pattern, a recipe for success that conservatives in the military wanted to leave unchanged.

Yet, as we have seen, empires in Asia and Africa were built on more than armies equipped with muskets. They were also formed by the complex interplay between European ambitions and rivalries in trade and politics, and the aspirations of indigenous societies at various stages of crisis and transition. Moreover, with the ongoing evolution of indigenous small arms technologies and manufacturing, reinforced by arms transfers from Europe, colonial troops did not always enjoy technological superiority or tactical advantage over indigenous armies and resistance groups. Traditional or new weaponry in the hands of native warriors could prove just as effective—if not more so—in the context of local geography, military culture, and modes of warfare.

This final chapter examines the impact of the arms trade on military balances and small wars, before and after the 'Breech-loader Revolution' of the mid-nineteenth century. It locates these arms transfers within the cross-cultural milieu of the Indian Ocean arena—a precursor of the modern Indo-Pacific regional system and context for much of today's illicit trafficking. Such arms transfers would prove instrumental in both the expansion and limitation of Western power and influence, against the changing temper of indigenous state formation, resistance, and warfare.

The impact of military diffusion on cross-cultural collision, *c.*1780–1850

Distinguished scholars have contended that the 'revolutionary' development of European arms and military organization from the early modern period conferred a decisive advantage over the comparatively static weapons technologies and deficient military skills of Asian and African societies.[1] Yet this is a view that must be carefully qualified, for it otherwise assumes teleological significance when indeed there was nothing inevitable about what Geoffrey Parker hailed as 'the triumph of the West' and Paul Kennedy labelled 'the eclipse of the non-European world'. As Peter Marshall has argued, the potential capacity for Europeans to wage war more effectively than Asians had perhaps existed since the 'military revolution' of the seventeenth century, but only from the long nineteenth century was the full force of such potential felt on Asian battlefields: it was then that arms production in Europe was accelerating; shipping costs to the East were falling; European governments were displacing the companies and taking a direct interest in Asia; and territorial possessions in India were providing the British with new resources of men and money.[2]

Even then, there remains a wider indigenous context to consider. First, the European colonial war machine could not depend solely on its technical prowess wherever it was up against Indian Ocean societies with comparable weapons technologies and production capabilities, and wherever arms transfers from Europe had placed 'Brown-Bess'-type muskets in the hands of natives. Second, the military objectives of European imperialism were thwarted wherever it could not overcome the military advantages possessed by indigenous societies: traditional weaponry in the hands of highly proficient warriors; local knowledge of physically difficult terrain; skirmishing tactics and irregular warfare.

In this light, Parker's thesis that eighteenth-century Western arms technology was innately superior while the Indian move to adopt Western military tactics and weaponry came too little, too late, must be revised.[3] By the dawn of the long nineteenth century, Indian firearms technology and production were assuming a new complexion, uniquely 'modern' among the indigenous societies of Eurasia, with muskets, cannon, shot, and gunpowder in some cases equal to European standards. By 1785, the Maratha warlord Mahadji Scindia had established ordnance factories near Agra, utilizing relatively sophisticated indigenous military technology— emphasizing adaptation rather than innovation—and involving the expertise of local artisans. State arsenals and magazines appeared across the former Mughal heartland, at Agra, Mattra, Delhi, Gwalior, Kalpi, and Gohand, producing munitions equal to (if not surpassing) European standards.[4] Indeed, the bulk of firearms used by the armies of the Mughal breakaway satrapies and successor states were manufactured in local factories, like those at Lucknow, Pondicherry, Hyderabad, and Lahore, even if their European mercenary officers advocated and used European-made weapons.[5] 'Military entrepreneur states' like Rohilkhand were emerging, too, while the more exceptional indigenous armies were using weaponry at least comparable to Western standards, with near-critical military repercussions.[6] For Mysore and the Maratha Confederation (*c*.1750–1818), and the Sikhs (*c*.1800–49), various improvements in indigenous arms technology and production, reinforced by weapons and techniques obtained from the Europeans, generated an almost 'revolutionary' impact on Indo-European relations, politics, and society. Before the advent of the breech-loader, their armies would actually achieve technological near-parity with their European counterparts. From the commencement of the First Anglo-Mysore War in 1767 to the conclusion of the Sikh Wars in 1849, these native polities developed or acquired sufficient armament to challenge British colonial authority, dealing a crippling if not mortal blow to the Company's war machine in the run-up to the great Indian Mutiny-Rebellion of 1857.

Witness the case of Tipu Sultan's Mysore. Whereas contemporary accounts written from the victor's perspective have sought to denigrate Tipu's military achievement, the weight of evidence is incontrovertible. Following the hard-won siege of Seringapatam in 1799, the British seized 99,000 flintlocks, 52,000 of which were European (15,000 'English' and 37,000 'French' muskets). Perhaps even more remarkable was the fact that they found only 320 matchlocks, all of which appeared unserviceable. This extremely small stock of what was then the Indian standard firearm attests to the modernity of Tipu's army: small arms which continued to be used generally in other parts of India well into the nineteenth century were being regarded as obsolescent by Mysore as early as the 1770s–80s. The manufacture of flintlocks was, however, discontinued after Tipu's downfall, for his European mercenaries and artificers were soon dislodged from Mysore and very likely deported. Nothing remains of the 'eleven armouries for making and finishing small arms'.[7] The Marathas, with their tactical and technological transition from irregular cavalry warfare to infantry battles, came close to defeating the British on several occasions. During the Second Anglo-Maratha War (1803–5), Arthur Wellesley's success at Assaye owed more to a bayonet charge than any advantage in firepower conferred by Western arms; in retrospect, the hero of Waterloo would acknowledge Assaye as the hardest-fought battle of his military career.[8] But finally, in order to defeat their opponents in the Sikh Wars (1845–49), the British had to deploy armies equal in size and superior in artillery firepower. This was no mean feat, as by the time of his death in 1839, the great 'Lion of the Punjab' Ranjit Singh had built up a 150,000-strong army, of whom about 65,000 were regulars 'trained by European soldiers of fortune and supported by ... guns of a type more modern than those used by the British'.[9]

Even in clashes with Indian Ocean societies whose weapons technologies and manufacturing were less advanced, European colonial forces had to grapple with other indigenous factors: traditional weaponry used by skilled warriors; local knowledge of intractable terrain; skirmishing tactics and guerrilla operations; anti-colonial religious protest and the sheer political will to resist. Here, above all, the cultural context of conflict would shape its outcome. In the Java War (1825–30), the Dutch could mobilize up to 12,000 troops (half European, half indigenous), yet they were no match for the forces (probably only 5,000 men) and fighting methods of Javanese resistance hero, Diponegoro, Prince of Yogyakarta. Whether through dealings with corrupt Dutch officials, British and American gun-runners, or local arms manufacturers, the Javanese had soon procured hundreds of British muskets, together with bayonets, ball cartridges, and gunpowder, which they combined with staves and slings.[10] But it was not this weaponry *per se* which brought defeat to the Dutch army in the early years of conflict. Rather, it was the mobility and use of traditional skirmishing tactics under Diponegoro's leadership that so skillfully exploited the weakness of the Dutch military position: in particular, their large, unmanageable forces and immobility. Only when the Dutch adapted themselves to the demands of indigenous warfare by deploying smaller fighting units with a similar hit-and-run approach, backed up by the construction of small fortified posts (*bentengs*), did they begin to pacify rural areas and turn the tide of the war.[11]

In peninsular Malaya, the campaigns of the Naning War (1831–32) likewise revealed a contest between two vastly different systems of war. Despite having a force of 1,200 seasoned redcoats from the Madras

Native Infantry, backed by British musketry and artillery, the first Naning campaign ended abysmally for the British. These same Madras redcoats had seen action in the First Anglo-Burmese War (1824–26), but had even then barely survived a logistics disaster in the swamps of the Irrawaddy.[12] Further south, the troops were non-acclimatized to the Malay Peninsula's humid jungles. Disciplined and trained for pitched battle, they were unaccustomed to the entanglements of jungle warfare. Wind and weather —especially 'Sumatras'—had a particularly ruinous effect on British fire-arms and ammunition.[13] The heavy artillery had to be jettisoned in the sea of green in the wake of the first campaign.[14] The small arms, as one British officer noted, scarcely made any difference; the troops were 'only armed with the Brown Bess, and consequently possessed weapons very little, if any, superior to those of the Malays, some of which were reported to have carried as far as twelve hundred yards'.[15] But inspired by their leader Dol Syed, the Malay raiding parties travelled light, equipped only with bare essentials: daggers, swords, throwing-spears, in addition to muskets which had been secured through marauding and smuggling activities.[16] Weaving in and out of Naning's thick primeval forests, the Malays expressed a cultural preference for guerrilla entanglements, whether sniping from hidden stockades or timing the precise moment for each offensive.[17] It was only after the British acknowledged their technological and military limitations, employing a strategy of divide-and-rule to exploit internal dissension among the Malays and turning their aptitude for 'commando' tactics against one another, that the British finally managed to win the second campaign, and with it, the war.[18]

Perhaps most infamous of all, however, was the cross-cultural collision of the First Anglo-Afghan War (1839–42). The British invasion of Afghanistan made initial progress but deteriorated precipitously from 1841, forcing an ignominious retreat: 4,500 British Indian troops (only 700 Europeans) and 12,000 camp-followers pulled back from Kabul to Jalalabad under the leadership of Major-General William Elphinstone, a veteran commander from Waterloo. Harassed all the way, the entire force was annihilated within a week, to cries of 'treachery'.[19] One survivor of the doomed garrison at Charikar lamented the limitations of British arms in inhospitable Afghan terrain: 'Our muskets were so foul from incessant use that the balls were forced down with difficulty, although separated from the paper of the cartridge which usually wraps them round.'[20] Conversely, having no regular army, the Afghan mode of warfare was based on skirmishing tactics and familiarity with the terrain, using the tribal long rifled-musket to good effect from behind natural cover. Swords and long-bladed knives were part of the indigenous arsenal, too, but the *jezail* undoubtedly outclassed the musket used by British Indian troops, enabling sniping tribesmen to harass the flanks of British forces from out of range. As such, there would be little need to import firearms from the West for decades to come. For their part, the British escaped by the skin of their teeth, and only by using Afghan skirmishing tactics to oust the Afghans from their positions on the Khyber Pass.[21]

H. L. Wesseling has reiterated that Western dominance hinged upon 'technical prowess', whether in the form of gunboats, artillery, cavalry, or small arms. But such superiority was not always usable, depending on the nature and degree of asymmetry in a given indigenous military context.[22] The deployment of 'Brown Bess' could not guarantee victory, given the kaleidoscopic interplay between rival groups of Europeans bearing the latest musketry; an evolving arms transfer system; and various indigenous forces wielding a combination of locally-made and imported weaponry on home ground. Over the first half of the nineteenth century, British military domination of the subcontinent was a closer-run thing than many have cared to admit, while the cross-cultural engagements of the Java War, the Naning War, and the First Anglo-Afghan War are salutary reminders of where Western technological dominance did, in fact, end.

The impact of military diffusion on cross-cultural collision, *c.*1850–1914

The dramatic advances in Western military technology over the second half of the nineteenth century would altogether transform the dynamics of cross-cultural collision. Between 1848 and 1878, the old flintlock muskets were effectively replaced when all the European powers rearmed their infantry with single-shot breech-loaders. Then, from 1878 through to the 1890s, a new generation of repeating firearms was spawned, which included magazine rifles and Maxim guns. This new weaponry far surpassed anything that had been achieved with muzzle-loading technology, whether in terms of accuracy, range, or rate of fire. In Paul Kennedy's view, such a 'firepower revolution' in the metropole had the potential to create a 'firepower gap' at the periphery, which 'quite eradicated the chances of a successful resistance by indigenous peoples reliant upon older weaponry'.[23] Combined with the proper training, a generous supply of spare parts, several light repair workshops, and (in the case of repeating firearms) huge amounts of ammunition, the new weaponry seemed virtually unstoppable. It was this new military configuration, showcased in Horatio Herbert Kitchener's conquest of the Sudan, which enabled the Westernized 'hybrid army' to overwhelm local resistance: perhaps the greatest disparity of all was seen at the battle of Omdurman (1898), when in one half-morning, the Lee-Enfields and Maxims of Kitchener's Anglo-Egyptian force wiped out 11,000 Dervishes for the loss of only 48 of their own men.[24]

Yet the image of impotent natives armed with arrows and spears being mown down by Europeans with machine-guns is simplistic and one-sided. It must be remembered that 'Brown Bess' had not merely been replaced but displaced, along with successive models of breech-loading precision-arms that were rendered obsolete by progressive waves of technological innovation. Each changeover immediately liberated vast quantities of still-lethal weapons for export, to the very places where the colonial powers were fighting to establish their authority. The new technology and weaponry certainly facilitated European expansion, summed up in the self-confidence of Hilaire Belloc's Captain Blood:

> Whatever happens, we have got
> The Maxim gun, and they have not.

But equally, if indigenous forces did not have the Maxim gun, they quickly acquired a host of fairly modern breech-loaders, and even the odd machine-gun, too. Moreover, where indigenous military culture continued to rely on superior field-craft, knowledge of the land, and political or religious motivation, the *assegai, jezail,* or *machete* could still be a more effective weapon than the modern firearm.

In Ethiopia, Western technological and military advantages were offset by indigenous military culture and procurement of the new weaponry in the context of imperial restoration and national reunification. Following the chaos of the 'age of princes' (1755–1855), the restored Emperor Tewodros II (r.1855–68) sought to recentralize political authority against the twin challenges of internal rebellion and European intervention. Although Sir Robert Napier, the British commander-in-chief, noted how the 'Brown Bess' muskets carried by elements of his Indian Army were 'hardly equal to the double barrelled percussion guns of the Abyssinians', the British soldiers armed with breech-loading rifles decisively outgunned the Ethiopian musketeers at Magdala (1868).[25] With the memory of Magdala firmly imprinted and the elevated threat of European incursions during the era of high imperialism, the possession and procurement of the new weaponry soon became a prerequisite for securing national leadership: Yohannis IV (r.1872–89) and Menelik II (r.1889–1913) both excelled in their mastery of the new technology, acquiring large quantities of quick-firing weapons, overawing their fellow warlords, and eventually crushing Italy's pretensions to exercise a protectorate over Ethiopia. By the 1890s, rifles and the occasional machine-gun were being deployed to devastating effect.[26] At Adowa (1896), General Oreste Baratieri's 'modern' Italian army was annihilated by an indigenous opponent fully aware that the clash would decide its destiny: an approximately 15,000-strong Italian force armed with just as many rifles was comprehensively defeated by an estimated 100,000-strong Ethiopian army in possession of superior topographical knowledge, tactics, and again just as many rifles, with cartridges by the crate, mostly acquired from the French at Jibuti.[27]

In South Africa, the Second Anglo-Boer War likewise illustrates the limitations of the British colonial war machine in the face of an alien military context and a burgeoning arms trade. British superiority in resources was frequently offset by the sheer tenacity of Boer offensives, organized resistance, and guerrilla warfare, backed up by modern firearms that included magazine rifles.[28] Until the arrival of British reinforcements, the Boers enjoyed both numerical and tactical superiority; no fewer than 147,000 Mauser rifles had passed through customs via Delagoa Bay in 1897, to be shared among the 30,000 able-bodied Boers in the Transvaal —a ratio of five Mausers per head. It was reported that these rifles engendered remarkable feats of marksmanship.[29] Only after heavy mobilization of British imperial forces—by war's end they had assembled more than 300,000 troops from Britain, Canada, Australia, and New Zealand—followed by brutal reprisals against Boer civilians, and a recourse to the irregular 'commando' warfare in which the Boers excelled, did the British finally break the back of the insurgency.[30]

Local geography, tactics, and transfers of the new weaponry also eroded Britain's technological and military advantage in Afghanistan and India's Northwest Frontier. Writing in 1911, one colonial official observed that,

> Mountain warfare, as carried on in the North-West Frontier, does not lend itself to impassioned descriptions of bloody fields of battle, of cavalry charges and heroic actions done in the limelight before the astounded gaze of two armies. The Afridi does not rush down into the open to certain death, but retires gracefully before a stronger force sent against him, skulking along the sky-line, and ready to take advantage of the smallest mistake on the part of his opponents, or to cut off any straggler. He has been described as the finest natural skirmisher in the world. ... Those who by repeated experience of this kind of warfare may be considered competent judges declare that nothing is more nerve-racking or more demoralizing to troops than to have to undergo night after night this constant sniping, coupled with the apprehension of a night attack at any moment.[31]

In this manner, the *jezail* continued to outclass the small arms employed by British Indian troops for years after the First Anglo-Afghan War, enabling resisting groups of Baluch and Pathan tribesmen to harass British forces from out of range, combining superior marksmanship with skirmishing tactics in guerrilla warfare. The progressive adoption of Snider, Martini-Henry, and Lee-Metford rifles by British forces redressed the imbalance for a time, but by the early 1890s, substantial numbers of modern precision-arms were beginning to equip and empower the frontier tribes, too, via Muscat, thus offsetting the British superiority in firepower. The intractability of the region's terrain, and the influx of even more lethal weapons, would be exploited to full advantage by the tactics and strategy of tribal warriors. Such developments ultimately jolted the British into formulating a specific training doctrine suited to military operations in the Northwest Frontier.[32]

There were also instances where tactics rather than transfers of technology and weaponry shaped the course of colonial warfare and expansion. The final phase of Dutch expansion in the East Indies—from the mid-nineteenth century up until 1910—did not depend on any superiority conveyed by modern firearms. In 1867, the old-fashioned standard muzzle-loader was converted into a more up-to-date breech-loading weapon, but even so, the incorporation of such new weaponry into the Dutch Colonial Army was very gradual. On the other hand, in addition to their traditional weapons (bows and arrows, lances, pikes, *kris*), indigenous resistance groups accessed a wide variety of firearms, old and new, home-made as well as the latest precision-arms imported from Singapore. In any case, neither the indigenous soldiers of the Colonial Army nor the resistance fighters seemed to be very skilled in the handling of their respective firearms. The 'Breech-loader Revolution' in the metropolis between the 1850s and 1880s did not thereby translate into either simultaneous rapid territorial expansion or effective indigenous resistance at the periphery: Dutch expansion in Aceh and other parts of the Archipelago ceased precisely during those years; conversely, the introduction of modern rifles in the Colonial Army had scarcely begun when the Dutch conquest was completed. The Colonial Army achieved their greatest measure of success when resorting to old-fashioned storming assaults in order to capture fortified villages, where even modern rifles would not have been of much use against the sheer solidity of indigenous fortifications.[33] As Wesseling has emphasized, 'the military breakthrough in Aceh eventually came when the *Corps Marechaussee* employed, not the Maxim gun, but that quintessentially primitive weapon, the naked sabre'.[34]

Again, as the example of Britain's epic engagement with the Zulus will bear out, the possession of sophisticated new firearms did not always guarantee success against indigenous weaponry and tactics. Given the disparity in military resources between the British (assisted by colonial volunteers and 'native levies') and the Zulus, the final conquest of Zululand was assured, depending largely upon problems of transport and supply being overcome. Yet in the Anglo-Zulu War of 1879, while the British defenders at Rorke's Drift repelled waves of Zulu warriors through the stopping power of their Gatling guns, Martini-Henry rifles, and bayonets, their compatriots at Isandlwana proved far less fortunate. The highly disciplined Zulu force caught Lord Chelmsford's main camp at Isandlwana by surprise, and launched wave upon wave of assaults inflicting huge losses. While there is evidence that the Zulus had used rifles for sniping and harassment, some of which had been old War Office surplus, their main weapons were the throwing-spear (*assegai*) and the stabbing-spear (*iklwa*), which they used to terrible effect at close range.[35] Conversely, the British supply system broke down at a critical point when those in charge of issuing ammunition demanded receipts; with the depletion of ammunition, Zulu superiority in numbers and tactics enabled them to slaughter their hapless opponents. The Zulus finally lost the war, but not before they had first offered ferocious resistance, challenging prevailing assumptions about the superiority of Western arms in the modern world.[36]

The future of arms transfers and war in the Indian Ocean world

Commenting on the changing military impact of European incursions into the Indian Ocean arena, Sir Michael Howard observed how,

> At the beginning of the sixteenth century, it was their monopoly of guns that enabled the Portuguese to break into and dominate the trading system of the Indian Ocean. But as the use of firearms became general throughout the world the advantage which Europeans gained from them disappeared. In the eighteenth century it was the professional qualities of drill and discipline, together with a careful attention to their supply system, that gave European armies such an ascendancy in, for example, India, rather than the weapons with which they were armed. But in the nineteenth century the balance swung decisively in favour of the technologically superior powers ... European artillery, breechloading rifles, and machine-guns made the outcome of any fighting almost a foregone conclusion.[37]

To be sure, rifling, the sliding bolt, and the magazine spring are mass-produced elements of high sophistication; and even more so, the chain-feed of a Maxim. The advent of precision-machined, mass-manufactured weapons and weaponry components was effectively superseding the craft-made article, however brilliantly executed. This helps to explain why societies without the requisite industrial infrastructure and other factor endowments could not keep up technologically, and lost out militarily in confrontations with societies that had those resources. From the perspective of military technology, the nineteenth-century disparity between the West and Indian Ocean societies could be explained in terms of the technological innovation occurring in the metropolis, with a growing technological gap and time-lag at the periphery. The margin of advantage narrowed at times but proved virtually unbridgeable after the 1850s–80s, when military technological 'take-off' in Europe was translating into a quantum leap forward at the Asian or African frontier. Furthermore, the system of rifling revolutionized military thought and organization in nineteenth-century Europe. Although the arms transfer system ensured that rifles reached the remotest frontiers, thereby triggering off limited 'military revolutions' in certain Asian and African societies, the corresponding changes in military organization usually also occurred after a time-lag. On balance, technological and organizational superiority remained with the West.

Still, there was nothing inevitable about Western military dominance—even after 1850—insofar as technological and organizational advantages were constrained by indigenous military contexts. The self-confidence of Hilaire Belloc's Captain Blood did not always ring true, as the British survivors of the Zulu victory at Isandlwana and the Italian survivors of the Ethiopian victory at Adowa could testify. The other cases of cross-cultural collision between 1850 and 1914—in South Africa, the Northwest Frontier, and Southeast Asia—further remind us that superiority in technology, organization, and weaponry did not always surmount obstacles presented by local geography, superior field-craft and tactics, knowledge of the terrain, and motivation, which sometimes made the use of traditional weapons more effective. Again, the arms transfer system could—and did—militarize many Indian Ocean societies, equipping innumerable resistance groups with the most modern rather than the most outmoded of weapons.

But if scholars today still marginalize or overlook these arguments, then is it surprising that the generations after 1914 should also fail, or find it difficult, to grasp these lessons in the long term? Recognition of the killing power of modern weapons and the efficacy of irregular 'commando' tactics in the Anglo-Boer War failed to influence the attitudes of British generals toward technological and tactical innovation in the First World War.[38] The British pacification of Kenya (1895–1912) had partly depended on the military contributions of indigenous recruits like the East African Rifles, a regiment that trained men previously involved in the region's gun, slave, and ivory trades. But adding a further twist of irony in the tragic narrative of the Mau Mau insurgency (1952–60), the ancient rifles of the triangular trade and British regimental life would survive—restored or modified—to be effective weapons against the colonial authorities.[39]

In France's pacification of Vietnam (1882–96) and Madagascar (1895–1914), the difficulties of fighting an enemy that limited themselves to guerrilla engagements became painfully apparent. Only through the 'Vietnamization' of French colonial troops and tactics was Indo-China finally subjugated.[40] And yet, these aspects of colonial warfare—the socio-political analysis of an enemy's nature, the formulation of means to win indigenous collaborators, and the increasing Vietnamization of colonial forces —not only characterized the colonial conflict of 1882–96, but also explained the catastrophic evolution of war in Indo-China from 1945–54 and 1960–75.[41]

Further progressing through the Gulf and across the troubled Northwest Frontier, large quantities of modern rifles from the West began to replace the tribal *jezails* with which Afghan warriors had hitherto been armed. The 'official' and 'unofficial' arming of nineteenth-century Afghanistan would lead to a third Anglo-Afghan war in 1919: 60 years before the onset of a costly Russian (Soviet) occupation; and 80 years before yet another world power would be drawn into the tribal politics of the militarized zone, with its globalized networks of terror. The British Raj would eventually recede into history, along with much of the Birmingham gun industry. But in the *mujahidin* resistance against the Soviet occupation of Afghanistan (1979–89), or Al-Qaeda's outmanoeuvring of US-led military operations (2001–2), the age-old constraints placed upon the sheer array of Western arms—incorporating sophisticated, 'smart' technology in the latter offensive—would again become apparent: geographical obstacles, such as the mountain ranges and the cave complex of Tora Bora; guerrilla tactics based on traditional Afghan warfare; and yet more guns (including many AK-47 and M-16 rifles) obtained from both Soviet and American sources by Islamic fundamentalist 'freedom fighters'.[42] From the dying stages of the global Cold War into the early phases of an international war against terrorism, the West was still learning—through its encounters with arms transfers and armed resistance across Indian Ocean societies—that the technological and organizational superiority of Western arms did not automatically translate into victory in warfare.

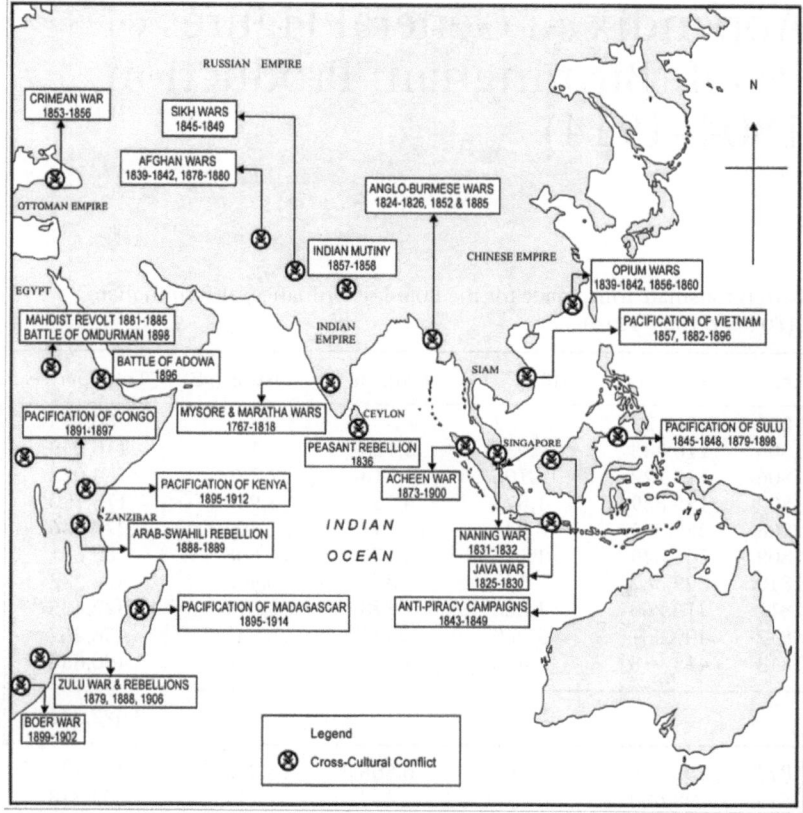

Cross-Cultural Collision in the Indian Ocean World

Appendix A: General Figures of Proof (Birmingham Production, 1804–1914)

Barrels for small arms, made for the Board of Ordnance, at Birmingham (1804–28, 1855–1914)

Year	Muskets	Rifles	Carbines	Pistols (×2)	Total per year
1804	80,123	62	–	–	80,185
1805	110,833	2,938	3,120	1,158	118,049
1806	112,222	15,106	4,418	2,276	134,022
1807	155,839	1,873	6,536	8,942	173,190
1808	229,355	6,334	15,245	21,402	272,336
1809	265,049	1,433	5,571	16,608	288,661
1810	299,382	133	313	6,405	306,233
1811	316,760	1,886	178	3,611	322,435
1812	409,961	2,260	7,694	16,347	436,262
1813	413,918	466	24,878	51,576	490,838
					2,622,211
1814	282,215	91	6,566	23,313	312,185
1815	98,689	–	442	4,117	103,248
1816	–	116,686	–	10,745	127,431
1817		87,217		9,456.5	96,673.5
1818		128,935		14,843.5	144,778.5
1819		102,499		18,118	120,617
1820		66,099		66,099	84,568
1821		63,754		10,825	74,579
1822		56,419		9,535	65,954
1823		97,926		10,521	108,447
					1,238,481
1824		104,713		8,611.5	113,324.5
1825		159,740		11,555	171,295
1826		126,745		11,425	137,170
1827		73,306		8,649	81,955
1828		68,615		8,362	76,977
					580,721.5

Year	Trade Proof House (Banbury Street)	Military Barrels	Government Proof House (The 'Tower', Bagot Street)	Pistols	Total per year
1855	264,477		131,869	28,776	425,122
1856	275,468		119,910	33,252	428,630
1857	302,670		136,247	35,593	474,510
1858	198,238		127,265	36,504	362,007
1859	250,922		133,823	42,775	427,520
1860	301,021		136,298	35,098	472,417
1861	380,781	142,819	101,241	32,637	657,478
1862	622,372	388,264	45,405	24,288	1,090,329
1863	460,140	210,078	19,263	38,522	728,003
1864	221,726	42,242	26,930	48,181	338,279
					5,396,295
1865	576,884				576,884
1866	582,127				582,127
1867	766,893				766,893
1868	961,459	(c.197,000			961,459
1869	693,572	21,000			693,572
1870	852,079	90,000			852,079
1871	891,228	119,000			891,228
1872	815,863	40,000			815,863
1873	756,056	19,000			756,056
1874	626,478	40,000			626,478
					7,522,639
1875	695,554	95,000			695,554
1876	466,748	10,000			466,748
1877	458,656	14,000			458,656
1878	559,815	13,000			559,815
1879	552,152	6,000			552,152
1880	638,070	16,000			638,070
1881	730,364	12,000			730,364
1882	771,597	7,000			771,597
1883	681,439	8,000			681,439
1884	694,035	7,000			694,035
					6,248,430
1885	501,634	21,000			501,634
1886	459,052	7,000			459,052
1887	440,334	5,000			440,334

Year	Trade Proof House (Banbury Street)	Military Barrels	Government Proof House (The 'Tower', Bagot Street)	Pistols	Total per year
1888	460,211	7,000			460,211
1889	529,082	12,000			529,082
1890	520,949	10,000)			520,949
1891	561,631				561,631
1892	379,086				379,086
1893	335,271				335,271
1894	299,273				299,273
					4,486,523
1895	328,791				328,791
1896	324,898				324,898
1897	402,115				402,115
1898	392,939				392,939
1899	375,513				375,513
1900	390,268				390,268
1901	355,270				355,270
1902	376,788				376,788
1903	427,474				427,474
1904	304,969				304,969
					3,679,025
1905	337,457				337,457
1906	370,528				370,528
1907	371,435				371,435
1908	326,697				326,697
1909	340,176				340,176
1910	420,239				420,239
1911	417,988				417,988
1912	514,681				514,681
1913	448,324				448,324
1914	361,011				361,011
					3,908,536

Note: The figures in italics are decadal totals, the only exception being the total for the five-year period, 1824–28.
Sources: *The Ironmonger*; *Arms and Explosives*; 'Artifex' and 'Opifex', *Causes of Decay in a British Industry*; and C. Harris, *History of the Birmingham Gun-Barrel Proof House*, pp. 152–5.

Appendix B: Monthly Fluctuations in British Small Arms Exports, *c.*1860 to *c.*1880

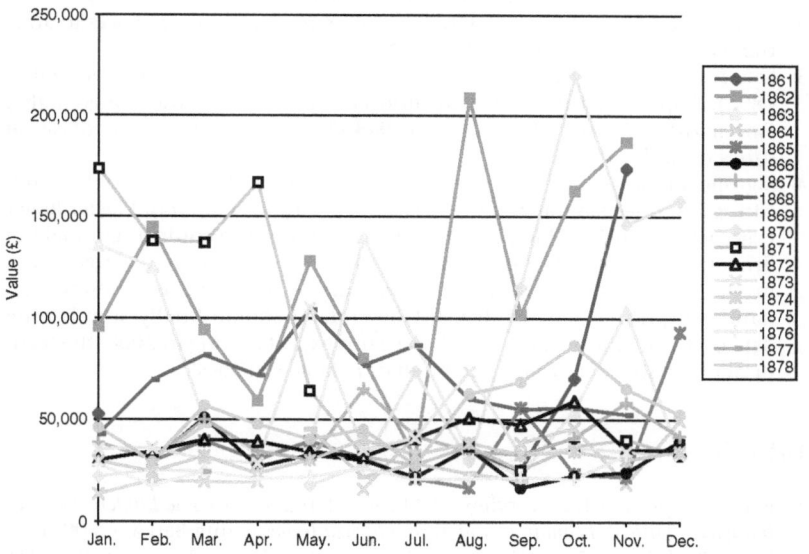

Sources: *The Ironmonger;* and Artifex and Opifex, *Causes of Decay*.

Notes

Preface

1 The Proof House archives held the informative Proceedings of the Guardians (PG), incorporating miscellaneous correspondence and specialist literature (both published and unpublished), plus newspaper clippings (including rare excerpts from the *Birmingham Daily Post*, 1864–69).

2 *Arms and Explosives* furnished Proof House returns and key references to arms trafficking in the Gulf region during the 1890s.

3 See, for example, L. Decle, 'The Murder in Africa', *New Review* (December 1895); 'Papers relating to the Execution of Mr. Stokes in the Congo State', *Parliamentary Papers*, Africa No. 8, C 944 (1896); A. Luck, *Charles Stokes in Africa* (Nairobi, 1972).

4 Significant gaps existed even at the better-documented metropolitan end. Citing just one example from conversations with former Proof Master Roger Hancock, a considerable portion of the commercial records of gunmaking firms located in the Birmingham Proof House Library were destroyed in a mysterious boiler fire during the last century. Against this background, it is hardly surprising that the British Government should finally stop issuing data on the financial value of Britain's arms exports, citing in 2008 'the technical difficulty of continuing to produce reliable statistics'.

Introduction

1 For a comprehensive overview, see R. A. Bitzinger (ed.) *The Modern Defence Industry: Political, Economic, and Technological Issues* (Santa Barbara, 2009).

2 See C. Trebilcock, *The Industrialization of the Continental Powers, 1780–1914* (London, 1981); I. T. Berend, *The European Periphery and Industrialization 1780–1914* (Paris, 1982); M. E. Chamberlain, *'Pax Britannica'?: British Foreign Policy, 1789–1914* (London, 1988); and D. Blackbourn, *Fontana History of Germany 1780–1918: The Long Nineteenth Century* (London, 1997). European historians justify their starting-point in the 1780s by arguing that the French Revolution and the armies it spawned changed forever the geopolitical and military configuration of Europe, even as the industrialization that began in Britain was also starting to work its socio-economic transformation in the rest of the Continent.

3 See V. T. Harlow, *The Founding of the Second British Empire 1763–1793*, 2: *New Continents and Changing Values* (London, 1964), pp. 482–594; and, more recently, C. A. Bayly, *The Birth of the Modern World 1780–1914: Global Connections and Comparisons* (Oxford, 2004). The imperial historian's case for starting from the late eighteenth century rests not only on Britain's loss of the American colonies and, as an extension of European great power rivalry, the growth of commercial and strategic interests in India and the Far East. As this book argues, indi-

genous societies in Asia and Africa were themselves at various stages of crisis and transition, which in turn shaped the course of colonial encounters throughout the century.

4   F. Braudel, *The Mediterranean and the Mediterranean World in the Age of Philip II*, trans. S. Reynolds, 2 vols (London, 1966). Fernand Braudel (1902–85) was a great oceanic historian, though known primarily as a leading light of the French Annales School of historical writing. The *longue durée* would be the expression denoting their distinctive approach to the study of history: historical events or movements in politics, economics, and society yield particular consequences in the short to medium term, but the seasonal cycles and structural crises of the shorter time-scale are regulated over the long term by underlying structures or forces of geography, climate, and culture. All but permanent or perennial in nature, these larger structures or forces predispose particular communities to characteristic patterns of thought and action, thereby shaping their identities and destinies in the long run.

5   K. N. Chaudhuri, *Asia before Europe: Economy and Civilization of the Indian Ocean from the Rise of Islam to 1750* (Cambridge, 1990), pp. 10–41, 147–8; cf. R. Kaplan, *Monsoon: The Indian Ocean and the Future of American Power* (New York, 2010). While occasional reference will be made to what some scholars and analysts have defined as the 'greater' Indian Ocean, extending further eastward across the South China Sea, this book focuses on the 'lesser' Indian Ocean, encompassing East Africa, the Gulf region, South Asia, and Southeast Asia.

6   See J. R. V. Prescott, *Political Frontiers and Political Boundaries* (London, 1987). In defining 'borders' for the modern world, Prescott makes a clear distinction between 'frontiers' and 'boundaries', in which the indistinct contours of frontier spaces have given way to distinct boundary lines, while 'borderland' refers to the transitional zone within which a border lies. From my study of the Indian Ocean 'periphery', this was at best an uneven process of transition, occurring at different times in different places. Such terms are therefore used more or less interchangeably throughout this book.

Chapter 1   The Arms Trade and Global Empire in the Indian Ocean

1   See *SIPRI Yearbook 1999* through to *2010* (Oxford, 1999 through to 2010); cf. *SIPRI Yearbook 2010*, especially pp. 177–283, 319–21. Estimated world military expenditure: $1,208 billion (1989); $834 billion (1998); $1,071 billion (2004); and $1,214 billion (2007). Figures are in US dollars at constant (2007) prices, the most recent year for which official data was available at the time of writing. Estimated total value of global arms exports was most probably below the true figure since a number of significant exporters, including China, released no official data on the financial value of their arms exports. From 2008, the British Government also stopped providing such data, given 'the technical difficulty of continuing to produce reliable statistics'.

2   J. Boutwell, M. T. Klare and L. W. Reed (eds) *Lethal Commerce: The Global Trade in Small Arms and Light Weapons* (Cambridge, Massachusetts, 1995); Bonn International Center for Conversion, *Conversion Survey 1997: Global Disarmament and Disposal of Surplus Weapons* (Oxford, 1997); and D. Capie,

Small Arms Production and Transfers in Southeast Asia (Canberra, 2002), pp. 4–5. Whereas 'white-market' arms transfers are legitimate inter-state transactions, 'grey-market' transfers are the channelling of arms surpluses from states (or private arms companies with the knowledge of 'state actors') to groups or individuals acting in a private capacity betwixt the interstices of state power ('non-state actors'). The grey market also encompasses illicit transfers from legal stockpiles to states under arms control restrictions (e.g. embargoes or sanctions). 'Black-market' transfers are totally illicit arms trafficking linked to organized criminal activity, where state-controlled arms stockpiles are accessed through corruption, coercion, or theft; and where organized criminal groups act as principal arms suppliers to insurgent and terrorist groups worldwide. Such participation in the illicit arms trade is usually related to wider smuggling activities, including trafficking in drugs, gemstones, and other commodities treated as contraband.

3   See *The Straits Times* (13–14 April 2006) and *Today* (14 April 2006). Singaporean, Indonesian, and British nationals were apprehended in the United States for conspiring to supply munitions to markets in Indonesia. As arms dealers in the black market of the international arms bazaar, they were regional participants in a worldwide web of commercial intrigue involving multi-million dollar purchases of large quantities of handguns, machine guns, sniper rifles, air-to-air missiles, and aviation radar equipment for shipment to Southeast Asia. Although militants or terrorists may not have been intended recipients, the subversive nature of the operations and the destructive potential of all such weaponry could nonetheless escalate threats to national security if allowed to interpenetrate the porous borders of nation-states.

4   T. Makarenko, 'Tracing the Dynamics of the Illicit Arms Trade', *Jane's Intelligence Review* (1 September 2003).

5   'Crime, Conflict, Corruption: Global Illicit Small Arms Transfers', *Small Arms Survey 2001* (Oxford, 2001). Groups supplied via the grey and black markets have included Chechen rebels in Central Asia and the Liberation Tigers of Tamil Eelam (LTTE) in South Asia.

6   Data derived from United States Arms Control and Disarmament Agency (ACDA), *World Military Expenditures and Arms Transfers, 1989, 1994* and *1996* (Washington, DC, 1990, 1995, and 1997); *SIPRI Yearbook 1999* and *2005*. For an overview, see 'Small Arms, Big Business: Products and Producers', *Small Arms Survey 2001* (Oxford, 2001).

7   See the SIPRI Annual Report, 13 June 1996; *SIPRI Yearbook 2010*, pp. 285–318; cf. *The Military Balance 1996* through to *2011* (London, 1996 through to 2011). In terms of global arms transfer agreements and deliveries (2005–9), the leading suppliers of conventional arms—apart from China and Brazil—were all European countries (Russia, Germany, France, and Britain) and the United States; the leading recipients—apart from Venezuela and Greece—were Asian and African countries.

8   M. Tully, 'The Arms Trade and Political Instability in South Asia', *Churchill Review* (1995), p. 17.

9   *SIPRI Yearbook 1999*, p. 425; *SIPRI Yearbook 2000*, p. 8.

10 P. Chalk, 'Light Arms Trading in SE Asia', *Jane's Intelligence Review* (1 March 2001); and 'Reaching Consensus in New York: The UN 2001 Small Arms Conference', *Small Arms Survey 2002* (Oxford, 2002).

11 *SIPRI Yearbook 1999*, pp. 1, 7, 10; cf. *SIPRI Yearbook 2005*.

12 See 'Making the Difference? Weapon Collection and Small Arms Availability in the Republic of Congo', *Small Arms Survey 2003* (Oxford, 2003); 'The Central African Republic: A Case Study of Small Arms and Conflict', *Small Arms Survey 2005* (Oxford, 2005); 'Fuelling Fear: The Lord's Resistance Army and Small Arms' and 'Stabilizing Cambodia: Small Arms Control and Security Sector Reform', in *Small Arms Survey 2006* (Oxford, 2006); I. Talbot, *India and Pakistan* (London, 2000), pp. 231–4; Chalk, 'Light Arms Trading in Southeast Asia'; A. Davis, 'Philippine Security Threatened by Small Arms Proliferation', *Jane's Intelligence Review* (1 August 2003); 'After the Smoke Clears: Assessing the Effects of Small Arms Availability' and 'Tackling the Small Arms Problem: Multilateral Measures and Initiatives', in *Small Arms Survey 2001* (Oxford, 2001); and 'Illicit Small Arms Trade in Africa Fuels Conflict, Contributes to Poverty, Stalls Development', United Nations Information Service, DC/3032 (28 June 2006). In Central Africa alone, militia in the Congo acquired approximately 74,000 weapons during the civil war of 1993–99. Despite disarmament and demobilization programmes implemented by the state between 2003 and 2005, only 17,371 weapons were collected. A comprehensive blueprint adopted on 14 February 2006 aimed to reintegrate 19,000 former combatants, and round up the equivalent of 10,000 weapons. In January 2006, the Congo and the International Monetary Fund (IMF) also signed a $17 million agreement to support further collaborative efforts aimed at disarming indigenous society and reintegrating local communities.

13 See 'Fuelling the Flames: Brokers and Transport Agents in the Illicit Arms Trade', *Small Arms Survey 2001*; and Makarenko, 'Tracing the Dynamics of the Illicit Arms Trade'.

14 See A. R. Markusen and S. S. Costigan (eds) *Arming the Future: A Defence Industry for the Twenty-first Century* (New York, 1999); N. Cooper, *The Business of Death: Britain's Arms Trade at Home and Abroad* (London, 1997); P. Cornish, *The Arms Trade and Europe* (London, 1995); E. A. Kolodziej, *Making and Marketing Arms: The French Experience and Its Implications for the International System* (Princeton, 1987); and E. J. Laurance, *The International Arms Trade* (New York, 1992).

15 Many excellent contemporary studies explore the interconnections between Islamic terrorist networks, arms transfers, and asymmetric conflict, including S. Ulph, 'Shifting Sands: Al-Qaeda and Tribal Gun-Running Along the Yemeni Frontier', *Terrorism Monitor*, 2:7 (8 April 2004); A. Rashid, *Taliban: Militant Islam, Oil and Fundamentalism in Central Asia* (London, 2000), pp. 13–30, 117–27; P. Marsden, *The Taliban: War and Religion in Afghanistan* (London, 2002), pp. 26–66, 146–56; S. Elegant, 'Eye of the Storm: Asia's War on Terrorism', *Time* (11 February 2002); R. Gunaratna (ed.) *Terrorism in the Asia-Pacific: Threat and Response* (Singapore, 2003), pp. 18, 76, 144; and K. Ramakrishna and S. S. Tan (eds) *After Bali: The Threat of Terrorism in Southeast Asia* (Singapore, 2003), p. 101.

16 A. Sampson, *The Arms Bazaar: The Companies, The Dealers, The Bribes: from Vickers to Lockheed* (London, 1977), p. 35.

17 See Thucydides, *The Peloponnesian War*, trans. R. Warner (Harmondsworth, 1972); J. P. Puype and M. van der Hoeven (eds) *The Arsenal of the World: The Dutch Arms Trade in the Seventeenth Century* (Amsterdam, 1996); C. A. Bayly,

Indian Society and the Making of the British Empire (Cambridge, 1988), pp. 47–8.

18 See, for example, D. R. Tahtinen, *Arms in the Indian Ocean* (Washington, DC, 1977). 'East of Suez' was a phrase used repeatedly in British military and political discussions following the end of the Second World War and the onset of the Cold War. It referred to imperial interests beyond the European theatre, held together strategically by sea-lanes of communication through the Mediterranean Sea via the Suez Canal, and round the Cape, to East Africa and the Middle East, South Asia, Southeast Asia, East Asia, and Australia. With decolonization, starting with Indian independence in 1947, there would be progressive dilution of the military presence 'east of Suez', culminating in a definitive decision to close major military bases and reduce naval commitments across Asia.

19 M. Kaldor, *The Baroque Arsenal* (London, 1982), p. 4.

20 M. R. Smith, *Harpers Ferry Armory and the New Technology: The Challenge of Change* (Ithaca, 1977), pp. 21–3. See also Trebilcock, *The Industrialization of the Continental Powers*, pp. 26–9.

21 Thucydides, *The Peloponnesian War*, pp. 608–9.

22 K. Krause, *Arms and the State: Patterns of Military Production and Trade* (Cambridge, 1992), pp. 1–2, 9–18.

23 Krause, *Arms and the State*, pp. 18–80, 205–11; G. Parker, *The Military Revolution: Military Innovation and the Rise of the West, 1500–1800* (Cambridge, 1988), pp. xiv–xvii, 1–5; J. Black, *European Warfare 1660–1815* (London, 1994), pp. 32–3.

24 See J. Needham, *Science and Civilization in China*, 5:7: *Military Technology: The Gunpowder Epic* (Cambridge, 1986); and N. Tarling (ed.) *The Cambridge History of Southeast Asia From Early Times to c.1800*, vol. 1: (Cambridge, 1992), pp. 379–80.

25 C. J. Rogers, 'The Efficacy of the English Longbow', *War in History*, 5 (1998), pp. 233–42.

26 A. E. Prince, 'The Importance of the Campaign of 1327', *English Historical Review*, 50 (1935), pp. 301–2; C. J. Rogers, 'Military Revolutions of the Hundred Years' War', in C. J. Rogers (ed.) *The Military Revolution Debate* (Boulder, 1995), pp. 64–5.

27 Tarling (ed.) *The Cambridge History of Southeast Asia*, vol. 1, pp. 380, 389–93.

28 P. Borschberg, *The Singapore and Melaka Straits: Violence, Security and Diplomacy in the Seventeenth Century* (Singapore, 2010); M. W. Charney, *Southeast Asian Warfare, 1300–1900* (Leiden, 2004), pp. 42–72.

29 P. A. Lorge, *The Asian Military Revolution: from Gunpowder to the Bomb* (Cambridge, 2008), pp. 4–10; cf. N. Perrin, *Giving Up the Gun: Japan's Reversion to the Sword, 1543–1879* (Boston, 1979).

30 A. Reid, 'Europe and Southeast Asia: The Military Balance', James Cook University of North Queensland, Occasional Paper 16 (Townsville, 1982), pp. 90–6.

31 Krause, *Arms and the State*, p. 80; G. Parker (ed.) *Cambridge Illustrated History of Warfare: The Triumph of the West* (Cambridge, 1995), p. 340.

32 L. Mathiak and L. Lumpe, 'Government Gun-Running to Guerrillas', in L. Lumpe (ed.) *Running Guns: The Global Black Market in Small Arms* (London, 2000); cf. Chalk, 'Light Arms Trading in Southeast Asia'. Other estimates are less conservative, setting the monetary value of weapons and ammunition flowing through the US-Pakistan-Afghanistan conduit at $6–8 billion.

33 Many proponents of this concept have also displayed an astonishing lack of historical consciousness. See, for instance, J. Blacker, 'Understanding the Revolution in Military Affairs: A Guide to America's Twenty-first Century Defence', *Defence Working Paper No. 3*, Progressive Policy Institute (Washington, DC, 1997). For a measured historical approach to the 'Revolution in Military Affairs' (RMA) debate, see M. Knox and W. Murray (eds) *The Dynamics of Military Revolution 1300–2050* (Cambridge, 2001).

34 'On the Equipment of the British Infantry', *United Service Magazine and Naval and Military Journal*, 2 (1831), p. 202. These 'Brown Bess' muskets were also popularly known as 'Tower' muskets, since they were proved with charges of powder by government inspectors at the Tower of London. Tried and tested, they were preferred by the Government for use in war.

35 See J. de Vries, 'The Industrious Revolution and the Industrial Revolution', *Journal of Economic History*, 54 (1994), pp. 249–70.

36 W. H. McNeill, *The Pursuit of Power: Technology, Armed Force, and Society since AD 1000* (Chicago, 1982), pp. 236–7, 262ff.

37 D. W. Bailey, *British Military Longarms, 1715–1865* (London, 1986), p. 85.

38 'The Fusil *Gras*, or French Model Rifle', *Colburn's United Service Magazine*, 3 (1875), pp. 88–99.

39 H. Strachan, 'Military Modernization 1789–1918', in T. C. W. Blanning (ed.) *The Oxford Illustrated History of Modern Europe* (Oxford, 1996), p. 75; A. Corvisier and J. Childs (eds) *A Dictionary of Military History and the Art of War* (Oxford, 1994), pp. 249–51, 674–5; R. W. Beachey, 'The Arms Trade in East Africa in the Late Nineteenth Century', *Journal of African History* (*JAH*), 3:3 (1962), p. 452; cf. C. H. Roads, *The British Soldier's Firearm, 1850–1864* (London, 1964).

40 'Arms and Ammunition', *The Ironmonger*, October 1878, p. 1106.

41 'A Terrible Weapon', *The Ironmonger*, April 1880, p. 429.

42 B. Collier, *Arms and the Men: The Arms Trade and Governments* (London, 1980), p. 65.

43 Corvisier and Childs (eds) *A Dictionary of Military History and the Art of War*, pp. 674–5. Cordite was made by dissolving nitrocellulose and nitroglycerine in acetone, and acquired its name because the mixture was then hardened into a long, thin rope or cord. Used in the 0.303 inch rifle, cordite was found to increase muzzle velocity from the 1,800 feet per second achieved with compressed black powder to 2,000 feet per second and later to 2,440 feet per second when used with a lighter, pointed bullet.

44 C. G. Slade, 'Modern Military Rifles and Fire Tactics', *Journal of the Royal United Service Institution*, 32:143 (1888), pp. 899–917; C. Trebilcock, '"Spin-off" in British Economic History: Armaments and Industry, 1760–1914', *Economic History Review* (*EcHR*), 22:3 (1969); C. Trebilcock, 'British Armaments and European Industrialization, 1890–1914', *EcHR*, 26:2 (1973), p. 254; and Trebilcock, *The Industrialization of the Continental Powers*, pp. 346–9.

45 D. R. Headrick, *The Tentacles of Progress: Technology Transfer in the Age of Imperialism* (New York, 1988), p. 13.

46 R. Gray, 'Portuguese Musketeers on the Zambezi', *JAH*, 12:4 (1971), pp. 531–3; S. Marks and A. Atmore, 'Firearms in Southern Africa: A Survey', *JAH*, 12:4 (1971), pp. 517–18; A. Das Gupta and M. N. Pearson (eds) *India and the Indian Ocean 1500–1800* (Delhi, 1987), pp. 218, 256; Puype and van der Hoeven (eds) *The Arsenal of the World*, pp. 9, 19, 47–51; D. F. Harding, *Small*

Arms of the East India Company 1600–1856, vol. 1: *Procurement and Design*, pp. 1–2, and vol. 2: *Catalogue of Patterns* (London, 1997), p. 529. Despatches to Bombay, Madras and Bengal record numerous gifts of guns and other munitions by the English East India Company to indigenous rulers: for instance, Dowlat Row Scindia in 1800 (India Office [IO], E/4/1015); the King of Johanna in 1810 (IO, E/4/1026); the Imam of Muscat in 1822 (IO, E/4/1041); the Ras of Tigre in 1834 (IO, E/4/1057); the Raja of Manipur in 1849 (IO, E/4/799); and, as late as 1856, Minié rifles to King Mongkut of Siam (IO, E/4/841). See also travelogues, including W. C. Barker, 'Narrative of a Journey to Shoa', in G. W. Forrest (ed.) *Travels and Journals Preserved in the Bombay Secretariat* (Bombay, 1902), describing the presentation in the 1840s of a three-pounder cannon and muskets to Sahle Selassie, the Ethiopian king in Shoa, by an English East India Company mission.

47 D. R. Headrick, *The Tools of Empire: Technology and European Imperialism in the Nineteenth Century* (New York, 1981), pp. 9–11.

48 M. Adas, *Machines as the Measure of Men: Science, Technology, and Ideologies of Western Dominance* (Ithaca, 1989), p. 268.

49 See Adas, *Machines as the Measure of Men*, especially pp. 38–40, 143–4; and J. S. Galbraith, 'The "Turbulent Frontier" as a Factor in British Expansion', *Comparative Studies in Society and History*, 2 (1960), pp. 150–68.

50 K. N. Chaudhuri, *Trade and Civilization in the Indian Ocean: An Economic History from the Rise of Islam to 1750* (Cambridge, 1985), pp. 15, 97, 182.

51 C. A. Bayly, *Imperial Meridian: The British Empire and the World 1780–1830* (London, 1989), pp. 14–15.

52 See D. E. Leach, *Arms for Empire: A Military History of the British Colonies in North America, 1607–1763* (New York, 1973); H. A. Gemery and J. S. Hogendorn, 'Technological Change, Slavery and the Slave Trade', in C. Dewey and A. G. Hopkins (eds) *The Imperial Impact: Studies in the Economic History of India and Africa* (London, 1978); J. E. Inikori, 'The Import of Firearms into West Africa, 1750–1807', *JAH*, 18:3 (1977); W. A. Richards, 'The Import of Firearms into West Africa in the Eighteenth Century', *JAH*, 21:1 (1980); R. Hyam, *Britain's Imperial Century* (London, 1993), pp. 221–4.

53 M. Legassick, 'Firearms, Horses and Samorian Army Organization 1870–98', *JAH*, 7:1 (1966), p. 100.

54 D. W. Bailey and D. A. Nie, *English Gunmakers: The Birmingham and Provincial Gun Trade in the Eighteenth and Nineteenth Century* (London, 1978), p. 15.

55 R. A. Kea, 'Firearms and Warfare on the Gold and Slave Coasts from the Sixteenth to the Nineteenth Centuries', *JAH*, 12:2 (1971), pp. 194–5; cf. R. Law, 'Horses, Firearms and Political Power', *Past and Present*, 72 (1976), p. 122; Richards, 'The Import of Firearms into West Africa', pp. 46, 57.

56 Inikori, 'The Import of Firearms into West Africa', pp. 339–43. The general pressure on manufacturers and shippers whenever the volume of English trade to West Africa was rising (1751–54, 1771–75, 1790–92) underscores the great demand for guns in the trade; see, for example, the private records of the gunmaking firm of Farmer and Galton of Birmingham, and the slave-trading firm of J. Rogers & Co. of Bristol.

57 B. Parsons (attributed), *Observations on the Manufacture of Firearms for Military Purposes, etc* (London, 1829), p. 45.

58 Parsons (attributed), *Observations on the Manufacture of Firearms*, p. 46.

59 Kea, 'Firearms and Warfare on the Gold and Slave Coasts', p. 200; Headrick, *The Tools of Empire*, p. 106.
60 W. D. McIntyre, *The Imperial Frontier in the Tropics, 1865–75* (London, 1967), p. 116.
61 A. Atmore, J. M. Chirenje and S. I. Mudenge, 'Firearms in South Central Africa', *JAH*, 12:4 (1971), pp. 549–50; C. F. Goodfellow, *Great Britain and the South African Confederation (1870–1881)* (Cape Town, 1966), pp. 45, 153–5. The massive instability generated by such massive arms transfers was a sufficiently powerful dynmic to persuade both of these Colonial Secretaries, the Liberal Earl of Kimberley as much as the Conservative Earl of Carnarvon, to propose formal intervention irrespective of differences in political ideology.
62 H. J. Fisher and V. Rowland, 'Firearms in the Central Sudan', *JAH*, 12:2 (1971), p. 223; J. J. Guy, 'A Note on Firearms in the Zulu Kingdom with Special Reference to the Anglo-Zulu War, 1879', *JAH*, 12:4 (1971), p. 559.
63 W. E. H. Lecky, cited in Hyam, *Britain's Imperial Century*, p. 221.
64 Atmore, Chirenje and Mudenge, 'Firearms in South Central Africa', p. 547.
65 Gray, 'Portuguese Musketeers on the Zambezi', pp. 531–3; Das Gupta and Pearson (eds) *India and the Indian Ocean*, p. 256; K. M. Swope, 'Crouching Tigers, Secret Weapons: Military Technology Employed during the Sino-Japanese-Korean War, 1592–1598', *The Journal of Military History*, 69 (January 2005), p. 19.
66 Das Gupta and Pearson (eds) *India and the Indian Ocean*, pp. 213, 218; W. O. Dijk, 'The VOC's Gunpowder Factory, c.1620–1660', *The International Institute for Asian Studies Newsletter Online, 26* (November 2001); cf. Puype and van der Hoeven (eds) *The Arsenal of the World*.
67 Col. H. Munro's evidence, *Reports from Committees of the House of Commons*, 12 vols (London, 1803–6), III, p. 169.
68 See V. T. Harlow, *The Founding of the Second British Empire 1763–1793*, 1: *Discovery and Revolution* (London, 1952). Harlow's notion of the political and commercial 'swing to the East', begun by the Anglo-French duel for empire, confirmed by the loss of the American colonies and then magnified by the Napoleonic Wars, was certainly reflected in small arms proliferation across Africa and Asia.
69 See H. A. Young, *The East India Company's Arsenals and Manufactories* (Oxford, 1937); and Harding, *Small Arms of the East India Company*, vol. 1. From the establishment of the first military depots, workshops, and armouries in India during the late 1760s to the end of Company rule in 1858, the arsenals of the English East India Company were capable of producing their own gunpowder, brass ordnance, gun-carriages, percussion caps, bullets and other munitions. The only important articles of military equipment the Company did not manufacture in India were small arms; some were purchased in India or produced in several Indian states, but most were procured by the Company from its own network of workshops and factories in Birmingham and London, which supplied weapons of high standard for service not only in India but elsewhere.
70 Bayly, *Imperial Meridian*, p. 130.
71 Public Record Office (PRO), CUST 9/3. For Africa and Asia respectively, this represented 4.7 and 36.5 per cent of total British firearms exports and 2.7 and 50.3 per cent of the total real value.
72 PRO, CUST 9/107. For West Africa, this amounted to 38.9 per cent of total British small arms exports but only 10 per cent of the total real value for 1897.

For the entire Indian Ocean arena, this constituted 28.5 per cent of total British small arms exports and 38.6 per cent of total real value. By the 1870s, one *Birmingham Post* correspondent reported that the Ashantis in West Africa were still being armed with trade flintlock muskets, exported at 7 shillings apiece. Nevertheless, the bulk of the trade had already shifted eastwards, with major exports of firearms (including the newer breech-loading rifles) reaching Delagoa Bay, Mozambique, Natal, and Cape Colony. See 'The African Musket', *The Ironmonger*, April 1878, p. 192; 'Arms and Ammunition', *The Ironmonger*, August 1873, p. 944; 'Trade Report', *The Ironmonger*, November 1873; 'Gun Trade at the Cape', *The Ironmonger*, January 1875, p. 49.

73 'Cheap Rifles for Africa', *Arms and Explosives*, July 1895, p. 175.

74 Hyam, *Britain's Imperial Century*, p. 222; Birmingham Proof House returns, in the trade journal *Arms and Explosives*, June 1893 to June 1911. They reflect the consolidated output of the 'old' Birmingham gun trade, but exclude the production figures for state-run and state-assisted enterprises—at Birmingham, London, and Enfield—which were increasingly meeting the demand for military weapons in conflicts such as the Anglo-Boer War. The latter establishments were, by this time, proving their own guns. See 'The Birmingham Gun Trade', *The Ironmonger*, May 1894, p. 165.

75 F. A. Vali, *Politics of the Indian Ocean Region* (New York, 1976), pp. 24–30; cf. V. L. Forbes, *The Maritime Boundaries of the Indian Ocean Region* (Singapore, 1995).

76 K. McPherson, *The Indian Ocean: A History of People and the Sea* (Delhi, 1993), pp. 1–15; Das Gupta and Pearson (eds) *India and the Indian Ocean*, pp. 8–13; S. Bose, *A Hundred Horizons: The Indian Ocean in the Age of Global Empire* (Cambridge, Massachusetts, 2006), pp. 4–22. The original meaning of 'monsoon' (Arabic *mawsim*) is 'season', a seafarer's term from antiquity denoting the sheer regularity and steady rhythms of the trade wind cycle, blowing half the year from the southwest (April to October), then alternating from the reverse northeast direction in the other half of the year (November to March). From the Arabic word, the Portuguese would derive *moncao*, the Dutch *monssoon*, and the English *monsoon*.

77 Vali, *Politics of the Indian Ocean Region*, p. 27.

78 See M. G. S. Hodgson, *The Venture of Islam: Conscience and History in a World Civilization*, 3 vols (Chicago, 1974); Chaudhuri, *Asia before Europe*, pp. 101–3.

79 See H. Inalcik, 'The Socio-Political Effects of the Diffusion of Firearms in the Middle East', in V. J. Parry and M. E. Yapp (eds) *War, Technology and Society in the Middle East* (London, 1975). Demonstrating their military effectiveness in the early modern period, Turkish armies equipped with high-quality siege cannon and small arms were able to strike terror into the heart of Central Europe, besieging the gates of Vienna in 1529 and 1683. But this prompted the European powers to supply their own arms to indigenous opponents of the Ottomans in Asia and Africa.

80 J. F. Cady, *A History of Modern Burma* (Ithaca, 1958), p. 35.

81 Tarling (ed.) *The Cambridge History of Southeast Asia*, vol. 1, pp. 380–5.

82 Ibid.

83 Tarling (ed.) *The Cambridge History of Southeast Asia*, vol. 1, pp. 380, 386–91; S. Gordon, *Marathas, Marauders, and State Formation in Eighteenth Century India* (Delhi, 1994), p. 188.

84 Bayly, *Imperial Meridian*, pp. 16–74.
85 For the earlier period, see Harding, *Small Arms of the East India Company*, vol. 1, pp. 1–9.
86 McPherson, *The Indian Ocean*, pp. 216–25.
87 Bayly, *Indian Society and the Making of the British Empire*, pp. 47–8.
88 IO, P/D/74/424-26, Bombay Pol. & Sec. Consultations: Bombay to Bengal, 29 October 1787; cf. P. Nightingale, *Trade and Empire in Western India 1784–1806* (Cambridge, 1970), pp. 34–6, 45, 61, 162, 240–2.
89 H. Furber, *Rival Empires of Trade in the Orient 1600–1800* (Minneapolis, 1976), p. 293; Nightingale, *Trade and Empire in Western India*, pp. 18, 23–5.
90 Bayly, *Indian Society and the Making of the British Empire*, pp. 98–103; R. G. S. Cooper, 'Cross-cultural Conflict Analysis: The "Reality" of British Victory in the Second Anglo-Maratha War 1803–1805' (Ph.D. thesis, University of Cambridge, 1992), pp. 73–6. See also R. G. S. Cooper, *The Anglo-Maratha Campaigns and the Contest for India: The Struggle for Control of the South Asian Military Economy* (Cambridge, 2003).
91 Bayly, *Imperial Meridian*, pp. 36–9.
92 J. B. Kelly, *Britain and the Persian Gulf 1795–1880* (Oxford, 1968), pp. 1–2; S. Subrahmanyam, *The Portuguese Empire in Asia, 1500–1700* (London, 1993), pp. 148–9; Das Gupta and Pearson (eds) *India and the Indian Ocean*, p. 139.
93 A. Sheriff, *Slaves, Spices and Ivory in Zanzibar: Integration of an East African Commercial Empire into the World Economy, 1770–1873* (Athens, Ohio, 1987), pp. 18–24; Das Gupta and Pearson (eds) *India and the Indian Ocean*, p. 219.
94 Brigadier-General H. H. Austin, 'Gun-running in the Gulf', *Blackwood's Magazine*, 208 (London, 1920), p. 170.
95 V. Lieberman, 'Local Integration and Eurasian Analogies: Structuring Southeast Asian History', *Modern Asian Studies (MAS)*, 27:3 (1993), pp. 488–95, 518–19, 548–51, 568–9.
96 B. W. Andaya and L. Y. Andaya, *A History of Malaysia* (London, 1982), pp. 77–113.
97 IO, F/4/1724 (69433), 'Report on Piracy in the Straits Settlements', E. Presgrave, Registrar of Imports and Exports, to K. Murchison, Resident Councillor at Singapore, 5 December 1828. See also J. F. Warren, *The Sulu Zone 1768–1898: The Dynamics of External Trade, Slavery, and Ethnicity in the Transformation of a Southeast Asian Maritime State* (Singapore, 1981), pp. xix–xxvi, 3–4, 147–8, 198; and by the same author, *Iranun and Balangingi: Globalization, Maritime Raiding and the Birth of Ethnicity* (Singapore, 2002).
98 Das Gupta and Pearson (eds) *India and the Indian Ocean*, p. 13.
99 Bose, *A Hundred Horizons*, pp. 13, 274; cf. R. K. Ray, 'Asian Capital in the Age of European Expansion: The Rise of the Bazaar, 1800–1914', *MAS*, 29:3 (1995).
100 W. G. Clarence-Smith (ed.) *The Economics of the Indian Ocean Slave Trade in the Nineteenth Century* (London, 1989), pp. 60–1.
101 Andaya and Andaya, *A History of Malaysia*, pp. 100–1, 133–6, 210–14; Sir T. S. Raffles, *The History of Java* (London, 1817; reprinted Kuala Lumpur, 1965), pp. 246–62; J. Anderson, *Mission to the East Coast of Sumatra in 1823* (London, 1826; reprinted Oxford, 1971), pp. 227–8; P. J. Begbie, *The Malayan Peninsula* (Madras, 1834; reprinted Kuala Lumpur, 1967), pp. 270–4; H. St. John, *The Indian Archipelago*, vol. 2 (London, 1835), pp. 160–2; T. J. Newbold, *Political and Statistical Account of the British Settlements in*

the Straits of Malacca (London, 1839; reprinted Kuala Lumpur, 1971), vol. 1, pp. 36–47; C. B. Buckley, *An Anecdotal History of Old Times in Singapore* (Singapore, 1902; reprinted, Kuala Lumpur, 1965), pp. 276–82; and the series of articles published under the title, 'The Piracy and Slave Trade of the Indian Archipelago', *Journal of the Indian Archipelago and Eastern Asia* (*JIAEA*), 3 (1849).

102 C. A. Bayly (ed.) *Atlas of the British Empire* (London, 1989), pp. 46–7, 49–50, 87.

103 E. A. Alpers, *Ivory and Slaves in East Central Africa: Changing Patterns of International Trade to the Later Nineteenth Century* (London, 1975), pp. 12, 18, 54.

104 Kelly, *Britain and the Persian Gulf*, pp. 23, 36–7, 44, 412–14, 624–35, 834; Sheriff, *Slaves, Spices and Ivory in Zanzibar*, pp. 1–5.

105 Clarence-Smith (ed.) *The Economics of the Indian Ocean Slave Trade*, pp. 1–13, 54, 172–9. See also G. Campbell, *An Economic History of Imperial Madagascar, 1750–1895* (Cambridge, 2005). Malagasy slave imports would sustain the East African slave export trade even after British suppression of the Swahili coast trade by the 1880s. Staple imports comprised not only slaves, but also cloth for local communities and arms for regional armies, which underpinned the economic and political power of indigenous rulers. These weapons empowered anti-colonial resistance against the French in Madagascar until 1897, when a major uprising was finally suppressed.

106 E. B. Martin and C. P. Martin, *Cargoes of the East: The Ports, Trade and Culture of the Arabian Seas and Western Indian Ocean* (London, 1978), pp. 29–37. The Qawasim, a leading nomadic confederacy, participated in the Gulf carrying trade along with commercial voyages to Africa and India.

107 IO, F/4/1724 (69433), 'Report on Piracy in the Straits Settlements', Presgrave to Murchison, 5 December 1828.

108 G. R. Mundy, *Narrative of Events in Borneo and Celebes*, vol. 2 (London, 1848), p. 17.

109 O. Rutter, *The Pirate Wind: Tales of the Sea-Robbers of Malaya* (London, 1930; reprinted Singapore, 1986), pp. 26–8.

110 Warren, *The Sulu Zone*, p. 208.

111 See B. Elleman, A. Forbes and D. Rosenberg (eds) *Piracy and Maritime Crime: Historical and Modern Case Studies* (Newport, Rhode Island, 2010); J. Kleinen and M. Osseweijer (eds) *Pirates, Ports, and Coasts in Asia: Historical and Contemporary Perspectives* (Singapore, 2010). Both adopt a comparative approach, surveying patterns of continuity and change in piratical activities from Asia to Africa. Intelligence networks and news agencies also deliver security updates on recent arms transfers and related armed conflicts in these regions. See, for example, A. Rathmell, 'Policing the Arabian Gulf', *International Police Review* (January/February 1999), pp. 23–5; and 'East Africa: Bin Laden's New Front', *Jane's Terrorism and Security Monitor* (4 August 2006). Reports from the International Maritime Bureau (IMB) detail the growing incidence of piracy in international waters, particularly 'hot spots' off the Somali Coast, around the Indonesian Archipelago, and along the Malacca–Singapore Straits. Journals for maritime decision-makers focus on the heavily armed piratical activities off the coast of Somalia. See, for instance, *MarEx Newsletter* (7 April 2006). D. Elliott, 'Pirates of the South China Sea', *Newsweek Magazine* (5 July 1999), offers another contemporary angle on armed

plunderers in Southeast Asian waters, from Burma through Singapore, to the Indonesian and the Philippine Archipelagos, and upwards as far as Hong Kong and southern China. Finally, there are current reports about munitions traffic at Jolo; once a historical centre for the gun-slave traffic, it is today the stronghold of the Abu Sayyaf, a militant Islamic group with alleged ties to foreign Muslim cells accused of perpetrating terrorism.

112 J. Bevan, 'Military Demand and Supply: Products and Producers', in *Small Arms Survey 2006*, p. 15.

Chapter 2 The Arms Trade in the Metropolis

1 See J. Brewer, *The Sinews of Power: War, Money and the English State, 1688–1783* (Cambridge, Massachusetts, 1988). Backed by a secure, sophisticated capital market, this capacity to organize long-term credit to fund wartime public borrowing would prove momentous: Brewer's 'sinews of power' thesis furnishes a useful framework for understanding how military-fiscalism enables nations to emerge as major international powers.

2 P. Kennedy, *The Rise and Fall of the Great Powers: Economic Change and Military Conflict from 1500 to 2000* (New York, 1987), pp. 20–3.

3 Bayly (ed.) *Atlas of the British Empire*, p. 18.

4 Bayly (ed.) *Atlas of the British Empire*, p. 18; Kennedy, *The Rise and Fall of the Great Powers*, pp. 76, 80–2; J. Black, *European Warfare 1660–1815* (London, 1994), pp. 109–10; Parker (ed.) *Cambridge Illustrated History of Warfare*, p. 8. As Parker indicates, 'mobilizing credit to finance wars rests not only upon the existence of extensive private credit, but also upon the convergence of interest between those who make money and those who make war, for public loans depend both on finding borrowers willing to lend as well as taxpayers willing and able to provide ultimate repayment'. In England, tax revenues increased six times over in the century after 1689. By 1783, when the unsuccessful American war had ended, Britain's national debt stood at £245 million, in which the Bank of England alone handled bills of exchange worth over £2 million annually—a stunning extension of the available monetary stock.

5 See, for example, the articles in *Arms and Explosives*, December 1892, p. 67; *Arms and Explosives*, January 1893, p. 92; *Arms and Explosives*, March 1893, p. 138.

6 See L. Stone (ed.) *An Imperial State at War: Britain from 1689 to 1815* (London, 1994); and H. Strachan, *From Waterloo to Balaclava: Tactics, Technology, and the British Army 1815–1854* (Cambridge, 1985).

7 See H. Strachan, 'Military Modernization 1789–1918', in T. C. W. Blanning (ed.) *The Oxford Illustrated History of Modern Europe* (Oxford, 1996), pp. 71ff; articles by W. H. McNeill and H. Bailes in R. Haycock and K. Neilson (eds) *Men, Machines and War* (Waterloo, Ontario, 1988); and D. Showalter, *Railroads and Rifles: Soldiers, Technology and the Unification of Germany* (Hamden, 1975).

8 L. F. Salzman, *English Industries of the Middle Ages* (1823; reprinted London, 1964), pp. 158–9; T. F. Tout, 'Firearms in England in the Fourteenth Century', *English Historical Review*, 26 (1911), p. 699. The use of *gunarium cum pulvere* ('a gun with powder') was implied in the medieval account of

a muster in Norwich in 1355. Nicholas Herbert, a fourteenth-century gun-maker, was reported to have supplied 51 guns—small mobile pieces from 15 inches to 24 inches in length—to the municipality of Norwich in 1384, for which various citizens were assessed for the city's defence. The term 'handgun' first occurs in English accounts in 1386, when three cannon-like 'handgunnes' were delivered out of the Tower of London to Berwick.

9   A. H. Burne, 'Cannons at Crécy', *English Historical Review*, 77 (1962), pp. 335–42; Rogers, 'The Efficacy of the English Longbow', pp. 233–42; R. Payne-Gallway, *The Handgun in Relation to the Crossbow* (London, 1903), p. 39.

10 D. W. Young, 'History of the Birmingham Gun Trade' (M.Com. thesis, University of Birmingham, 1936), pp. 4–5.

11 See, for instance, [33.Hen.VIII.c6] in D. Pickering (ed.) *The Statutes at Large: from Magna Charta to ... 1761*, vol. 5 (Cambridge, 1763), pp. 71–8.

12 'Artifex' and 'Opifex' (pseud.), *The Causes of Decay in a British Industry* (London, 1907), p. 1.

13 J. Goodwin, *The Newdigates of Arbury: Early Memorials of the Birmingham Gun Trade* (Birmingham, 1869), p. 293. A good contemporary analysis of English military development during the Tudor period—a reluctant revolution that continued to combine old and new weapons out of 'sentiment and inertia'—is found in D. Eltis, *The Military Revolution in Sixteenth-Century Europe* (London, 1995), pp. 101–3.

14 Quoted in H. B. C. Pollard, *A History of Firearms* (London, 1926).

15 H. L. Blackmore, *Dictionary of London Gunmakers* (Oxford, 1986), pp. 11–21; Bailey, *British Military Longarms*, p. 19.

16 C. E. Hanson, *The Northwest Gun* (Lincoln, Nebraska, 1956).

17 PRO, T 70/95.

18 Blackmore, *Dictionary of London Gunmakers*, pp. 21, 27.

19 'Artifex' and 'Opifex' (pseud.), *The Causes of Decay*, p. 56.

20 J. D. Goodman, 'The Birmingham Gun Trade', in S. Timmins (ed.) *The Resources, Products, and Industrial History of Birmingham and the Midland Hardware District* (London, 1866), pp. 388–90; W. West, *The History, Topography and Directory of Warwickshire* (Birmingham, 1830), pp. 126, 582; Bailey and Nie, *English Gunmakers*, p. 17.

21 Bailey and Nie, *English Gunmakers*, p. 15.

22 Local historical examples cited in 'Artifex' and 'Opifex', *Causes of Decay*, p. 2.

23 J. Morfitt, 'Sketches of Birmingham Industries', in S. J. Pratt, *Harvest-Home*, vol. 1 (London, 1805), p. 331; J. Morrill, *The Nature of the English Revolution* (London, 1993), p. 218.

24 W. Hutton, *An History of Birmingham*, *6th edition* (Birmingham, 1835), p. 59.

25 Goodwin, *The Newdigates of Arbury*, p. 296.

26 Goodman, 'The Birmingham Gun Trade', p. 409.

27 Quoted in Goodman, 'The Birmingham Gun Trade', p. 410.

28 A. Edwards, 'Centenary of the Birmingham Proof House, 1813–1913', p. 16, cited in Young, 'History of the Birmingham Gun Trade', p. 17.

29 Goodwin, *The Newdigates of Arbury*, pp. 298, 300–1.

30 *Journal of the House of Commons*, 15 (1707), p. 550.

31 J. D. Goodman, 'The Small Arms Trade', *The Ironmonger*, September 1865; Young, 'History of the Birmingham Gun Trade', p. 24. According to Goodman's calculations, the total number of military barrels produced for the British Government between 1804 and 1815 was 3,037,644 (an average of 253,137 per year); the total number of arms supplied to the East India Company amounted to over four million.

32 Goodman, 'The Birmingham Gun Trade', pp. 412–14.

33 Ibid., pp. 414–19.

34 Bailey and Nie, *English Gunmakers*, p. 22.

35 J. Sketchley, *The Directory of Birmingham, Wolverhampton and Walsall* (Birmingham, 1767), pp. 31–3. Sketchley's Directory goes on to credit Wolverhampton with one gunsmith, and Walsall with one gunsmith and one gun locksmith. But these figures are approximate, as firms in the outworking system rarely knew the number of hands actually doing their work.

36 'Artifex' and 'Opifex', *Causes of Decay*, p. 21.

37 Goodman, 'The Birmingham Gun Trade', pp. 392–3.

38 The doctrine of free trade was advocated by statesman-manufacturer Richard Cobden and others from the 'Manchester School', the free trade movement of nineteenth-century Britain. In 1860–61, an important Anglo-French commercial agreement—the Cobden-Chevalier Treaty—inaugurated 30 years of lower tariffs across Europe.

39 Newspaper cuttings on Birmingham Industries, vols. 1 and 2 (1863–80), p. 10, cited in Young, 'History of the Birmingham Gun Trade', p. 116; and 'Artifex' and 'Opifex', *Causes of Decay*, p. 21.

40 'Bill to Insure the Proper and Careful Manufacturing of Firearms', *Parliamentary Papers (PP)*, 2 (1813), p. 853.

41 W. Greener, *The Present Proof Company: The Bane of the Gun Trade: A Letter Addressed to the Masters, and Journeymen Gun Makers of the Kingdom* (Birmingham, 1845), p. 4. While there was general improvement in the quality of manufacturing after 1813, there were times when quality control was not enforced or achieved. Three decades later, the patriarchal William Greener expressed his criticism of the 'legalized hypocrisy' of the Proof House in continuing to pass guns 'made from iron unfit to make firearms', the result being 'horribly dangerous'.

42 Goodman, 'The Birmingham Gun Trade', p. 427.

43 Edwards, 'Centenary of the Birmingham Proof House', pp. 21–9.

44 S. Shaw, *History of Staffordshire*, vol. 2 (London, 1801), pp. 126–34; Bailey and Nie, *English Gunmakers*, pp. 17, 23–4; D. M. Smith, 'Birmingham's Gun Quarter and its Workshops', in *The Journal of Industrial Archeology*, 1 (1964–65), pp. 106–19; W. A. Richards, 'The Birmingham Gun Manufactory of Farmer and Galton and the Slave Trade' (M.A. thesis, University of Birmingham, 1972), pp. 87–108.

45 Bailey, *British Military Longarms*, p. 8.

46 Blackmore, *Dictionary of London Gunmakers*, pp. 10–14; H. L. Blackmore, *British Military Firearms, 1650–1850* (London, 1994), pp. 257–8.

47 Collier, *Arms and the Men*, p. 28; Kennedy, *Rise and Fall of the Great Powers*, pp. 80–2.

48 Bailey, *British Military Longarms*, p. 11.

49 See D. O. Pam, *The Royal Small Arms Factory, Enfield and Its Workers* (Enfield, 1998). The Royal Small Arms Factory at Enfield continued to produce rifles for the British Government well into the twentieth century. In 1984, however, Enfield was privatized along with other Royal Ordnance Factories to become part of Royal Ordnance PLC. British Aerospace acquired the company in 1987, but closed the Enfield site in 1988 and auctioned off all the machinery.

50 C. Trebilcock, *The Vickers Brothers: Armaments and Enterprise 1854–1914* (London, 1977), p. 19.

51 M. J. Bastable, 'Arms and the State: The History of Sir William G. Armstrong & Co., 1854–1914' (Ph.D. thesis, University of Toronto, 1990), pp. 242–3; see also pp. 260–1, 269–71, 288–90, 350–2.

52 See, for instance, W. Anderson, 'The Capabilities of Private Firms to Manufacture Heavy Ordnance for Her Majesty's Service', *Journal of the Royal United Service Institution*, 29 (1885–86), pp. 497–520.

53 'War Office (Ordnance Department)—The Royal Commission on Warlike Stores—Observations', *Hansard's Parliamentary Debates*, House of Lords, 25 July 1887, Vol. 317, cc.1834–45.

54 'Artifex' and 'Opifex', *Causes of Decay*, p. 4; Goodman, 'The Birmingham Gun Trade', pp. 411, 415; W. J. Spurrier, 'Degenerates Regenerated: The Guns and Rifles of Birmingham', *The Birmingham Magazine of Arts and Industries*, 4:3 (1902), p. 103.

55 Goodman, 'The Birmingham Gun Trade', pp. 412–13; Bailey and Nie, *English Gunmakers*, p. 21. The detailed breakdown is as follows: 1,743,382 completed weapons (muskets, rifles, carbines, and pairs of pistols); plus a further 3,037,644 military barrels; and 2,879,203 military gun-locks. By comparison, Napoleonic France—with its control of the Belgian, Italian, and French manufactories—produced only 2,456,257 small arms. A full annual survey of barrels manufactured for the Board of Ordnance at Birmingham is displayed in Appendix A. It should be noted, however, that these figures refer to the barrels only, and not to other components. They also exclude sporting-arms, and the materials for guns—manufactured in Birmingham and made up into finished arms by the London trade—for the English East India Company. For the period 1804–15, this amounted to almost 500,000 fowling pieces, and around one million military firearms made for the Company.

56 Goodman, 'The Birmingham Gun Trade', p. 418.

57 Bailey and Nie, *English Gunmakers*, p. 20.

58 'Birmingham: From Our Own Correspondent', *The Ironmonger*, November 1870, p. 1031.

59 'Arms and Ammunition', *The Ironmonger*, April 1871, p. 312.

60 Bailey and Nie, *English Gunmakers*, p. 20.

61 B. M. D. Smith, 'The Galtons of Birmingham: Quaker Gun Merchants and Bankers, 1702–1831', *Business History* (1967), pp. 106–50; Richards, 'The Birmingham Gun Manufactory of Farmer and Galton and the Slave Trade', p. 181.

62 Birmingham Proof House Library, PG 87, G. Nicholls to R. Elsey, 14 October 1830.

63 Launched originally as *Morgan's Monthly Circular, and Metal Trades Advertiser* in June 1859, this trade magazine was later renamed and continued as *The Ironmonger*, arguably the most valuable resource for metropolitan studies on the small arms trade during Britain's Industrial Age.

64 'Report from the Select Committee on the Cheapest, Most Expeditious and Most Efficient Means of Providing Small Arms for Her Majesty's Service', *PP*, 18 (1854), pp. 10–11.

65 G. Markham, *Guns of the Empire: Firearms of the British Soldier, 1837–1987* (London, 1990), pp. 21, 23; Collier, *Arms and the Men*, p. 47.

66 *PP*, 18 (1854), pp. 10–11.

67 Markham, *Guns of the Empire*, p. 21.

68 R. I. Fries, 'British Response to the American System: The Case of the Small-Arms Industry after 1850', *Technology and Culture*, 16 (1975), pp. 386–7.

69 'Birmingham...', *The Ironmonger*, April 1865, p. 62; Goodman, 'The Birmingham Gun Trade', pp. 403–4, 430–1; and excerpts from the *Birmingham Post* on the history of BSA, *Arms and Explosives*, January 1898, p. 56.

70 'Birmingham...', *The Ironmonger*, April 1866, p. 64.

71 Ibid., May 1866, p. 82.

72 Ibid., June 1866, p. 96.

73 Ibid., July and October 1866. The administration of the new policy was, unfortunately, bungled. While the offer drew a response of 50 entries, only eight were shortlisted, of which five were subjected to trials. Only one rifle from Westley-Richards was found to actually comply with the conditions, but was rejected in favour of a Snider that required further modification by government officials. In Viscount Lifford's view, no system was 'better calculated to deter gunmakers from competing on future occasions'. See 'Breech Loading Small Arms', *Hansard's Parliamentary Debates*, House of Lords, 19 April 1866, Vol. 182, cc.1638–40.

74 Spurrier, 'Degenerates Regenerated', pp. 108–9; 'Trade Notes', *The Ironmonger*, September 1894, p. 471; R. P. T. Davenport-Hines, *Dudley Docker: The Life and Times of a Trade Warrior* (Cambridge, 1984), p. 48; S. B. Saul, 'The Market and the Development of the Mechanical Engineering Industries in Britain, 1860–1914', *EcHR*, 20:1 (1967), pp. 123, 125. BSA's net profit margins (1891 – £60,459; 1892 – £54,902; 1893 – £24,660; 1899 – £51,984; 1901 – £85,500; 1902 – £93,049) may be found in *Arms and Explosives*, October 1892, p. 20; September 1893, p. 273; October 1899, p. 154; October 1901, p. 157; and November 1902, p. 174.

75 PRO, Board of Trade 41-378/2138 and 31/177, The London Armoury Company, Ltd; and 31/30754, The London Small Arms Co., Ltd.

76 G. C. Allen, *The Industrial Development of Birmingham and the Black Country* (1929; reprinted London, 1966), p. 219.

77 PRO, Director Army Contracts (DAC) Reports (1889–99).

78 Spurrier, 'Degenerates Regenerated', p. 107; 'The Government and the Small Arms Factories', *Arms and Explosives*, April 1893, p. 144; 'Bagot Street Factory', *Arms and Explosives*, March 1894, p. 96.

79 'Birmingham and London Small Arms Trade', *The Ironmonger*, April 1861, p. 111.

80 'Birmingham...', *The Ironmonger*, January 1866, p. 18.

81 Young, 'History of the Birmingham Gun Trade', p. 77.

82 'The Birmingham Gun Trade', *The Ironmonger*, March 1869, p. 205.

83 'Birmingham...', *The Ironmonger*, March 1869, p. 214.

84 *Arms and Explosives*, June 1893 to June 1911; PRO, CUST 9/107.

85 'South Staffordshire', *The Ironmonger*, October 1870, p. 936.

86 'Birmingham...', *The Ironmonger*, April 1867, p. 61 and October 1867, p. 162; 'Ammunition and Small Arms', *The Ironmonger*, January 1871, p. 15 and 'Arms and Ammunition', *The Ironmonger*, April 1877, p. 151; 'A Large Contract', *The Ironmonger*, October 1872, p. 1121; 'Arms and Ammunition', *The Ironmonger*, June 1878, p. 466; and 'Birmingham...', *The Ironmonger*, October 1879, p. 582, March 1884, p. 456 and May 1885, p. 738.

87 'Arms and Ammunition', *The Ironmonger*, April 1879, p. 512; cf. 'Birmingham...', *The Ironmonger*, May 1880, p. 718; 'Birmingham...', *The Ironmonger*, July 1880, p. 77. BSA, LSA, and NAA had to share by the late 1870s one-third each of a government order for 10,000 Martini-Henry rifles, compared to the contract of 30,000 given directly to Enfield.

88 'Birmingham...', *The Ironmonger*, January 1883, p. 92.

89 PRO, CUST 9, Exports by Article.

90 Trade production annual averages are derived from Birmingham Proof House returns in 'Artifex' and 'Opifex', *Causes of Decay*, p. 279; and *Arms and Explosives*, June 1893 to June 1911. 'Military barrels' in these returns refer to the vestiges of military production by the old gun trade, excluding the output of the state or private factories. For the latter, see DAC Reports (1899–1910). For general employment figures, see 'Artifex' and 'Opifex', *Causes of Decay*, pp. 20–1.

91 F. W. Hackwood, *Wednesbury Workshops* (Birmingham, 1889), p. 49.

92 'Artifex' and 'Opifex', *Causes of Decay*, pp. 56–7.

93 'Birmingham...', *The Ironmonger*, September 1878, p. 975; cf. 'The Government and the Gun Trade', *The Ironmonger*, May 1880, p. 700.

94 'Artifex' and 'Opifex', *Causes of Decay*, pp. 32–3.

95 Ibid., p. 42.

96 'The Prohibition of Arms Act', *Arms and Explosives*, August 1900, pp. 117–18.

97 'Artifex' and 'Opifex', *Causes of Decay*, pp. 42, 46, 49.

98 Smith, *Harpers Ferry Armory and the New Technology*, pp. 38, 65.

99 Smith, *Harpers Ferry Armory and the New Technology*, pp. 323–35; Fries, 'British Response to the American System', pp. 382–4.

100 Goodman, 'The Birmingham Gun Trade', pp. 384–5, 388; R. B. Prosser, *Birmingham Inventions and Inventors, Being a Contribution to the Industrial History of Birmingham* (Birmingham, 1881), pp. 183, 191.

101 Fries, 'British Response to the American System', p. 377; cf. Smith, *Harpers Ferry Armory and the New Technology*, p. 19. Fries notes that Samuel Colt operated a gun factory in London from 1852–57; this dovetailed with the publication of the 'Report from the Select Committee on ... Small Arms for Her Majesty's Service', *PP*, 18 (1854) and 'Report of the Committee on the Machinery of the United States', *PP*, 50 (1854–55).

102 Fries, 'British Response to the American System', p. 385.

103 *PP*, 18 (1854), pp. 8–9.

104 The essential diversification of the industry in the production of bicycles and motorized vehicles came much later, only after the bigger firms—like BSA—had learnt their costly lesson, and their expectation of the Government and state-led enterprise had matured in the 1880s–90s.

105 'Birmingham...', *The Ironmonger*, October 1867, p. 162.

106 'The Hardware Trades of South Staffordshire', *The Ironmonger*, August 1867, p. 131.

107 'Birmingham...', *The Ironmonger*, March 1865, p. 44.

108 Allen, *Industrial Development of Birmingham*, p. 189.
109 Bailey and Nie, *English Gunmakers*, pp. 22–3.
110 Cuttings from the *Birmingham Daily Post*, published between 5 January 1864 and 1 May 1869.
111 N. Rosenberg (ed.) *The American System of Manufactures* (Edinburgh, 1969), pp. 180–92; Smith, *Harpers Ferry Armory and the New Technology*, p. 19.
112 'Artifex' and 'Opifex', *Causes of Decay*, p. 17.
113 *Arms and Explosives*, August 1893, p. 252; April 1894, p. 123; and August 1900, p. 123; 'Artifex' and 'Opifex', *Causes of Decay*, p. 12; Young, 'History of the Birmingham Gun Trade', p. 80.
114 'The Manufacture of Damascus Barrels in Belgium', *The Sporting Goods Review*, 15 May 1896; 'Artifex' and 'Opifex', *Causes of Decay*, pp. 9–11; 'The Arms Industry in Belgium', *Arms and Explosives*, October 1904, p. 129; Bailey and Nie, *English Gunmakers*, p. 25.
115 Greener, *The Present Proof Company*, p. 7.
116 Goodman, 'The Birmingham Gun Trade', p. 427.
117 Greener, *The Present Proof Company*, pp. 8–9; Goodman, 'The Birmingham Gun Trade', pp. 394–6.
118 Goodman, 'The Birmingham Gun Trade', pp. 419, 426; cf. Birmingham Proof House returns in *Arms and Explosives*, June 1893 to June 1900.
119 'Artifex' and 'Opifex', *Causes of Decay*, p. 130, citing figures provided by S. B. Allport, presiding at the 1891 annual meeting of the Birmingham Gun Trade.
120 E. Landers, *The Birmingham Gun Trade* (Birmingham, 1869), p. 23; Bailey and Nie, *English Gunmakers*, p. 25.
121 Allen, *Industrial Development of Birmingham*, p. 266.
122 'Artifex' and 'Opifex', *Causes of Decay*, pp. 25, 294; 'Belgian Competition in the Gun Trade', *The Ironmonger*, September 1903, p. 379; 'The Birmingham Gun Trade', *The Ironmonger*, May 1894, p. 159.
123 *The Sporting Goods Review*, 39 (1913), p. 222.
124 'Artifex' and 'Opifex', *Causes of Decay*, pp. 11–12.
125 Ibid.
126 British Commissioner's Report (1857), cited in 'Artifex' and 'Opifex', *Causes of Decay*, p. 12.
127 PRO, Diplomatic and Consular Reports, No. 650, 'The Arms Industry of Liège' (1906), pp. 6–7.
128 'Belgian Small-Arms Factory', *Arms and Explosives* (July 1893), p. 220.
129 'Artifex' and 'Opifex', *Causes of Decay*, p. 58.
130 PRO, Diplomatic and Consular Reports, No. 650, 'The Arms Industry of Liège', p. 9.
131 Figures are derived from the 'Statistical Abstract for the United Kingdom', *PP*, and Secretary of Legation's report from Brussels, cited in Goodman, 'The Birmingham Gun Trade', pp. 417, 427–8; and 'Artifex' and 'Opifex', *Causes of Decay*, pp. 17–18.
132 Figures are derived from Appendices in Bailey and Nie, *English Gunmakers*, p. 109; 'Artifex' and 'Opifex', *Causes of Decay*, pp. 278–9.
133 'Belgian Small-Arms Factory', *Arms and Explosives*, July 1893; 'The Arms Industry in Belgium', *Arms and Explosives*, October 1904, p. 130.
134 'The Marking of Guns', *The Ironmonger*, August 1903, p. 256.
135 'Artifex' and 'Opifex', *Causes of Decay*, pp. 18–21, 30–1, 280–7, 293.

136 A. L. Levine, *Industrial Retardation in Britain 1880–1914* (London, 1967), pp. 7–8, 14–15, 21, 31–6, 43–54.

137 Levine, *Industrial Retardation in Britain*, pp. 57–78, 118, 145–50; D. C. Coleman, 'Gentlemen and Players', *EcHR*, 26:1 (1973), pp. 92–116; B. Elbaum and W. Lazonick, *The Decline of the British Economy* (Oxford, 1986), p. 162; B. Collins and K. Robbins (eds) *British Culture and Economic Decline* (London, 1990), p. 61.

138 Elbaum and Lazonick, *The Decline of the British Economy*, p. 2.

139 Ibid., pp. 18–20, 40, 75–6.

140 A. L. Friedberg, *The Weary Titan: Britain and the Experience of Relative Decline* (Princeton, 1988), pp. 30–6, 45–51; B. Supple, 'Fear of Failing: Economic History and the Decline of Britain', *EcHR*, 47:3 (1994), pp. 445–6, 452.

141 'The Birmingham Gun Trade', *Arms and Explosives*, June 1894, p. 151.

142 Quoted in front matter of Friedberg, *The Weary Titan*.

143 Bastable, 'Arms and the State', p. 354.

144 Collier, *Arms and the Men*, pp. 65–72; cf. W. J. Reader, *Imperial Chemical Industries: A History*, vol. 1 (London, 1970).

145 Young, 'History of the Birmingham Gun Trade', p. 126.

146 For more on the Lee-Enfield rifle, see I. D. Skennerton, *The Lee-Enfield Story* (London, 1993). The cartridge-loaded Snider-Enfield breech-loading rifles illustrate both the 'baroque' continuities and the innovative changes inherent in small arms production. They also show something of the resourcefulness of Enfield in adapting its production to the needs of the British state as well as the wider empire. Essentially, the Snider-Enfields were converted 1853 Enfield pattern muzzle-loading rifled-muskets, used throughout the British Empire during the period 1853–67, but which now incorporated the higher accuracy and rapid-firing ability of the Snider breech-loading system.

147 'Belgian Small-Arms Factory', *Arms and Explosives*, July 1893; 'The Arms Industry in Belgium', *Arms and Explosives*, October 1904.

148 'The Sale of Old Rifles', *The Ironmonger*, June 1882, pp. 832–3.

149 'Birmingham…', *The Ironmonger*, February 1879, pp. 209–10; 'The Sale of Unserviceable Arms, etc.', *The Ironmonger*, May 1881, pp. 739–40; 'The Importation of Guns into India', *Arms and Explosives*, November 1892, p. 29. Of course, there were colonial outposts as close to home as Ireland, where in February 1880, some 5,000 to 8,000 converted Snider-Enfields and many thousands of English and Belgian revolvers were despatched by dealers that included Birmingham gunmakers buying old government stocks, with at least one consignment intended for the arming of the Protestant population in the North.

150 'Arms and Ammunition', *The Ironmonger*, March 1873, p. 303, and May 1878, p. 367; cf. 'The Sale of Old Rifles', *The Ironmonger*, June 1882, pp. 832–3.

151 C. A. Kemball, British Resident's Report, quoted in *The Times of India*, 24 August 1901.

152 'Artifex' and 'Opifex', *Causes of Decay*, p. 31.

Chapter 3   The Arms Trade in the Western Indian Ocean

1 A. Keppel, *Gun-running and the Indian North-West Frontier* (London, 1911), pp. 124–5. Arnold Joost William Keppel (1884–1964), second son of the 8th Earl of Albemarle, served as the British Government's Honorary Attaché in Bucharest and Tehran. He was appointed special correspondent for the

government expedition to Persian Makran in 1911, and became a correspondent for *The Times* newspaper from 1912–14, with a special interest in the affairs of the Middle East and South Asia. His narrative on the arms traffic in this region continues to yield valuable primary source material.

2   R. Coupland, *The Exploitation of East Africa 1856–1890*, 2nd edition (London, 1968), p. 320; R. Oliver and G. Mathew, *History of East Africa*, vol. 1 (Oxford, 1963), p. 277; R. Coupland, *East Africa and its Invaders* (Oxford, 1938), pp. 316–17. In 1859, small arms worth some £18,840 were imported into Zanzibar. By 1879, this figure had risen to £70,000, of which some 50 per cent came from Belgium, 40 per cent from Britain, and 10 per cent from Germany. Although the old 'Tower' musket remained the most characteristic weapon in the 1880s, the muzzle-loaders were giving way steadily to single-shot breech-loaders, followed by repeating rifles.

3   Beachey, 'The Arms Trade in East Africa', p. 452; Foreign Office (FO) 881/6497/8, Rodd to Anton, 6 September 1893.

4   FO 881/6403/97, Enclosure 3, Lugard to Kitchener, 10 February 1893. Henry Morton Stanley (born John Rowlands, 1841–1904) was a Welsh journalist and explorer noted for his much-vaunted exploits in Africa, such as 'finding' David Livingstone (1871–72) and 'rescuing' Emin Pasha (1886–90). But he was also one of most theatrical agents of sub-imperialism, unwittingly securing for the Belgian King Leopold II various concessionary rights to the Congo (1874–77).

5   FO 881/5732/52, Euan-Smith to Rosebery, 28 June 1888. Archibald Primrose (1847–1929), the 5th Earl of Rosebery, served as Foreign Secretary in Gladstone's Liberal administrations (1886, 1892–94) before himself becoming Prime Minister (1894–95). Rosebery's views on foreign policy, however, differed substantially from those of Gladstone. Inspired by the political legacy of the two Pitts and the intellectual legacy of British imperial historian Sir John Seeley, Rosebery was motivated to visit the colonies in person to become an advocate of the imperial federation he termed a 'Commonwealth of Nations'. He predicted that war would not, as was commonly held, dissolve the Empire, but rather unite it if colonial opinion were taken seriously. Acknowledged as the leader of the Liberal Imperialist faction of the Liberal Party, Rosebery's designs in defence, foreign, and colonial policy included proposals for British naval expansion, support for the Anglo-Boer War, and opposition to Irish Home Rule. In 1894, Rosebery established a protectorate over Uganda, from which Gladstone had wished to withdraw all British influence. The extension of British colonial authority even under a Liberal Government reflected the high imperialism of the times; it created further opportunities for colonial competition and crisis at the imperial periphery, as well as cross-border arms control in the face of rampant arms trafficking.

6   FO 881/5732/52, Euan-Smith to Rosebery, 28 June 1888. For the period 1 January to 23 June 1888 alone, these were 37,441 rifles; 188 pistols; one million bullets; three million caps; 70,650 cartridges; and 69,350 lbs of gunpowder.

7   FO 881/5732/52, Mackay to Euan-Smith, 18 April 1888.

8   Hyam, *Britain's Imperial Century*, p. 223; Beachey, 'The Arms Trade in East Africa', pp. 455–62, 467.

9   IO, L/P&S/18, Memorandum D 182, Appendix T, 'The Arms and Ammunition Traffic in the Gulfs of Persia and Oman', Government of India (GoI) Foreign Department, 1907.

10 'Importation of Arms in the Tirah Valley', *Arms and Explosives*, November 1900, p. 171.

11 'Arms in the Persian Gulf', *Arms and Explosives*, October 1901, p. 156.

12 IO, L/P&S/10, Subject Files 13/06, 1792, Major Sir Henry McMahon (Baluchistan) to Foreign Secretary (GoI), 26 July 1907; Annual Trade Reports, *Diplomatic and Consular Reports* (London, 1905–16); IO, L/P&S/10, Home Correspondence 4011/08, Memorandum C 120, 'The Illicit Trade in Arms and Ammunition between the Persian Gulf and Afghanistan', GoI, Division of the Chief of Staff, Simla, 1908.

13 'I had a little nut tree, nothing would it bear, but a silver nutmeg and a golden pear. The king of Spain's daughter came to visit me, and all for the sake of my little nut tree.' The imagery of the nut tree, the silver nutmeg, and the golden pear are drawn from this traditional children's nursery rhyme. They are regarded as metaphors for the early voyages of European exploration and discovery, which came to encompass the wealth of both the East and West Indies.

14 C. A. Trocki, *Opium, Empire and the Global Political Economy: A Study of the Asian Opium Trade* (London, 1999), p. 8.

15 Subrahmanyam, *The Portuguese Empire in Asia*, pp. 30–54, 270–7.

16 Gray, 'Portuguese Musketeers on the Zambezi', pp. 531–3.

17 S. Marks and A. Atmore, 'Firearms in Southern Africa: A Survey', *JAH*, 12:4 (1971), p. 517.

18 Sheriff, *Slaves, Spices and Ivory in Zanzibar*, pp. 41–2; Coupland, *East Africa and its Invaders*, pp. 77, 80. See also Campbell, *An Economic History of Imperial Madagascar*, which demonstrates the centrality of Madagascar's role in the proposed French commercial empire, stretching from the Mascarenes to the Swahili and Mozambique coasts. Madagascar was a vital source of provisions and plantation workers for the Mascarenes. During the nineteenth century, Madagascar played a pivotal role in the East African slave trade, and was integrated progressively—alongside other countries in the region—into the burgeoning world economy.

19 B. Nicolini, 'Religion and Trade in the Indian Ocean: Zanzibar in the 1800s', *Newsletter of the International Institute for the Study of Islam in the Modern World*, July 1999, p. 28; Coupland, *The Exploitation of East Africa*, p. 7.

20 A. M. F. Al-Wuhaibi, 'Oman under Sultans Taimur and Sa'id 1913–1970' (Ph.D. thesis, University of Cambridge, 1995), p. 12.

21 Bayly, *Indian Society and the Making of the British Empire*, pp. 50–3, 84–7; cf. D. M. Peers, *Between Mars and Mammon: Colonial Armies and the Garrison State in India 1819–1835* (London, 1995).

22 R. Kumar, *India and the Persian Gulf Region 1858–1907* (London, 1965), pp. 20, 245; Kelly, *Britain and the Persian Gulf*, pp. 2, 50–98, 167–92; G. H. Mungeam, *British Rule in Kenya 1895–1912: The Establishment of Administration in the East Africa Protectorate* (Oxford, 1966), pp. 5–7. Careful to maintain largely informal influence in the region, the British Admiralty ended their temporary occupation of Qishm as a naval base for the suppression of piracy in 1823, and the Foreign Office refused the request of local Arabs to establish a British Protectorate at Mombasa in

1826. The cautious appointment of a British Agent and Consul at Zanzibar in 1841 came mostly as a concession to safeguard British interests in India and cultivate good relations with the Sultan, since many British Indian traders had established operations in Zanzibar and coastal East Africa.

23 D. Gillard, *The Struggle for Asia, 1828–1914* (London, 1977), pp. 26, 34, 43.

24 P. J. Brobst, *The Future of the Great Game* (Akron, Ohio, 2005), adopts a useful comparative approach that explores both the historical context and the contemporary resonances of the 'Great Game'.

25 Kumar, *India and the Persian Gulf Region*, p. 250; Kelly, *Britain and the Persian Gulf*, pp. 260–410, 500–716.

26 See H. P. Merritt, 'Bismarck and the German Interest in East Africa', *Historical Journal*, 21 (1978).

27 Mungeam, *British Rule in Kenya*, pp. 1–2.

28 B. C. Busch, *Britain and the Persian Gulf, 1894–1914* (Berkeley, 1967), pp. 6–7.

29 Busch, *Britain and the Persian Gulf*, pp. 19, 106, 241.

30 Ibid., pp. 1–2.

31 Lieutenant-Colonel Sir A. T. Wilson, *South-west Persia: A Political Officer's Diary 1907–1914* (London, 1941), p. x.

32 FO 84/24, 6 February 1879. In early 1879, a British vessel, the *Vortigern*, was wrecked in the Gulf of Aden. On board were found 'forty tons of gunpowder and 1,000 stand of arms', destined for the Delagoa Bay area.

33 Colonial Office (CO) 879/16/200, Memorandum on 'The South African arms question', 21 July 1879; Beachey, 'The Arms Trade in East Africa', p. 454.

34 FO 5867/48, Memorandum by Hon. Guy Dawnay, 12 December 1888.

35 *Berliner Tageblatt*, 12 October 1888; *Nationale Zeitung*, 24 October 1888, cited in Beachey, 'The Arms Trade in East Africa', p. 455.

36 FO 5770/215, Universities Mission Report, 7 November 1888.

37 FO 6025/1, Admiralty, 30 September 1889.

38 FO 6039/279, Memorandum by von Wissman, 28 January 1890.

39 FO 6127/12, Sultan's Proclamation, February 1889. The scope of prohibition did not, however, extend to flintlocks.

40 FO 5867/314, I.B.E.A.C., 24 February 1890.

41 CO 879/31/381, Correspondence about trade in arms and liquor 1881–89 (1890); cf. Sir E. Hertslet, *The Map of Africa by Treaty*, vol. 3, p. 60, cited in Beachey, 'The Arms Trade in East Africa', p. 457. The full title of the Brussels Conference Act was 'Convention Relative to the Slave Trade and Importation into Africa of Firearms, Ammunition, and Spiritous Liquors'.

42 FO 6340/206, Memorandum by Portal (Zanzibar), 20 May 1892; FO 6454/265, Rodd to Rosebery, 12 May 1893; FO 6341/264, Proclamation of Von Soden, 1 December 1891; FO 6341/132, Proclamation of Von Soden, 9 July 1892.

43 FO 6341/91, Portal to Lord Salisbury, 4 July 1892.

44 Ibid.

45 FO 6340/206, Memorandum by Portal, 20 May 1892.

46 FO 6490/121, Captain Williams to Rodd, 3 July 1893.

47 FO 7401/100, Report on Toru Confederacy by Major Sitwell, 13 May 1899.

48 *PP*, Africa No. 4, Lugard to the IBEAC, 13 August 1891. British Army Captain Frederick Dealtry Lugard (1858–1945) was an ambitious military adventurer turned soldier-administrator, later ennobled as Baron Lugard of Abinger. Starting off as an explorer for the Imperial British East Africa

Company, Lugard became Military Administrator of Uganda (1890–92) before leading the West African Frontier Force in expeditions along the Niger Delta. Lugard was subsequently appointed High Commissioner of Northern Nigeria (1900–6), Governor of Hong Kong (1907–12), Governor of Northern and Southern Nigeria (1912–14), and Governor-General of Nigeria (1914–1918).

49   Ibid.
50   F. D. Lugard, *Diaries* (London, 1959), vol. 2, p. 118.
51   FO 6454/231, Count Hatzfeldt to Rosebery, 23 May 1893.
52   Beachey, 'The Arms Trade in East Africa', p. 461.
53   FO 7868/48, Notes on Uganda by Intelligence Officer, E. Knox, 12 October 1901; FO 7401/100, Report on Toru Confederacy by Major Sitwell, 13 May 1899; FO 7946/145, Intelligence Report, Uganda, February 1902.
54   FO 7868/48, Notes on Uganda by Intelligence Officer, E. Knox, 12 October 1901.
55   FO 7024/25 and 79, Wilson to Salisbury, 9 November and 7 December 1897.
56   IO, L/P&S/18, Memorandum D 182, Appendix T, 'The Arms and Ammunition Traffic in the Gulfs of Persia and Oman'. As early as 1888, permission to tranship 1,477 guns, 44 pistols, and 32,050 bullets at Zanzibar for Bahrein was denied by the British Indian authorities at Bombay.
57   FO 7401/102, Foreign Office to India Office, 23 May 1899.
58   FO 8040/165, Resident, Aden to Bombay government, March 1901.
59   FO 6617/11, Enclosure 1, Captain Dugmore to Mr Cracknell, 23 May 1894.
60   FO 8040/165, Enclosure, Resident, Aden, 8 March 1901.
61   FO 6043/294, Rosebery to Monson, 20 March 1893; FO 7954/75, Monson to M. Delcasse, 14 August 1902; FO 7954/60, Lansdowne to Monson, 23 April 1902.
62   FO 7953/222, Cordeaux to Lansdowne, 7 June 1902; FO 8040/165, Enclosure 5, Intelligence Division to War Office, 27 August 1902. These 'pigs' of lead were crude lead that had been cast in blocks or ingots, typically weighing around 70 lbs each, and supplying sufficient quantities of raw material for the mass production of bullets.
63   FO 8040/166, Enclosure 4, Lieutenant-Colonel Abud to Bombay Government, 13 September 1902. These figures of arms and ammunition were typical of the cargoes of captured dhows.
64   'The Persian Arms Trade', *Arms and Explosives*, July 1898, p. 157.
65   FO 8040/165, Enclosure 5, Major Cox to Lieutenant-Colonel Kemball, Muscat, 27 August 1902.
66   IO, L/P&S/10, Subject Files 13/06, 1995a/07, Grey telegram to Foreign Secretary (GoI), 26 October 1907.
67   Austin, 'Gun-running in the Gulf', pp. 171–2.
68   IO, L/P&S/7/94, Lord Elgin to Lord George Hamilton (India Office), 4 August 1897.
69   IO, L/P&S/3/361/vol. 174, Memorandum by W. Lee-Warner, 27 October 1897; IO, L/P&S/18, Memorandum C 88, 'The Trade in Arms with the Persian Gulf', W. Lee-Warner, 3 June 1898.
70   *Kilbracken MSS*, private letter, Elgin to Godley, 11 November 1897; cf. Elgin (private) to Godley, 18 April 1894. Elgin actually passed over Lee-Warner in favour of Cunningham when appointing a Foreign Secretary.

71 *Curzon MSS*, Meade (private) to Curzon, 14 January 1898.
72 *The Times of India*, 18 December 1897.
73 *Curzon MSS*, Curzon (private) to Godley, and to Hamilton, 12 July and 23 August 1899.
74 Busch, *Britain and the Persian Gulf*, pp. 52–4, 92.
75 IO, L/P&S/18, Memorandum D 182, Appendix T, 'The Arms and Ammunition Traffic in the Gulfs of Persia and Oman'; Busch, *Britain and the Persian Gulf*, p. 281.
76 FO 7946/68, Lord Currie to Lansdowne, 4 February 1902.
77 IO, L/P&S/18, Memorandum D 182, Appendix T, 'The Arms and Ammunition Traffic in the Gulfs of Persia and Oman'.
78 FO 7954/70, Commander Pears to Admiralty, 11 June 1902.
79 FO 8040/70, Enclosure 1, Commander Pears to Admiralty, 12 September 1902.
80 FO 8040/218, GoI to Lord G. Hamilton, 9 October 1902; FO 8040/180, Commander Cartwright to Rear-Admiral Bosanquet, July 1902.
81 FO 7823/168, Macdougall to Eliot, 6 May 1901; FO 8040/24, Enclosure 3, Commander Swayne to Lansdowne, 4 July 1902.
82 J. R. P. Postlethwaite, *I Look Back* (London, 1947), pp. 51–2.
83 Busch, *Britain and the Persian Gulf*, pp. 278–81.
84 Admiralty (ADM) 116/1046, Warrinder to Secretary of Admiralty, 30 January 1908.
85 Busch, *Britain and the Persian Gulf*, pp. 282–3, 288.
86 Austin, 'Gun-running in the Gulf', pp. 173, 177–88, 324–53.
87 IO, L/P&S/10, Subject File 239, 'Compensation for French Dealers at Muscat, 1913–14'.
88 Busch, *Britain and the Persian Gulf*, p. 302; see also pp. 286–7, 301.
89 IO, L/P&S/10, Subject File 115, 'Dubai Incident, 1910–11'; Wilson, *Southwest Persia*, p. 130.
90 See *The Times*, 30 December 1910; *The Times of India*, 31 December 1910; and IO, L/P&S/10, Subject Files 7/07, 262/11, Commander H.M.S. *Hyacinth* to CIC H. M. Ships and Vessels, East Indies, 31 December 1910; Cox to Foreign Secretary (GoI), 8 January 1911; and Admiral Slade to Admiralty, 12 January 1911.
91 IO, L/P&S/10, Subject Files 7/07, 151 and 205/11, Viceroy telegrams to Secretary of State for India (SSI), 2 and 23 January 1911; and SSI telegram to Viceroy, 3 January 1911.
92 See F. Cooper, *Plantation Slavery on the East Coast of Africa* (New Haven, 1977).
93 P. J. Cain and A. G. Hopkins, *British Imperialism: Innovation and Expansion 1688–1914* (London, 1993), p. 387.
94 IO, Home Correspondence 280/92, SNO to Commander-in-Chief, East Indies Station, 1 January 1892.
95 IO, L/P&S/18, Memorandum D 182, Appendix T, 'The Arms and Ammunition Traffic in the Gulfs of Persia and Oman'.
96 Ibid.
97 IO, L/P&S/18, B400, 'Muscat 1908–1928'.
98 Al-Wuhaibi, 'Oman under Sultans Taimur and Sa'id', p. 13.

99   Bayly, *Imperial Meridian*, pp. 14–15; cf. I. Wallerstein, *The Modern World System, 2: The Consolidation of the European World Economy, 1600–1750* (New York, 1980); Chaudhuri, *Asia before Europe*.

100 Bayly, *Imperial Meridian*, p. 18.

101 Chaudhuri, *Asia before Europe*, p. 384; McPherson, *The Indian Ocean*, pp. 5, 37–64; Bayly, *Imperial Meridian*, pp. 16, 19, 34; cf. S. Chander, 'From a Pre-Colonial Order to a Princely State: Hyderabad in Transition' (Ph.D. thesis, University of Cambridge, 1987), pp. 194–5.

102 Sheriff, *Slaves, Spices and Ivory in Zanzibar*, p. 8; Alpers, *Ivory and Slaves in East Central Africa*, pp. 2–8; Martin, *Cargoes of the East*, pp. 16–19.

103 Kelly, *Britain and the Persian Gulf*, pp. 1–2; Subrahmanyam, *The Portuguese Empire in Asia*, pp. 148–9; Das Gupta and Pearson (eds) *India and the Indian Ocean*, p. 139.

104 Kelly, *Britain and the Persian Gulf*, pp. 1–2; Das Gupta and Pearson (eds) *India and the Indian Ocean*, p. 139; Nicolini, 'Religion and Trade in the Indian Ocean', p. 28.

105 T. L. Pennell, *Among the Wild Tribes of the Afghan Frontier: A Record of Sixteen Years' Close Intercourse with the Natives of the Indian Marches* (London, 1909), p. 62.

106 Keppel, *Gun-running and the Indian North-West Frontier*, pp. 57–61. The blood-feuding and factiousness persisted into the twentieth century; among the Khyber Afridis, for instance, there was a dynastic alliance between the Malikdin Khel and the Khambar Khel, while the Zakka Khel and the Kuki Khel were mortal enemies.

107 Das Gupta and Pearson (eds) *India and the Indian Ocean*, pp. 12–15.

108 Coupland, *The Exploitation of East Africa*, pp. 135–7.

109 Nicolini, 'Religion and Trade in the Indian Ocean', p. 28.

110 Clarence-Smith (ed.) *The Economics of the Indian Ocean Slave Trade*, pp. 1–3; Oliver and Mathew, *History of East Africa*, vol. 1, p. 273.

111 Alpers, *Ivory and Slaves in East Central Africa*, p. 266; cf. Oliver and Mathew, *History of East Africa*, vol. 1, p. 219; Coupland, *East Africa and Its Invaders*, pp. 316–17; Coupland, *The Exploitation of East Africa*, pp. 147, 320; 'Ivory', *The Ironmonger*, March 1891, p. 544. We need only look at Zanzibar's trade profits in 1859 and 1879 to see how lucrative the triangular trade had become by the second half of the nineteenth century. In 1859, the value of trade was as follows: ivory (£146,666); cloves (£55,666); and small arms (£18,840). In 1879: cloves (£170,000); ivory (£160,000); and small arms (£70,000). At the height of his prosperous reign, Sultan Seyyid Said of Zanzibar obtained between £10,000 and £20,000 a year from the duty on slaves. Finally, there was a clear 'swing to the east' in terms of the quantity of ivory exports; by 1891, West Africa yielded only two tons, compared to 18 tons of East African and East Indian ivory.

112 Fisher and Rowland, 'Firearms in the Central Sudan', p. 233.

113 FO 97/334, no. 3 of 1864, Baikie to Russell, 20 January 1864; cf. J. P. Smaldone, 'Firearms in the Central Sudan: A Revaluation', *JAH*, 13:4 (1972), p. 594; D. Birmingham, 'The Forest and Savanna of Central Africa', in J. E. Flint (ed.), *Cambridge History of Africa, 5: From c.1790 to c.1870* (Cambridge, 1976), p. 262. In this regard, the observation of William B. Baikie, the British consul at Lokoja, is instructive. Writing to the Foreign Office in 1864 on behalf

of Emir Masaba of Nupe, Baikie requested a shipment of Enfield rifles as well as old 'Tower' muskets; the former for firepower, but the latter 'for noise, a very important element in warfare here'.

114 H. A. Gemery and J. S. Hogendorn, 'Technological Change, Slavery and the Slave Trade', in C. Dewey and A. G. Hopkins (eds) *The Imperial Impact: Studies in the Economic History of India and Africa* (London, 1978), pp. 248–50.

115 Guy, 'A Note on Firearms in the Zulu Kingdom', p. 562; R. A. Caulk, 'Firearms and Princely Power in Ethiopia in the Nineteenth Century', *JAH*, 13:4 (1972), pp. 609–19.

116 G. L. Sullivan, *Dhow Chasing in Zanzibar Waters* (London, 1873), p. 125; W. A. Chanler, *Through Jungle and Desert: Travels in Eastern Africa* (London, 1896), p. 217; D. Birmingham, *Central Africa to 1870: Zambezia, Zaire and the South Atlantic* (Cambridge, 1981), p. 131; Fisher and Rowland, 'Firearms in the Central Sudan', pp. 234–5.

117 Fisher and Rowland, 'Firearms in the Central Sudan', p. 232. Despite the greater elephant-stopping power of a rifle as compared with spear or sword, this was an advantage not immediately obvious to wary natives unreceptive to the new technology, especially in circles where traditional modes of elephant-hunting remained effective and thereby persisted.

118 Young, 'History of the Birmingham Gun Trade', p. 126; Atmore, Chirenje and Mudenge, 'Firearms in South Central Africa', pp. 548–9; D. Birmingham and P. M. Martin, *History of Central Africa* (London, 1983), vol. 1, pp. 273–4 and vol. 2, p. 4. As Birmingham and Martin point out in the case of Ndebele ivory procurement in southern Central Africa as late as the 1880s, a gun originally bought with a tusk became a profitable investment in later elephant hunts, supplying meat and providing security in various ways (payment for consumer goods, like imported blankets or clothes; cattle; and traditional dowries or bride-price).

119 R. Hall, *Empires of the Monsoon: A History of the Indian Ocean and its Invaders* (London, 1996), p. 440.

120 'The Ivory Trade of the East Coast of Africa', *The Ironmonger*, November 1887, p. 210.

121 Ibid., p. 211.

122 Oliver and Mathew, *History of East Africa*, vol. 1, pp. 270, 273–4. Other local products were sometimes included as part of the exchange; indigenous copper was one of the chief lubricants of this interior trade.

123 Birmingham, *Central Africa to 1870*, pp. 130–1; Beachey, 'The Arms Trade in East Africa', p. 458; cf. D. A. Low, 'Uganda: The Establishment of the Protectorate, 1894–1919', in V. Harlow and E. M. Chilver (eds) *History of East Africa*, vol. 2 (Oxford, 1965), pp. 106–7; A. Roberts, 'The Nyamwezi', in A. Roberts (ed.) *Tanzania before 1900* (Nairobi, 1968), pp. 117–50; and S. J. Rockel, '"A Nation of Porters": The Nyamwezi and the Labour Market in Nineteenth-Century Tanzania', *JAH*, 41:2 (2000), pp. 173–95.

124 Oliver and Mathew, *History of East Africa*, vol. 1, pp. 323–7, 334–5; cf. R. Reid, *Political Power in Pre-Colonial Buganda: Economy, Society and Warfare in the Nineteenth Century* (Oxford, 2002). Whereas Mutesa had artfully managed the rival groups (along with the colonial powers that supported them), Mwanga singlehandedly alienated them all through acts of violence perpetrated against his own people as well as the Europeans. The

unfolding civil war prompted the British to back a rebellion by Christian and Muslim groups who supported Mwanga's brother and deposed Mwanga (1888). Although he regained the throne after negotiations with the British (1889), further agreeing to cede some of his sovereignty to Frederick Lugard of the Imperial British East Africa Company (1890), it merely delayed the final eclipse of Buganda's political and military power. Mwanga accepted Buganda's absorption into the British protectorate of Uganda (1894) but his subsequent war against the British led to consecutive military defeats (1897, 1898) and deportation to the Seychelles (1899), alongside fellow African ex-ruler, Kabarega of Bunyoro.

125 D. Livingstone, *Narrative of an Expedition to the Zambesi and Its Tributaries and of the Discovery of the Lakes Shirwa and Nyassa, 1858–1864* (London, 1865), p. 360; D. Livingstone, *The Last Journals of David Livingstone in Central Africa from 1865 to His Death* (London, 1874), vol. 1, pp. 78, 126; R. F. Burton, *The Lake Regions of Central Africa* (1860, reprinted New York, 1962), vol. 1, p. 89 and vol. 2, p. 308; J. H. Speke, *What Led to the Discovery of the Source of the Nile* (1863, reprinted New York, 1967), pp. 235; R. Stanley and A. Neame (eds) *The Exploration Diaries of H. M. Stanley* (London, 1961), pp. 117, 119.

126 J. L. Krapf, *Travels, Researches and Missionary Labours, during an Eighteen Years' Residence in Eastern Africa* (London, 1860), p. 364.

127 Speke, *What Led to the Discovery of the Source of the Nile*, pp. 170, 245.

128 V. L. Cameron, *Across Africa* (London, 1877), vol. 1, pp. 201, 373; D. Livingstone, *The Last Journals of David Livingstone in Central Africa*, vol. 2, p. 73.

129 'Large Ivory Exports from the Congo', *The Ironmonger*, August 1889, p. 383.

130 'An Ivory Famine', *The Ironmonger*, August 1889, p. 395.

131 FO 6454/261, Rodd to Rosebery, 11 May 1893.

132 *The Times*, 9 November 1888.

133 Beachey, 'The Arms Trade in East Africa', p. 455.

134 Busch, *Britain and the Persian Gulf*, pp. 55–61, 302.

135 R. G. Landen, *Oman since 1856: Disruptive Modernization in a Traditional Arab Society* (Princeton, 1967), p. 392.

136 See, for instance, M. Adas, *Prophets of Rebellion: Millenarian Protest Movements Against the European Colonial Order* (Cambridge, 1987); and J. A. Clancy-Smith, *Rebel and Saint: Muslim Notables, Popular Protest, Colonial Encounters (Algeria and Tunisia, 1800–1904)* (Berkeley, 1994), pp. 4–6, 92–124, 153–7.

137 Austin, 'Gun-running in the Gulf', p. 324. British Army Major (later Brigadier-General) Herbert Henry Austin (1868–1937) commanded two surveys in the Anglo-Egyptian Sudan and Lake Rudolf at the twilight of the Victorian era (1899–1901), along with other explorations in East Africa and the Middle East. Austin was assigned special duty in the prevention of gun-running in the Persian Gulf (1909–10), and later became the British officer commanding the Bakuba Refugee Camp, located 33 miles northeast of Baghdad (1918–20). During his tenure in the Gulf, Austin made a detailed personal study of arms trafficking in the region, including the measures undertaken by various officials to curb the illicit trade.

138 Keppel, *Gun-running and the Indian North-West Frontier*, pp. 56–7, 68–70, 72–5.

139 Bayly, *Imperial Meridian*, pp. 179–84; Beachey, 'The Arms Trade in East Africa', p. 454; Al-Wuhaibi, 'Oman under Sultans Taimur and Sa'id', p. 16.

140 N. Harman, *Bwana Stokesi and His African Conquests* (London, 1986), pp. 28–33. Mutesa was fascinated with Mackay's printing press, set up to produce samples of Christian devotional literature, though somewhat less impressed with muskets presented by the White Fathers. From their determined efforts to outdo one another with competing gifts that revealed different grades of Western technology, the Kabaka quickly learnt that rifles were superior and might be obtained if he managed the Europeans with diplomatic cunning. By playing off Catholics, Protestants, and Muslims against one another, Mutesa successfully balanced the colonial powers that supported each group.

141 Ibid., pp. 72–3.

142 Beachey, 'The Arms Trade in East Africa', p. 460.

143 *PP*, Africa No. 4 (1892), Lugard to IBEAC, 13 August 1891; *PP*, Africa No. 8 (1896), Enclosure in No. 10, Annex 8, Extract from Captain F. D. Lugard's *The Rise of our East African Empire* (Edinburgh, 1893), vol. 2, p. 63.

144 *PP*, Africa No. 8 (1896), Enclosure in No. 10, Annex 1 (translation), Commandant Lothaire, President of the Court-martial of the Arab zone, 14 January 1895.

145 See reports on the 'Stokes Affair' and the wider 'Scramble for Africa', in *Daily Telegraph*, 6 September 1895; *Pall Mall Gazette*, 19, 23 & 28 November 1895, 11 & 16 June 1896, and 8 August 1896.

146 W. R. Louis, 'The Stokes Affair and the Origins of the Anti-Congo Campaign, 1895–1896', *Revue Belge de Philologie et d'Histoire*, 43:2 (1965), pp. 574, 579; cf. R. P. F. Ceulemans, *La Question Arabe et le Congo, 1883–1892* (Brussels, 1959).

147 FO 10/716, Salisbury's notation on Hill's minute of 2 November 1896.

148 *PP*, Africa No. 8 (1896), No. 26, Memorandum by Major W. H. Williams, 25 September 1895.

149 *PP*, Africa No. 8 (1896), Enclosure in No. 10, Annex 1 (translation), Commandant Lothaire, President of the Court-martial of the Arab zone, 14 January 1895. The German authorities actually favoured Stokes' arming of the native as a way of building up the power of Mwanga against their British rivals.

150 *PP*, Africa No. 8 (1896), Enclosure in No. 10, Annex 1 (translation), Commandant Lothaire, President of the Court-martial of the Arab zone, 14 January 1895; *PP*, Africa No. 8 (1896), No. 6, Sir Edward Malet to Salisbury, 23 August 1895.

151 FO 539/79/9, Memorandum by General T. E. Gordon on the export of arms to Persia (undated).

152 IO, L/P&S/3/361/2370a/97, Memorandum by W. Lee Warner on the export of arms and ammunition to the Persian Gulf, 27 October 1897; IO, L/P&S/18, Memorandum D 182, Appendix T, 'The Arms and Ammunition Traffic in the Gulfs of Persia and Oman'; FO 539/79/22, Elgin to Hamilton, 4 August 1897; FO 539/79/81, Sir F. Plunkett to Salisbury, 2 January 1898; and FO 539/79/224 and 230, for the role of Lloyds in financing the shipment of arms from the metropolis.

153 'The Persian Gulf Trade in Fire-Arms' and 'The Kynoch Co. Annual Meeting and Our Foreign Trade Relations', *Arms and Explosives*, July 1898, pp. 159–61;

cf. 'The Persian Gulf Fire-Arms Trade', *Arms and Explosives*, November 1898, p. 225. For further perspective, the value of Birmingham arms and ammunition exports to the Gulf in 1897 (£105,000) was a major share of the total value of all such imports into the Gulf that year (£181,977).

154 Austin, 'Gun-running in the Gulf', pp. 170–1.
155 IO, L/P&S/18, Memorandum D 182, Appendix T, 'The Arms and Ammunition Traffic in the Gulfs of Persia and Oman'.
156 FO 539/79/26, Horace Walpole (India Office) to Foreign Office, 1 December 1897; IO, L/P&S/18, Memorandum C 88, 'The Trade in Arms with the Persian Gulf'. By way of comparison, see 'Armaments for the Boers', *Arms and Explosives*, February 1901, p. 19, for Lord Salisbury's historic complaint that the weapons with which the Boer forces inflicted major losses on British troops were smuggled through to their destination 'under the unassuming guise of boilers, locomotives, and pianos'. In this manner, no fewer than 147,000 Mauser rifles had passed through customs via Delagoa Bay in 1897, to be shared among the 30,000 able-bodied Boers in the Transvaal—a ratio of five Mausers per head.
157 FO 881/6557/73, Colvile to Cracknell, 10 December 1893; FO 6557/6, Owen to Colvile, 8 March 1894.
158 FO 881/6557/3, Owen to Colvile, 29 March 1894.
159 *PP*, Africa No. 8 (1896), Enclosure in No. 3, Memorandum (translation), Brussels, 11 August 1895.
160 *PP*, Africa No. 8 (1896), Enclosure in No. 10, Annex 1, Stokes to Lothaire, January 1895.
161 *PP*, Africa No. 8 (1896), Enclosure in No. 128, Report by Lord Vaux on the Trial of Captain Lothaire before the Conseil Supérieur at Brussels.
162 Austin, 'Gun-running in the Gulf', p. 172.
163 Ibid., p. 173.
164 IO, L/P&S/18, Memorandum D 182, Appendix T, 'The Arms and Ammunition Traffic in the Gulfs of Persia and Oman'.
165 Keppel, *Gun-running and the Indian North-West Frontier*, pp. 142–4.
166 *PP*, Africa No. 8 (1896), Enclosure in No. 10, Annex 1 (translation), Commandant Lothaire, President of the Court-martial of the Arab zone, 14 January 1895; Harman, *Bwana Stokesi*, p. 77. Stokes' marriage to Limi, a Nyamwezi woman, ensured that he could muster her Nyamwezi clansmen as porters on his gun-running expeditions to and from the coast.
167 Oral history of the Nyamwezi, cited in Hall, *Empires of the Monsoon*, p. 444.
168 Burton, *The Lake Regions of Central Africa*, vol. 2, p. 308; Oliver and Mathew, *History of East Africa*, vol. 1, pp. 274, 276.
169 Stanley and Neame (eds) *The Exploration Diaries of H. M. Stanley*, pp. 117–19; Oliver and Mathew, *History of East Africa*, vol. 1, pp. 280–1. On 21 April 1876, Stanley mentions the roar of 'gemeh-gumeh'—'Tower' muskets—deployed by the 'notorious Chief' Mirambo to signal his near arrival for a meeting Serombo in Central Africa. On 3 May 1876, Stanley presented Mirambo with a pistol and 100 cartridges, after the latter had asked him for 'a gun or pistol' and prevented Stanley's departure until they had thus 'made friends'.
170 See F. Renault, *Tippo Tip* (Paris, 1987); and, for a wider survey of contemporaneous partnerships and personalities, B. G. Martin, *Muslim Brotherhoods in Nineteenth-Century Africa* (Cambridge, 2003). His real name was Hamid bin

Muhammad, but he was also known as 'Kingugwa' (The Leopard), from his fearsome reputation as a hunter-gatherer of slaves and ivory, and 'Tippu Tib', an even more familiar nickname derived from the sound of his guns.

171 Oliver and Mathew, *History of East Africa*, vol. 1, p. 274; Beachey, 'The Arms Trade in East Africa', p. 460.

172 FO 6557/73, Colvile to Cracknall, 10 December 1893.

173 FO 6261/50, Lugard to the IBEAC, 4 April 1891.

174 IO, L/P&S/18, Memorandum C 88, 'The Trade in Arms with the Persian Gulf'; Al-Wuhaibi, 'Oman under Sultans Taimur and Sa'id', pp. 1–16.

175 Coupland, *East Africa and Its Invaders*, p. 486.

176 Oliver and Mathew, *History of East Africa*, vol. 1, p. 275.

177 'The Ivory Trade of the East Coast of Africa', *The Ironmonger*, November 1887, pp. 210–11.

178 FO 8040/165, Enclosures from various Residents, 1902.

179 ADM 116/1089, Case 11115, Commander H.M.S. *Sphinx* and SNO Persian Gulf Division, to CIC H.M. Ships and Vessels, East Indies, 19 October 1908.

180 Austin, 'Gun-running in the Gulf', pp. 174, 326–7.

181 Ibid., p. 174.

182 Ibid., pp. 326–7.

183 IO, L/P&S/18, Memorandum D 182, Appendix T, 'The Arms and Ammunition Traffic in the Gulfs of Persia and Oman'.

184 ADM 116/1089, Case 11115, Commander H.M.S. *Sphinx* and SNO Persian Gulf Division, to CIC H.M. Ships and Vessels, East Indies, 19 October 1908, 25 September 1909 and 19 October 1909; Austin, 'Gun-running in the Gulf', pp. 171–4, 177; Keppel, *Gun-running and the Indian North-West Frontier*, pp. 144–7, 160–2. The large *Powindah* (warrior-merchant) tribe of the Ghilzais possessed no fewer than 50,000 camels, a tremendous asset for gun-running in the desert economy, given the fact that each Arabian camel was capable of carrying up to 380 lbs. of munitions (compared to the mule's 160 lbs) and making progress across desert terrain at the rate of two miles per hour (when fully loaded), without need of water or fodder for considerable lengths of time.

185 Austin, 'Gun-running in the Gulf', p. 173.

186 Ibid., p. 173.

187 Coupland, *The Exploitation of East Africa*, pp. 146–7; FO 881/5732/52, Euan-Smith to Rosebery, 28 June 1888. At this time, the price of an adult slave in the Zanzibar market was £4–5, although in Muscat and ports along the Arabian coast, this was as high as £13–20. One *frasillah* (36 lbs) of ivory—typically the quantity yielded by one tusk—was valued at £15 (up to £17 for a top-quality tusk). At Zanzibar, a first-class Snider rifle retailed at 13 shillings, while Snider carbines were sold from 9–10 shillings apiece.

188 FO 539/79/1, India Office to Foreign Office, 7 May 1880; IO, L/P&S/18, Memorandum D 182, Appendix T, 'The Arms and Ammunition Traffic in the Gulfs of Persia and Oman'.

189 FO 539/79/6, Mirza Said Khan, Persian Minister of Foreign Affairs, to R. Thompson, British agent in Tehran, 3 July 1881.

190 'The Persian Gulf Trade in Fire-Arms', *Arms and Explosives*, July 1898, p. 159.

191 IO, L/P&S/18, Memorandum D 182, Appendix T, 'The Arms and Ammunition Traffic in the Gulfs of Persia and Oman'.

192 FO 539/79/22, Elgin to Hamilton, 4 August 1897.

193   G. N. Curzon, then Under-Secretary for Foreign Affairs, and C. E. Mathews, solicitor acting for Birmingham gunmakers, quoted in 'The Persian Gulf Trade in Fire-Arms', *Arms and Explosives*, July 1898, p. 159.

194   'The Persian Trade in Fire-Arms', *Arms and Explosives*, August 1898, pp. 180–1.

195   IO, L/P&S/7/67, Sultan Faisal to Political Agent, Muscat, 22 June 1892.

196   IO, L/P&S/18, Memorandum D 182, Appendix T, 'The Arms and Ammunition Traffic in the Gulfs of Persia and Oman'; cf. Wilson, *South-west Persia*, pp. 176–7.

197   IO, L/P&S/18, Memorandum C 88, 'The Trade in Arms with the Persian Gulf'.

198   T. R. Moreman, 'The British and Indian Armies and North West Frontier Warfare, 1849–1914', *Journal of Imperial and Commonwealth History*, 20:1 (1992), p. 36.

199   Keppel, *Gun-running and the Indian North-West Frontier*, pp. 49–50.

200   'The Persian Gulf Trade in Fire-Arms', *Arms and Explosives*, July 1898, p. 159; 'The Persian Trade in Fire-Arms', *Arms and Explosives*, August 1898, p. 179.

201   FO 539/79/52, Anglo-Arabian and Persian Steamship Co. to Foreign Office, 23 December 1897.

202   FO 539/79/38, Enclosure, Hamilton to GoI, 17 December 1897; FO 539/79/ 62, Consul Perceval to Salisbury, 27 December 1897; FO 539/79/77, Salisbury to C. Hardinge, British Chargé d'Affaires, Tehran, 1 January 1898; FO 539/79/93, Foreign Office to India Office, 8 January 1898; FO 539/79/224 and 230, cf. FO 539/79/110, 176–84, 208–9; IO, L/P&S/18, Memorandum D 182, Appendix T, 'The Arms and Ammunition Traffic in the Gulfs of Persia and Oman'; 'The Muscat Seizure', *The Sporting Goods Review*, 15 July 1899, pp. 142–3.

203   'The Birmingham Gun Trade Annual Meeting', *Arms and Explosives*, June 1898, p. 144.

204   'The Persian Arms Trade', *Arms and Explosives*, July 1898, p. 157.

205   FO 539/79/224 and 230, Lloyds Bank to Foreign Office, 15 and 21 April 1898.

206   'The Gun Trade of the Persian Gulf', *The Sporting Goods Review*, 15 March 1898, p. 55.

207   'The Muscat Seizure', *The Sporting Goods Review*, 15 July 1899, p. 143.

208   'The Persian Gulf Trade in Fire-Arms', *Arms and Explosives*, July 1898, p. 160; 'The Prohibited Persian Arms Trade: Important Insurance Action', *The Sporting Goods Review*, 15 July 1898, p. 147.

209   'The Seizure of Ammunition in the Persian Gulf: Judgment in the Court of Appeal', *Arms and Explosives*, June 1900, pp. 96–7; 'The Seizure of Ammunition in the Persian Gulf', *Arms and Explosives*, April 1901, pp. 59–60; 'Judgment regarding Fracis, Times & Co. versus Meade', *Arms and Explosives*, June 1901, pp. 93–4; 'Appeal in the House of Lords', *Arms and Explosives*, August 1901, pp. 123–4; 'The Persian Gulf Trade in Fire-Arms', *Arms and Explosives*, July 1898, p. 160; 'The Prohibited Persian Arms Trade: Important Insurance Action', *The Sporting Goods Review*, 15 July 1898, p. 146.

210   IO, L/P&S/18, Memorandum D 182, Appendix T, 'The Arms and Ammunition Traffic in the Gulfs of Persia and Oman'.

211   'The Persian Arms Trade: Deputation to Mr. Curzon', *The Sporting Goods Review*, 15 June 1898, p. 121.

212   'The Persian Trade in Fire-Arms', *Arms and Explosives*, August 1898, p. 181.

213   Ibid., p. 180.

214   'The Persian Gulf Trade', *Arms and Explosives*, August 1898, p. 169.

215 'The Firearms Trade of the Persian Gulf', *The Sporting Goods Review*, 15 December 1897, p. 261.

216 'The Gun Trade of the Persian Gulf', *The Sporting Goods Review*, 15 March 1898, p. 55; 'Arms on the Indian Frontier', *The Sporting Goods Review*, 15 January 1898. The 'special report', published in the *Army and Navy Gazette*, indicated that among the 120 rifles captured from the tribesmen were 61 made from pieces sold as old iron, 47 weapons stolen from British troops, and 12 Winchesters and sporting weapons (including one of Russian provenance). There was no obviously incriminating proof of Birmingham origin.

217 'The Persian Gulf Fire-Arms Trade', *Arms and Explosives*, November 1898, p. 224; 'Rifle Thefts in India', *Arms and Explosives*, November 1899, p. 172; 'Rifle-Thieving on the Indian Frontier', *Arms and Explosives*, November 1900, p. 164.

218 *The Sporting Goods Review*, 15 June 1898, p. 116.

219 'The Persian Gulf Firearms Seizure in Court', *Arms and Explosives*, July 1899, p. 106.

220 Lieutenant-Colonel Sir A. T. Wilson, *The Persian Gulf: An Historical Sketch from the Earliest Times to the Beginning of the Twentieth Century* (London, 1928; reprinted 1954), p. 269; C. Collin Davies, *The Problem of the North-West Frontier, 1890–1908* (Cambridge, 1932), p. 175; Sir P. Sykes, *A History of Afghanistan*, vol. 2 (London, 1940), p. 244.

221 IO, L/P&S/18, Memorandum D 182, Appendix T, 'The Arms and Ammunition Traffic in the Gulfs of Persia and Oman'.

222 Ibid.

223 'The Prohibited Persian Arms and Ammunition Trade: Proceedings in Birmingham and at Westminster', *The Sporting Goods Review*, 15 August 1898, p. 177.

224 'The Persian Gulf Trade in Fire-Arms', *Arms and Explosives*, July 1898, p. 160.

225 'The Persian Trade in Fire-Arms', *Arms and Explosives*, August 1898, p. 181.

226 IO, L/P&S/18, Memorandum D 182, Appendix T, 'The Arms and Ammunition Traffic in the Gulfs of Persia and Oman'.

227 'The Kynoch Co. Annual Meeting and Our Foreign Relations', *Arms and Explosives*, July 1899, p. 106.

228 'The Persian Trade', *The Sporting Goods Review*, 15 August 1898, p. 167.

229 IO, L/P&S/18, Memorandum D 182, Appendix T, 'The Arms and Ammunition Traffic in the Gulfs of Persia and Oman'; cf. Lord Hardinge of Penshurst, *Old Diplomacy* (London, 1947), p. 65.

230 'The Persian Gulf Firearms Seizure in Court', *Arms and Explosives*, July 1899, p. 107; 'The Seizure of Ammunition in the Persian Gulf', *Arms and Explosives*, June 1900, p. 97; 'The Seizure of Ammunition in the Persian Gulf', *Arms and Explosives*, August 1901, pp. 123–4.

231 IO, L/P&S/18, Memorandum C 88, 'The Trade in Arms with the Persian Gulf'.

232 IO, L/P&S/18, Memorandum D 182, Appendix T, 'The Arms and Ammunition Traffic in the Gulfs of Persia and Oman'.

233 IO, L/P&S/7, Letters from India, No. 257 of 1902, Curzon to Hamilton, 23 January 1902.

234 'The New Indian Arms Order', *Arms and Explosives*, October 1899, p. 147; cf. 'Rifles in India', *Arms and Explosives*, September 1899, p. 135; 'Armaments

in Afghanistan' and 'Re-Arming of Indian Troops', *Arms and Explosives,* July 1901, p. 100; 'Importation of Arms into India', *Arms and Explosives,* June 1902, p. 95; Keppel, *Gun-running and the Indian North-West Frontier,* pp. xii–xiii, 41–8.

235   See Martin, *Cargoes of the East,* for fascinating impressions of life aboard such vessels in the Arabian Sea, well into the twentieth century; and D. Cordingly (ed.) *Pirates: Terror on the High Seas—from the Caribbean to the South China Sea* (Atlanta, 1996), p. 240, for contemporary piratical scenarios. In 1994, for instance, the vessel *Bonsella* was hijacked by 24 armed men on a dhow and held for five days, being repeatedly used for attacks on other ships off the coast of Somalia.

236   R. Orenstein (ed.) *Elephants: The Deciding Decade* (New York, 1997), pp. 12–16; 'A System of Extinction: The African Elephant Disaster', *Report of the Environmental Investigation Agency* (1989); 'Made in China: How China's Illegal Ivory Trade is Causing a Twenty-first Century African Elephant Disaster', *Report of the Environmental Investigation Agency* (2007). The colonial programmes of wildlife conservation, based as they were on European—not African—values and privileges, came under attack as newly independent states began to forge their own destinies. Demand for ivory surpassed pre-1914 levels, particularly in a now affluent Japan where ivory name-seals, or *hanko,* had become a status symbol. The price, hovering around $5.45 per kilo (kg) for years, now skyrocketed: from $7.44 per kg (1970) to $74.42 per kg (1978) to $150 per kg (1989) in Africa (and $300 to $400 in Japan at the height of the ivory crisis). Simultaneously, the means to acquire this fabulous source of wealth fell increasingly into the hands of Africans, even as the anti-colonial struggles and civil wars that afflicted Africa throughout the 1960s attracted international arms dealers. Thousands of semi-automatic weapons—AK-47s, Kalashnikovs, and other models—inundated Africa once more. According to the US Arms Control and Disarmament Agency, arms imports into Africa increased (in real terms) from US $500 million (1971) to $4,500 million (1980). When law and order broke down—as it did in Uganda and elsewhere—it was easy for many of these small arms to get into the hands of modern-day elephant-hunters, the ivory poachers. These poachers formed the base of a new economic pyramid dominated by millionaire dealers, notably the 'kingpins' of Hong Kong, who used their wealth and ingenuity to bribe officials, forge permits, circumvent anti-poaching laws, smuggle the poached tusks out of Africa, have them carved in Hong Kong, and funnel the bulk of their hefty profits into Swiss bank accounts. In 1975, the Asian elephant was placed on Appendix I of the Convention on International Trade in Endangered Species (CITES) that bans international trade between member countries; the African elephant was placed on Appendix I in 1990. Between 1975 and 1990, investigators from the Environmental Investigation Agency (EIA) discovered that CITES sales of stockpiles from Singapore and Burundi (270,000 kg and 89,500 kg, respectively) had generated a system which boosted the value of ivory worldwide, rewarded international smugglers, and enabled them to continue smuggling new ivory. Although the ivory trade was outlawed internationally in 1990, global demand for ivory—especially in Chinese markets—has continued to fuel elephant poaching.

237   See, for instance, *SIPRI Yearbook 2000,* pp. 8, 291–8. Africa would be the only region of the world to experience a significant increase in major

armed conflicts after 1995, most notably the escalation of war in the Democratic Republic of Congo, involving Zimbabwe, Namibia, Rwanda, and Uganda.

Chapter 4   The Arms Trade in the Eastern Indian Ocean

1 Straits Settlements Records (SSR), A68, Governor's Minute, 2 March 1830. The Scotsman Robert Fullerton (1773–1831) was appointed by the English East India Company to serve as the first Governor of the Straits Settlements (1826–30). Against the Company's official policy of non-intervention, Fullerton was decidedly proactive in his approach to the trade and politics of the Malay Peninsula. He countered Siamese aggression by threatening war, sanctioning the use of diplomatic pressure (via envoys like Henry Burney and James Low) as well as military force in the Northern Malay States, Perak, Selangor, and Naning (on the border with Malacca), to protect and promote British commercial interests.

2 E. Tagliacozzo, *Secret Trades, Porous Borders: Smuggling and States Along a Southeast Asian Frontier, 1865–1915* (New Haven, 2005), p. 273.

3 For evidence of Singapore's rapid development as an arms transhipment centre, see C. D. Cowan (ed.) 'Early Penang and the Rise of Singapore 1805–1832: Documents from the Manuscript Records of the East India Company', *JMBRAS*, 23:2 (1950), Document no. 136, The Acting Master Attendant, Singapore, to the Resident Councillor, Singapore, 21 February 1826; and L. K. Wong, 'The Trade of Singapore, 1819–69', *Journal of the Malayan Branch, Royal Asiatic Society (JMBRAS)*, 33:4 (1960), pp. 160–1. For examples of the usage of 'Eastern Seas' or 'Eastern Archipelago', see G. W. Earl, *The Eastern Seas or Voyages and Adventures in the Indian Archipelago in 1832, 1833, 1834* (London, 1837); A. R. Wallace, 'On the Trade between the Eastern Archipelago and New Guinea and Its Islands', *Proceedings of the Royal Geographical Society of London*, 6:2 (1862); C. N. Parkinson, *Trade in the Eastern Seas, 1793–1813* (Cambridge, 1937) and by the same author, *War in the Eastern Seas, 1793–1815* (London, 1954). The historical term 'Eastern Archipelago' or 'Eastern Seas' was used broadly to describe the eastern Indian Ocean zone, the geographical region between India and China, encompassing coastal areas of mainland Southeast Asia as well as all of maritime Southeast Asia. At the heart of this maritime region were the 'East Indies', the islands of the Malay Archipelago.

4 SSR, A68, 27 February 1830, pp. 45–62. Official correspondence about the 'James Fraser Gunpowder controversy' affords early glimpses of smuggling activities and the illegal storage of munitions in godowns along the Singapore River.

5 Tagliacozzo, *Secret Trades, Porous Borders*, pp. 298–9.

6 CO 273/209/21045, 'Arms for Arabia', 25 November 1895.

7 CO 144/71/7308, The Joló Protocol, 7 April 1897.

8 C. R. Boxer, *The Portuguese Seaborne Empire 1415–1825* (New York, 1969), pp. 46–53.

9 Das Gupta and Pearson (eds) *India and the Indian Ocean*, p. 256. See also Swope, 'Crouching Tigers, Secret Weapons', pp. 19–28, for a discussion of cross-cultural arms technology transfers even further eastward during the early modern period. According to local tradition, the Japanese first received

firearms technology from the Portuguese who landed off the island of Tanegashima in 1543.

10 J. Israel, *The Dutch Republic: Its Rise, Greatness, and Fall 1477–1806* (Oxford, 1995), pp. 936–7. For a classic overview of the early phases of Dutch expansion, see C. R. Boxer, *The Dutch Seaborne Empire 1600–1800* (New York, 1965), pp. 84–112, 187–294.

11 Dijk, 'The VOC's Gunpowder Factory'; 'The Colonization of the Eastern Archipelago', *United Service Magazine and Naval and Military Journal*, 2 (1837), pp. 330–2.

12 Charney, *Southeast Asian Warfare*, pp. 55–8. Even arms terminology was transferred: the earliest example of flintlock musket, the Dutch *schnapphahn* (also *snaphaen* or *snaphaan*) came to known in English as *snaphaunce* (or *snaphance*) and in Malay as *senapang*.

13 P. J. Marshall, 'Western Arms in Maritime Asia in the Early Phases of Expansion', *MAS*, 14:1 (1980), p. 20; Tagliacozzo, *Secret Trades, Porous Borders*, p. 10. The commercial and military operations of the Dutch VOC were far more sophisticated and powerful than those of the Portuguese. Dutch shipping was more numerous and better gunned; by 1626, there were 29 Dutch ships in Asia; and, by 1635, 76 of them. The *hongi-tochten* were specialized warships stationed permanently in the East to enforce the VOC's spice trade monopoly, whether by policing offshore areas to prevent smuggling, or engaging in slash-and-burn tactics to destroy the clove and nutmeg crops of rival cultivators. In the Moluccas (Maluku) that many regarded as the definitive 'Spice Islands', such enforcement would extend to the deportation or massacre of indigenous populations on certain clove- and nutmeg-producing islands, and armed surveillance over spice gardens in order to prevent the commodities from being smuggled out to the financial detriment of the Dutch monopoly.

14 Cowan (ed.), 'Early Penang and the Rise of Singapore', Document no. 112, 'Treaty between His Brittanick Majesty and the King of the Netherlands, respecting Territory and Commerce in the East Indies', London, 17 March 1824. The British had initially established bases at Penang, Malacca, and parts of the East Indies in order to counter French influence in the region, but their focus would again shift to countering Dutch influence in the Eastern Seas after the defeat of Napoleonic France.

15 Bayly (ed.) *Atlas of the British Empire*, pp. 101–5.

16 M. Kuitenbrouwer, *The Netherlands and the Rise of Modern Imperialism: Colonies and Foreign Policy, 1870–1902* (New York, 1991), pp. 34–7, 113–22.

17 See J. F. Cady, *The Roots of French Imperialism in Eastern Asia* (Ithaca, 1954); and M. E. Osborne, *The French Presence in Cochin China and Cambodia; Rule and Response, 1859–1905* (Ithaca, 1969). Laos was also incorporated into French Indochina after the Franco-Siamese War of 1893.

18 Warren, *The Sulu Zone*, pp. 64, 104. The Spanish would be driven out altogether, however, when the Americans decided to join the fray later, taking up the 'white man's burden' by 'liberating' the Philippines in 1898.

19 'Sultanate of Sulu: Act of Incorporation into the Spanish Monarchy', 30 April 1851; text of treaty cited in Warren, *The Sulu Zone*, pp. 285–7.

20 A. Dalrymple, *Reprint From Dalrymple's Oriental Repertory, 1791–7, of Portions Relating to Burma* (Rangoon, 1926), pp. 52, 82, 107, 125; W. Milburn, *Oriental Commerce*, vol. 2 (London, 1813), p. 425.

21 IO, Home Miscellaneous Series (H/Misc.), 771/2, Dalrymple to the Court of Directors, 8 August 1770.

22 K. H. Lee, *The Sultanate of Aceh: Relations with the British, 1760–1824* (Kuala Lumpur, 1995), pp. 8, 29, 53, 62–4, 140–5; Furber, *Rival Empires of Trade in the Orient*, pp. 294–5.

23 IO, L/P&S/6, Bengal Political Consultations, 119/42, Enclosure 19, Lord Minto to Sultan of Aceh, 6 February 1811.

24 Lee, *The Sultanate of Aceh*, pp. 159–60, 164; SSR, I9, W. A. Clubley, Acting Secretary to the Government (Fort Cornwallis), to John Hall, Deputy Collector, 23 September 1811.

25 Lee, *The Sultanate of Aceh*, pp. 160–9.

26 Cowan (ed.) 'Early Penang and the Rise of Singapore', Document No. 33, The Governor, Prince of Wales Island, to the King of Acheen (Aceh), 24 June 1813.

27 Ibid.

28 Cowan (ed.) 'Early Penang and the Rise of Singapore', Document No. 34, The Governor, Prince of Wales Island, to the Court of Directors, 30 June 1813; and SSFR, Vol. 39, Penang to Sultan of Aceh, undated (Fort Cornwallis Council Proceedings, 24 June 1813).

29 Lee, *The Sultanate of Aceh*, pp. 166–206.

30 SSR, K5, Regulation III 'to prevent the Exportation of Arms, Ammunition and Warlike Stores for the purposes of Trade', promulgated on 26 December 1816 and re-enacted on 1 January 1825; cf. Cowan (ed.) 'Early Penang and the Rise of Singapore', Document No. 88, Governor, Prince of Wales Island, to the Court of Directors, 30 November 1820.

31 Cowan (ed.) 'Early Penang and the Rise of Singapore', Document No. 119, Minute by the Malay Translator on the Trade of Acheen, Fort Cornwallis, 15 March 1825.

32 Lee, *The Sultanate of Aceh*, pp. 206–97.

33 For a classic exposition, see J. Gallagher and R. Robinson, 'The Imperialism of Free Trade', *EcHR*, 6:1 (1953); and J. Bastin, *The Native Policies of Sir Stamford Raffles in Sumatra and Java* (Oxford, 1957).

34 C. E. Wurtzburg, *Raffles of the Eastern Isles* (London, 1954), pp. 450–74; Lee, *The Sultanate of Aceh*, pp. 273–88; N. Tarling, *Anglo-Dutch Rivalry in the Malay World, 1780–1824* (Cambridge, 1962), pp. 84–94.

35 Wurtzburg, *Raffles of the Eastern Isles*, pp. 475–542, 606–56; E. C. T. Chew and E. Lee (eds) *A History of Singapore* (Singapore, 1991), pp. 21–65.

36 Wong, 'The Trade of Singapore, 1819–69', pp. 86–105, 203; C. M. Turnbull, *The Straits Settlements 1826–67: Indian Presidency to Crown Colony* (London, 1972), pp. 162–5, 176–9, 188–9; Tagliacozzo, *Secret Trades, Porous Borders*, p. 260.

37 Tagliacozzo, *Secret Trades, Porous Borders*, pp. 5–6.

38 Ibid., p. 273.

39 SSR, A39, R. Fullerton, Governor of the Straits Settlements, to the Court of Directors, 21 October 1827.

40 Wong, 'The Trade of Singapore, 1819–69', p. 203; P. Auber, *An Analysis of the Constitution of the East India Company, and of the Laws Passed by Parliament for the Government of Their Affairs, at Home and Abroad* (London, 1826), p. 506; J. Crawfurd, *Journal of an Embassy from the Governor-General of India to the Courts of Siam and Cochin-China*, vol. 2 (London, 1830), p. 372.

41 J. R. Logan (ed.) 'Notices of Singapore', *JIAEA*, 9, p. 474; Crawfurd, *Journal of an Embassy*, pp. 372–3.

42 SSR, B10, 30 April 1830; SSR, A68, 2 March 1830.

43 *PP*, No. 1 and Enclosure, and No. 2 in Appendix 19, in *Paper 735, Part II*, 10:2 (1831–32).

44 *PP*, No. 6, in *Paper 735, Part II*, 10:2 (1831–32).

45 Act XXXIII of 1857; SSR, W25, No. 405; SSR, S25, No. 275.

46 SSR, R31, E. A. Blundell, Governor of the Straits Settlements, to the GoI, 25 March 1857, pp. 95–7; SSR, R32, pp. 177–81.

47 SSR, S26, 'Sale of Old Arms to Powers Hostile to Britain', William Leach (India Board) to Sir James Melville (Court of Directors), 9 February 1858; J. D. Dickinson (East India House) to C. Beadon, Secretary to the GoI, 17 March 1858; and Principal Military Storekeeper's report and Customs returns, 22 January and 1 February 1858. Along with the availability of newer rifles, the influx of decades-old War Office surplus was no doubt still an internally destabilizing factor, given the relative backwardness of Chinese military technology. The situation was fraught with security implications for British strategic and commercial interests wherever the writ of British political authority and military control did not run directly.

48 SSR, R31, Blundell to the GoI, 25 March 1857.

49 Government Proclamation, 7 August 1863. The colonial authorities in Singapore, headed by Blundell's successor Colonel Cavenagh, were granted urgent new powers and responsibilities to regulate that trade.

50 *PP*, C 3104, No. 42 and No. 53, Annual Trade Report of Singapore from the Resident Councillor (for 1852 and 1854), 73, 1863.

51 'Wolverhampton and Birmingham Trades', *The Ironmonger*, December 1863, p. 355. The article also indicates that much of the weaponry imported into Singapore was not exclusively British-made, but had wider European (and especially Belgian) origins.

52 Singapore Chamber of Commerce to the Straits Government, 24 August 1863, reported in the *Singapore Free Press (SFP)*, 10 September 1863; and the Secretary to the Straits Government to the Singapore Chamber of Commerce, 31 August 1863, *SFP*, 10 September 1863.

53 Singapore Chamber of Commerce to the Straits Government, *SFP*, 12 October 1863, *SFP*, 8 October 1863; Singapore Chamber of Commerce to the Straits Government, 4 February 1863, *SFP*, 11 February 1864; Straits Government to the Singapore Chamber of Commerce, 28 September 1863, *SFP*, 1 October 1863; Straits Government to the Singapore Chamber of Commerce, 5 February 1864, *SFP*, 11 February 1864; cf. SSR, R41, pp. 119–21; SSR, W47, Nos. 132, 163; SSR, W48, Nos. 184, 193, 207, 220; SSR, S32, No. 29.

54 CO 144/27, Pope Hennessey to the Duke of Buckingham and Chandos, 16 February 1868.

55 CO 144/24, Governor T. J. Callaghan to E. Cardwell, Secretary of State for the Colonies, 22 April 1865.
56 CO 144/24, Callaghan to Cardwell, 22 April 1865.
57 CO 144/24/6329, Callaghan to Cardwell, 23 April 1865.
58 Ibid.
59 CO 144/42, Bulwer to the Earl of Carnarvon, 27 July 1874.
60 Warren, *The Sulu Zone*, pp. 110, 129–31.
61 CO 144/56/18592, 'The Export of Arms and Ammunition to the Territories of the British North Borneo Company', P. Leys, Administrator of Labuan, to the Earl of Kimberley, Secretary of State for the Colonies, 4 September 1883, pp. 5, 9–12.
62 Tagliacozzo, *Secret Trades, Porous Borders*, p. 269.
63 Ibid.
64 CO 273/73/9450, J. Hammond, Under-Secretary of State for Foreign Affairs, to the Colonial Office, 9 September 1873; cf. PRO/Admiralty/China Station: Correspondence [No. 140: The Straits of Malacca and Siam], CO to Governor Straits Settlements, 16 May 1873 and 23 September 1873.
65 'Proclamation, 31 March 1873' and 'Government Notification, 13 June 1873, No. 125', in *Straits Settlements Government Gazette* (1873).
66 CO 273/100/20345, 'Proclamation Prohibiting Export of Arms and Munitions of War to Dutch Possessions', 19 November 1879; and CO 144/56/22522, 'Export of Arms etc.: Regulations in Dutch Borneo', 18 November 1882.
67 Tagliacozzo, *Secret Trades, Porous Borders*, pp. 270, 294. Ordinance Nos. 1 and 23 of 1899 both listed and regulated various explosive substances that included nitro-glycerine and fulminates of mercury, in addition to explosives such as blasting gelatins, ammonite, picric powders, and teutonite.
68 Ibid., p. 270.
69 FO 220, Oleh Oleh Consulate, 1882–85 (vol. 2), 'Memorandum on the Alleged Smuggling of War Stores into Acheen by the British Consul in Oleh Oleh', 6 August 1883, cited in Tagliacozzo, *Secret Trades, Porous Borders*, p. 270.
70 Tagliacozzo, *Secret Trades, Porous Borders*, pp. 270–1.
71 'Bond entered by Datu Klana Abdulrahman and Datu Muda Lingie', 21 April 1874, in W. G. Maxwell and W. S. Gibson (eds) *Treaties and Engagements Affecting the Malay States and Borneo* (London, 1924), p. 38.
72 Loopholes in Straits Settlements arms control legislation would later convince the Federated Malay States to introduce their own regulations. See, for instance, Federated Malay States Ordinance No. 13, 1915, cited in Tagliacozzo, *Secret Trades, Porous Borders*, p. 270.
73 CO 144/23, Colonial Office to the Governor of Labuan, 10 January 1865; CO 144/24, Governor of Labuan to the Colonial Office, 1 April 1865 and 22 April 1865. See also the later debate over the Arms Exportation Bill in *Straits Settlements Legislative Council Proceedings*, 21 December 1887.
74 Tagliacozzo, *Secret Trades, Porous Borders*, pp. 307–8.
75 'Export Statistics on Arms and Ammunition from Singapore', *Straits Settlements Government Gazette* (1873–74).
76 Tagliacozzo, *Secret Trades, Porous Borders*, p. 304.
77 Ibid., p. 264.

78 Ibid., pp. 266–8. The rapid diffusion of steam navigation throughout the Indian Ocean following the opening of the Suez Canal 'revolutionized' traditional commercial networks by opening new avenues for trade, while altering or displacing old transportation routes governed by the monsoons. Distances and travel times between metropolis and periphery were reduced dramatically, paving the way for expanded and sophisticated port facilities at strategically-positioned colonial settlements like Singapore, which could either facilitate the growing arms traffic or check it by assisting maritime policing.

79 Ibid.

80 Ibid., p. 269.

81 Ibid., p. 293.

82 CO 273/100/20345, 'Proclamation prohibiting export of arms and munitions of war to Dutch possessions', 19 November 1879; Tagliacozzo, *Secret Trades, Porous Borders*, pp. 275, 292–3.

83 SSR, R33, Blundell to the Secretary to the GoI, 24 September 1858; SSR, V26, Blundell to the Temenggong of Johor, 2 December 1858; SSR, X35, Governor's Diary, 11 December 1858.

84 Tagliacozzo, *Secret Trades, Porous Borders*, pp. 291–2.

85 Ibid., pp. 298–9.

86 FO 72/1283, Ricketts to the Earl Granville, 1 November 1871; CO 144/35, Bulwer to Granville, 29 December 1871; CO 144/37, Bulwer to Granville, 12 June 1872; CO 144/45, Commander Buckle to Admiral Ryder, 28 February 1875.

87 CO 144/71, Article 4, 'Protocol Relative to the Sulu Archipelago, Signed at Madrid by the Representatives of Great Britain, Germany, and Spain', 7 March 1885.

88 Tagliacozzo, *Secret Trades, Porous Borders*, p. 275.

89 CO 273/198/20624, 'Lombok export of arms from Singapore', 31 October 1894.

90 CO 273/209/21045, 'Arms for Arabia', 25 November 1895.

91 CO 144/71/7308, The Joló Protocol, 7 April 1897.

92 Tagliacozzo, *Secret Trades, Porous Borders*, pp. 308–9.

93 For more on 'archaic globalization', see C. A. Bayly, '"Archaic" and "Modern" Globalization in the Eurasian and African Arena, c. 1750–1850', in A. G. Hopkins (ed.) *Globalization in World History* (London, 2002), pp. 47–73.

94 McPherson, *The Indian Ocean*, pp. 56–7.

95 Ibid., pp. 31–7. Note, however, that just as the term 'dhow' does not do justice to the sheer range of Middle Eastern vessels, so the term 'prahu'—simply meaning 'ship'—obscures the variety of Southeast Asian shipping types.

96 *The Suma Oriental of Tomé Pires: An Account of the East, from the Red Sea to Japan*, trans. A. Cortesão (London, 1944), vol. 1, p. 115; Charney, *Southeast Asian Warfare*, pp. 52–3, 67.

97 Andaya and Andaya, *A History of Malaysia*, pp. 12–13, 57–62, 76–99.

98 C. A. Trocki, *Prince of Pirates: The Temenggongs and the Development of Johor and Singapore 1784–1885* (Singapore, 1979), pp. 8–27, 54–5.

99 Crawfurd to Calcutta, 1 October 1824, quoted in Buckley, *An Anecdotal History*, p. 178.

100 Trocki, *Opium, Empire and the Global Political Economy*, p. 8. Gunpowder and cordite are also chemicals, of course, and cordite quite an exotic one— capable, as it turned out, of shifting the balance of military power between the West and the Indian Ocean world.

101 Ibid., p. 9.

102 Bayly, *Indian Society and the Making of the British Empire*, p. 77.

103 Trocki, *Opium, Empire and the Global Political Economy*, pp. 56, 188. 'Straits produce' was the nineteenth-century generic term used to describe an assortment of items from the rainforests and seas of Southeast Asia. Along with the items already mentioned, it included cinnamon, camphor, benzoin, gambier, betel, coconuts, rattans, sappanwood, beeswax, birds' feathers, hides, buffalo horn, rhinoceros horn, shark's fin, tortoise shell, tapioca, and sago.

104 S. B. Singh, *European Agency Houses in Bengal, 1783–1833* (Calcutta, 1966), p. 145; *Egremont MSS* 30/47/20/1, 'Memoir of the Sooloogannan Dominions and Commerce', 26 February 1761; G. Watt, *The Commercial Products of India* (London, 1908), pp. 432–3, 444–7. In his magisterial survey of nineteenth-century India's economic products, Sir George Watt noted: 'Saltpetre engaged the attention of the mercantile governors of India for fully a hundred years. In the Proceedings of the Hon. East India Company from 1784 to 1820, frequent mention is made of it, the exports being then viewed as the most profitable.' East Asia, in particular, was recognized as the principal market for saltpetre from Bengal.

105 See C. A. Trocki, *Opium and Empire: Chinese Society in Colonial Singapore 1800–1910* (New York, 1990); Trocki, *Opium, Empire and the Global Political Economy*, pp. 61–159.

106 Trocki, *Prince of Pirates*, pp. 62–5; Warren, *The Sulu Zone*, pp. 147–8; cf. A. L. Reber, 'The Sulu World in the Eighteenth and Early Nineteenth Centuries: A Historiographical Problem in British Writings on Malay Piracy' (M.A. dissertation, Cornell University, 1966). For a more Euro-centric perspective, see N. Tarling, *Piracy and Politics in the Malay World: A Study of British Imperialism in Nineteenth-Century South-East Asia* (Melbourne, 1963), pp. 20, 146; and L. A. Mills, *British Malaya 1824–1867* (Kuala Lumpur, 1966), pp. 323–4, 328–9. Primary sources positing the 'decay theory' of Malay piracy include a personal memorandum from Raffles to Crawfurd, 7 June 1823, quoted in Buckley, *An Anecdotal History*, pp. 116–29; and IO, F/4/1724 (69433), 'Report on Piracy in the Straits Settlements', E. Presgrave, Registrar of Imports and Exports, to K. Murchison, Resident Councillor at Singapore, 5 December 1828.

107 J. L. Anderson, 'Piracy in the Eastern Seas, 1750–1850: Some Economic Implications', in D. Starkey, E. S. van Eyck van Heslinga and J. A. de Moor (eds) *Pirates and Privateers: New Perspectives on the War on Trade in the Eighteenth and Nineteenth Centuries* (Exeter, 1997), p. 88.

108 L. Blussé, 'Chinese Century: The Eighteenth Century in the China Sea Region', *Archipel* 58 (Paris, 1999), pp. 107–29.

109 Capt. E. Belcher, *Voyage of H.M.S. 'Samarang'* (London, 1848), vol. 1, pp. 209–70; Warren, *The Sulu Zone*, pp. 149–211.

110 'Oriental Pirates', *United Service Magazine and Naval and Military Journal*, 2 (1835), pp. 34–42; Logan (ed.) 'The Piracy and Slave Trade of the Indian

Archipelago', *JIAEA*, 4 (1850), pp. 619, 626. Just as the Wahhabi raiders of the western Indian Ocean zone in the early 1800s had been stigmatized as 'pirates', so the Sulu marauders of the East were castigated as 'sea-banditti'. See also Rutter, *The Pirate Wind*, which illustrates how marauding in the Eastern Seas provoked reprisals from European cruisers and provided rationale for the extension of European 'protection' to areas such as Sarawak and Labuan. In the Malay Archipelago, the long-drawn crisis in Kedah and the Malay 'piratical system' operating around Penang—'from whence ... they were recruiting their scattered forces and obtaining supplies of arms and ammunition without the possibility of prevention by our police'—prompted Robert Ibbetson, the Resident at Singapore, to propose a new naval blockade in consultation with Governor-General Lord William Bentinck and naval commander Rear-Admiral Sir Edward Owen. [IO, F/4/ 1331 (53245), Bentinck to Owen, 18 January 1832; Ibbetson to Swinton (Bentinck's Secretary), 25 April 1832, pp. 45, 93.]

111   Rutter, *The Pirate Wind*, pp. 27–8; cf. D. G. E. Hall, *A History of South-East Asia* (London, 1964), pp. 498ff.
112   Mills, *British Malaya*, p. 231; P. Gosse, *The History of Piracy* (New York, 1934).
113   Anderson, 'Piracy in the Eastern Seas, 1750–1850', p. 89.
114   Begbie, *The Malayan Peninsula*, pp. 271–73; Trocki, *Prince of Pirates*, pp. 56–7.
115   L. Y. Andaya, 'The Structure of Power in Seventeenth Century Johore', in A. Reid and L. Castles (eds) *Pre-Colonial State Systems in Southeast Asia: the Malay Peninsula, Sumatra, Bali-Lombok, South Celebes* (Kuala Lumpur, 1975), p. 7; Andaya and Andaya, *A History of Malaysia*, pp. 42, 50, 109.
116   Abdullah bin Abdul Kadir, *The Hikayat Abdullah*, trans. A. H. Hill (Kuala Lumpur, 1970), p. 163.
117   IO, F/4/1331 (52585), Owen to Pridham, 31 July 1830.
118   See A. P. Rubin, *Piracy, Paramountcy and Protectorates* (Kuala Lumpur, 1974).
119   Mundy, *Narrative of Events in Borneo and Celebes*, vol. 2, p. 17.
120   See, for example, 'Oriental Pirates', *United Service Magazine and Naval and Military Journal*, 2 (1835), pp. 34–42; 'The Malay Pirates, with a Sketch of Their System and Territory', *United Service Magazine and Naval and Military Journal*, 1 (1837), p. 459; and 'Piracy on the Coast of Borneo', *United Service Magazine and Naval and Military Journal*, 1 (1852), p. 338. Of course, colonial or cross-cultural stereotyping extends to Western depictions of the use of traditional weaponry in indigenous military culture and warfare, thus forming part of a larger debate. See P. Porter, *Military Orientalism: Eastern War Through Western Eyes* (New York, 2009).
121   Rutter, *The Pirate Wind*, pp. 34–5.
122   Tagliacozzo, *Secret Trades, Porous Borders*, pp. 261, 304–5; Charney, *Southeast Asian Warfare*, pp. 64–6.
123   Rutter, *The Pirate Wind*, pp. 32–4, 52, 118, 150; Capt. H. Keppel, *Expedition to Borneo of H.M.S. 'Dido'*, vol. 2 (London, 1847), p. 90; CO 144/24/4668, Callaghan to Cardwell, 1 April 1865; cf. Tarling (ed.) *The Cambridge History of Southeast Asia*, vol. 1, pp. 388–9. The barrel of the swivel-gun was typically 180 centimetres long. In 1865, a *pikul* (133.33 lbs) of brass guns was

worth $30. The finest guns were cast in Brunei and were often accumulated in the houses of indigenous chiefs and traders as the 'most easily convertible form of wealth'. They were also used as payment for a bride-price (*berian*) that amounted to two *pikul* of guns.

124 Tarling (ed.) *The Cambridge History of Southeast Asia*, vol. 1, pp. 383–5, 388; J. M. Gullick, *Indigenous Political Systems of Western Malaya* (London, 1988), pp. 120–2.

125 S. St. John, *The Life of Sir James Brooke* (Edinburgh, 1879), pp. 160–3; Capt. H. Keppel, *A Visit to the Indian Archipelago in H.M.S. 'Maeander'*, vol. 1 (London, 1853), pp. 126–33.

126 Governor Captain-General (Manila) to the *Secretario de Estado*, 16 January 1772, cited in Warren, *The Sulu Zone*, p. 170.

127 Turnbull, *The Straits Settlements*, p. 245.

128 Raffles, *History of Java*, vol. 1, p. 296; Anon., 'Journal of an Excursion to the Native Provinces of Java', *JIAEA*, 9 (1854), p. 81; P. B. R. Carey, *Babad Dipanagara: An Account of the Outbreak of the Java War 1825–1830* (Kuala Lumpur, 1981), pp. 274–6.

129 CO 144/24/6329, Callaghan to Cardwell, 23 April 1865.

130 CO 144/56/18592, 'The Export of Arms and Ammunition to the Territories of the British North Borneo Company', P. Leys, Administrator of Labuan, to Lord Kimberley, Secretary of State for the Colonies, 4 September 1883, pp. 5, 9–12.

131 CO 144/71/5245, 'Importation of arms from Straits', T. H. Sanderson (Foreign Office) to the Under Secretary of State, Colonial Office, 13 March 1897.

132 Capt. C. C. Saxton, 'The War in Malacca', *Colburn's United Service Magazine*, 1 (1874), pp. 354–5.

133 E. M. Chew, 'The Naning War 1831–1832: Colonial Authority and Malay Resistance in the Early Period of British Expansion', *MAS*, 32:2 (1998).

134 SSR, V41, Colonel Protheroe to Resident Councillor, Malacca, 23 August 1865.

135 CO 273/67/8353, Governor Sir H. Ord to Lord Kimberley, Secretary of State for the Colonies, 24 July 1873, and 'Proclamation prohibiting export of arms to the west coast of Malay Peninsula', 30 July 1873; CO 273/81/581, 'Ordinance 11/1875 for Prohibiting the Sale of Arms and Ammunition', 17 December 1875; cf. A. Harfield, *British and Indian Armies in the East Indies 1685–1935* (Chippenham, 1984), pp. 228–69.

136 ADM 125/133, Extract from *Singapore Free Press*, April 1847; Warren, *The Sulu Zone*, pp. 41, 74–6, 151–3, 182–6.

137 Logan (ed.) 'The Piracy and Slave Trade of the Indian Archipelago', *JIAEA*, 3 (1849), pp. 585–7 and 4 (1850), p. 194; J. T. Thomson, 'Description of the Eastern Coast of Johore and Pahang, and Adjacent Islands', *JIAEA*, 5 (1851), p. 137; CO 144/24/6329, Callaghan to Cardwell, 23 April 1865; cf. Warren, *The Sulu Zone*, pp. 156–8.

138 IO, F/4/1724 (69433), 'Report on Piracy in the Straits Settlements', Presgrave to Murchison, 5 December 1828; SSR, R3, Bonham to Prinsep, 23 April 1835; H. St. John, *The Indian Archipelago* (London, 1835), vol. 2, pp. 160–2; Logan (ed.) 'The Piracy and Slave Trade of the Indian Archipelago', *JIAEA*, 3 (1849), p. 585; *SFP*, 12 November 1835; *SFP*, 25 May 1843; Rutter, *The Pirate Wind*, p. 231.

139 ADM 125/133, Presgrave to Murchison, 5 December 1828; J. Anderson, *Acheen and the Ports on the North and East Coasts of Sumatra* (London, 1840; reprinted Kuala Lumpur, 1971), pp. 54, 171–2; J. H. Moor, *Notices of the Indian Archipelago and Adjacent Countries* (Singapore, 1837; reprinted London, 1968), pp. 15, 73–4; Logan (ed.) 'The Piracy and Slave Trade of the Indian Archipelago', *JIAEA*, 4 (1850), p. 144; *Hikayat Abdullah*, trans. Hill, pp. 161–2. By the mid-1820s, large fleets of Bugis prahus were arriving in Singapore, laden with cargoes of sarongs, gold dust, beeswax, bird's nests, agar-agar and other Straits produce, individual consignments being worth as much as $30,000. Munshi Abdullah also depicts the Bugis as herding droves of slaves through the streets of Singapore in the early days, selling them to Malay, Chinese, and Indian clients from their prahus moored in the harbour. This trade conducted by the Bugis remained, for many years, one of the mainstays of Singapore's prosperity and its major indigenous commercial link with the islands of the Archipelago.

140 CO 144/71/5245, L. P. Beaufort, Governor of British North Borneo Territory, to Sir C. B. H. Mitchell, High Commissioner for Borneo, 8 January 1897.

141 CO 144/71/5245, Report by F. O. Maxwell, Acting Consul for Borneo, 4 January 1897.

142 In 1865, $500 (or £36) worth of Straits produce could buy: 17 *pikuls* of brass guns; one *pikul* of opium; or five adult slaves. See Trocki, *Opium, Empire and the Global Political Economy*, p. 82; and Warren, *The Sulu Zone*, p. 114.

143 Tarling (ed.) *The Cambridge History of Southeast Asia*, vol. 1, pp. 389–92.

144 See Trocki, *Opium, Empire and the Global Political Economy*, pp. 56–7, 138–9. Mainland Southeast Asian states had banned the import and use of opium at the beginning of the nineteenth century. Following a period of upheaval, Burma, Siam, and Vietnam had emerged under strong, self-confident, and somewhat traditional rulers, who saw the eradication of opium as necessary for the expulsion of foreigners, especially Europeans. Initially, these states managed to resist the pressures of British commercial and military expansion. Ultimately, whether by outright annexation, diplomatic pressure, or the corruption of their own subjects and officials, all were compelled to accept European colonial authority, and with it, opium.

145 V. B. Lieberman, *Burmese Administrative Cycles: Anarchy and Conquest, c.1580–1760* (Princeton, 1987), pp. 126–7, 221.

146 *The Royal Orders of Burma, 1598–1885: Part 6, 1807–1810*, trans. Than Tun (Kyoto, 1987); Charney, *Southeast Asian Warfare*, p. 246. These traders obtained their guns presumably through trade with Chinese and European gun-runners, since the only gun foundry in the country was in royal hands and did not produce small arms until the mid-nineteenth century.

147 Charney, *Southeast Asian Warfare*, pp. 67, 247.

148 M. Thant, *The Making of Modern Burma* (Cambridge, 2001), pp. 104–13. The Kanaung Prince was himself a senior military commander who had experienced the superiority of Western arms in the Second Anglo-Burmese War. Apart from small arms, heavy armaments were imported along with ten steamers, which became instrumental in maintaining internal security.

149 *Elgin MSS*, Elgin to Secretary of State for India, 9 December 1862.
150 D. R. SarDesai, *British Trade and Expansion in Southeast Asia 1830–1914* (Delhi, 1977), pp. 135–7, 181, 207; D. K. Fieldhouse, *Economics and Empire 1830–1914* (London, 1973), p. 184.
151 'Arms and Ammunition: Smuggling Arms into Upper Burmah', *The Ironmonger*, September 1879, p. 402; *PP*, C 4614 (1886), Telegram from Dufferin to Churchill, 4 August 1885. The 'arms' eventually seized as 'prizes of war' from the Mandalay palace in 1885 included hundreds of muzzle-loading and breech-loading weapons, and one German artillery piece manufactured by Krupp.
152 See M. Htin Aung, *Lord Randolf Churchill and the Dancing Peacock: British Conquest of Burma 1885* (Delhi, 1990).
153 Tarling (ed.) *The Cambridge History of Southeast Asia*, vol. 1, pp. 382–3. When the Thai capital Ayuthia finally capitulated to the Burmese in 1767, the city's armoury contained 1,000 muskets inlaid with gold and silver tracery, more than 10,000 regular muskets, and a whole range of larger guns suitable for mounting on elephants, carriages, and war-boats. Such weaponry had been manufactured locally as well as procured from China, Laos, India, and Europe.
154 SSR, A11, President's Minute (W. E. Phillips), 30 November 1816; Lady S. Raffles, *Memoir of the Life and Public Services of Sir Thomas Stamford Raffles* (London, 1830; reprinted Singapore, 1991), p. 50.
155 SSR, A48, Penang Council to Superintendent of Police, 17 and 20 September 1828; SSR, I38, J. Anderson, Secretary to Government, to Superintendent of Police, 20 September 1828; SSR, U1, R. Ibbetson, Resident (Singapore), to Deputy Resident (Prince of Wales Island), 22 December 1831.
156 IO, E/4/841, Minié Rifles presented to the King of Siam, August–September 1856.
157 Tagliacozzo, *Secret Trades, Porous Borders*, pp. 298–9.
158 SSR, GD/C 1, C. Clementi-Smith, Governor of the Straits Settlements, to the Earl of Derby, 5 January 1885.
159 CO 273/133/3246 and 273/137/11205, Derby and Granville to Clementi-Smith (undated). Lord Derby had been Colonial Secretary in Lord Salisbury's Conservative Government (1885–86), while Lord Granville was Gladstone's appointee in his third Liberal Ministry (1886). Ideological differences did not fundamentally alter policy; realism dictated the need to preserve British influence and defend British interests.
160 Trocki, *Opium, Empire and the Global Political Economy*, pp. 48–51.
161 See, for example, CO 273/2, 'Report on the Supposed Sale of Arms to Hostile Powers', 21 March 1859, providing a list of such firms, both European and local, quantities of arms, ships and shipping destinations.
162 C. M. Turnbull, *A History of Singapore 1819–1975* (Kuala Lumpur, 1977), p. 43. According to Turnbull, Sir Jose D'Almeida was a prominent merchant who ran 'one of the oldest and most respected of Singapore business houses'. The firm of José D'Almeida and Sons went bankrupt in 1864, however, as a result of the depression of the 1860s.
163 For instance, see *The Straits Times*, 6 July 1888, and *Bintang Timor*, 19 November 1894, cited in Tagliacozzo, *Secret Trades, Porous Borders*, p. 291.
164 Raffles, *Memoir of the Life and Public Services of Sir Thomas Stamford Raffles*, p. 74.

165 J. Dalton, 'On the Present State of Piracy, Amongst These Islands, and the Best Method of Its Suppression', in Moor, *Notices of the Indian Archipelago*, p. 26.

166 G. W. Earl, *The Eastern Seas, or Voyages and Adventures in the Indian Archipelago in 1832, 1833, 1834* (London, 1837), p. 444.

167 S. St. John, *Life in the Forests of the Far East* (London, 1862), vol. 2, p. 203; Keppel, *A Visit to the Indian Archipelago*, vol. 1, p. 59.

168 Warren, *The Sulu Zone*, pp. 51–3.

169 CO 144/42, Bulwer to Carnarvon, 4 July 1874; and Statement of John Dill Ross, Enclosure 2 in No. 1, Bulwer to Kimberley, January 1874; cf. CO 144/31, Pope Hennessey to Granville, 23 January 1870. It was a lucrative business: Ross had begun trading as part owner of a small, heavily-mortgaged schooner, but had soon acquired sufficient funds to purchase the schooner outright, before replacing it with a brig and then a barque. Ross was also able to reinvest the proceeds of the trade in several brick shops and money-lending establishments at Labuan, where he provided capital—a few hundred dollars at a time, at high rates of interest—to Chinese merchants. At a time when piratical attacks were rife, Ross' ability and courage earned the respect of indigenous sea-captains (*nakhodas*).

170 CO 144/42, Bulwer to Carnarvon, 25 June, 14 July, and 17 July 1874; cf. Warren, *The Sulu Zone*, pp. 114–17. Kampong German, far up Sandakan Bay, served as a centre for the running of rifles, muskets, and gunpowder, the transhipment of other goods, and direct trade with Sulu. Consignments of arms, opium, textiles, and tobacco were shipped directly from Singapore to Tawi-Tawi, for example, and exchanged for slaves at the rate of 30 pesos per person. Schuck then shipped the slaves and remaining trade goods to Jolo and bartered them for mother-of-pearl shell.

171 CO 273/156/5810, 5880, 8626, and 163/5315, 'Export of Arms from Portuguese Possessions in Timor Islands', 23 March 1888 to 12 May 1889.

172 Turnbull, *A History of Singapore*, pp. 94, 102. Choon Bock & Co. was established by a Malacca *Peranakan* family with shipping interests and long-standing ties with the European companies. Tan Keong Saik inherited the business on the death of his uncle Choon Bock in 1880; along with two other influential *Peranakan* entrepreneurs, Tan Jiak Kim and Lee Cheng Yam, and the directors of Mansfields, he founded the Straits Steamship Company (1890) that was to dominate coastal trade between the British and Dutch colonial spheres into the early twentieth century. Tan Kim Ching, the son of local merchant-philanthropist Tan Tock Seng, together with the European agency house Guthries, set up the Tanjong Pagar Dock Company in 1864. This was to be the precursor of the Tanjong Pagar Dock Board (1905) and then the Singapore Harbour Board (1913), a corporate statutory board that was for decades the most important public utility in Singapore.

173 U.S. National Archives, Consular Despatches, Manila (1817–40), G. W. Hubbell to John Quincy Adams, Secretary of State, 31 December 1823, cited in Warren, *The Sulu Zone*, p. 52. John Quincy Adams (1757–1848) served as Secretary of State (1817–25) in the cabinet of U.S. President James Monroe, where he is credited for formulating the Monroe Doctrine. The son of the 2nd U.S. President John Adams, John Quincy Adams was himself elected the 6th President of the United States (1825–29) before deciding, unusually,

to continue in public office as an elected Congressman (1831–48) noted for his implacable stance against slavery.

174 Tagliacozzo, *Secret Trades, Porous Borders*, pp. 295ff.
175 Ibid.
176 Ibid., pp. 280–1.
177 'Arms and Ammunition: Smuggling Arms into Upper Burmah', *The Iron-monger*, September 1879, p. 402. The figures were taken from the *Rangoon Gazette*, 22 July 1879.
178 Letter from Lavino to the Governor-General, Netherlands East Indies, 21 June 1876, cited in Tagliacozzo, *Secret Trades, Porous Borders*, p. 317.
179 CO 144/24/4668, Callaghan to Cardwell, 1 April 1865.
180 Tagliacozzo, *Secret Trades, Porous Borders*, pp. 296–7.
181 Ibid., p. 277.
182 McPherson, *The Indian Ocean*, p. 222; Trocki, *Opium, Empire and the Global Political Economy*, pp. 83–5; J. S. Mangat, *A History of the Asians in East Africa c.1886 to 1945* (Oxford, 1969), pp. 2–26, 49–62; A. Farooqui, 'Opium Enterprise and Colonial Intervention in Malwa and Western India, 1800–24', *Indian Economic and Social History Review*, 32:4 (1995), pp. 447–73. In Malwa, for instance, groups of independent merchants, revenue-farmers, money-lenders, and other financial leaders in the major provincial towns rose to prominence through the opium trade. They were able to succeed because they already controlled networks of indebtedness extending into hundreds of villages, which then gave them leverage over the rulers of the states in which they operated. With such economic and political clout, they could invest in opium cultivation and marketing to the East, and reinvest the hefty, multi-million dollar profits from opium sales in wider business ventures in western India. In so doing, they galvanized the forces of capitalism on both sides of India and the Indian Ocean.
183 FO 84/1391, Frere to Granville, 7 May 1873.
184 Mangat, *A History of the Asians in East Africa*, p. 14.
185 FO, Confidential Prints No. 6454, Rodd to Rosebery, 2 May 1893, plus Enclosure 1, Sewa Hajee to Berkeley, 3 December 1891, and Enclosure 2, pp. 201, 207, 212; and No. 7946, pp. 28, 197; Mangat, *A History of the Asians in East Africa*, pp. 50–3; Harman, *Bwana Stokesi*, pp. 155, 179–80.
186 For cases of resurgent piracy in contemporary Southeast Asia, see Cordingly (ed.) *Pirates*, pp. 237–40; and Elliott, 'Pirates of the South China Sea', in *Newsweek Magazine*. For comparative analysis of parallels between past and present scenarios, see J. F. Warren, 'A Tale of Two Centuries: The Global-ization of Maritime Raiding and Piracy in Southeast Asia at the End of the Eighteenth and Twentieth Centuries', in P. Boomgard (ed.) *A World of Water: Rain, Rivers and Seas in Southeast Asian Histories* (Singapore, 2007), pp. 125–52. In Warren's view, 'Just as maritime raiders and slavers became generally active due to global economic development and disruption(s) in Asia in the 1790s, the incidence of piracy, or crime and terrorism on the high seas in Southeast Asia has steadily increased in a time of desperation at the end of the twentieth century; the final decade marked by widespread ethnic and political conflict and the near total collapse of global finan-cial systems and associated regional trade by the late 1990s. ... Nowadays, as the world contracts through ever-increasing connected ventures, a

somewhat different mirror image has appeared on the horizon once again, as new-wave pirates and ship thieves rule the seas of Asia.' One might add that the same holds true for new-age traffickers in drugs, guns, and humans.

Chapter 5   The Arms Trade and War in the Indian Ocean

1 See Parker, *The Military Revolution*, pp. 115–45; Parker (ed.) *Cambridge Illustrated History of Warfare*, p. 9; Kennedy, *The Rise and Fall of the Great Powers*, pp. 147–50; W. H. McNeill, 'European Expansion, Power and Warfare Since 1500', in J. A. de Moor and H. L. Wesseling (eds) *Imperialism and War* (Leiden, 1989), pp. 12–21; and C. Cipolla, *Guns, Sails and Empires: Technological Innovation in the Early Phases of European Expansion 1400–1700* (New York, 1965). Notions of Western technological and military superiority were reinforced by stereotypical images of the tradition-bound, 'superstitious' East, where even the role of firearms and gunpowder in indigenous military culture could be imbued with spiritual significance. For more on the religio-cultural symbolism ascribed to arms and ammunition in 'militarized' Asian and African societies, see G. M. Berg, 'The Sacred Musket: Tactics, Technology, and Power in 18th-Century Madagascar', *Comparative Studies in Society and History*, 27:2 (1985), pp. 261–79; Clancy-Smith, *Rebel and Saint: Muslim Notables, Popular Protest, Colonial Encounters*, p. 154; Fisher and Rowland, 'Firearms in the Central Sudan', pp. 231–5; V. G. Kiernan, *European Empires from Conquest to Collapse, 1815–1960* (London, 1982), p. 143; and Tarling (ed.) *The Cambridge History of Southeast Asia*, vol. 1, pp. 392–5.
2 Marshall, 'Western Arms in Maritime Asia in the Early Phases of Expansion', p. 28.
3 Parker, *The Military Revolution*, pp. 115, 128–9, 136; cf. B. P. Lenman, 'The Weapons of War in 18th-Century India', *Journal of the Society for Army Historical Research*, 46 (1968). Lenman concludes that 'there is enough evidence to suggest that a margin of superiority in military technology in military technology was at least one of the factors contributing to the rise of British supremacy in the subcontinent', but this, too, has to be qualified.
4 H. Compton, *A Particular Account of the European Military Adventurers of Hindustan from 1784–1803* (London, 1893), pp. 385–7.
5 Bayly, *Indian Society and the Making of the British Empire*, p. 100; Cooper, 'The Second Anglo-Maratha War', pp. 12, 63–6, 70; S. Bidwell, *Swords for Hire: European Mercenaries in Eighteenth-century India* (London, 1971), p. 55; P. Barua, 'Military Developments in India 1750–1850', *Journal of Military History*, 58:4 (1994), pp. 607, 611.
6 S. Alavi, *The Sepoys and the Company* (Delhi, 1995), pp. 23–4, 202–20.
7 Lieutenant E. Moor, *A Narrative of the Operations of Captain Little's Detachment during the Late Confederacy in India, against the Nawab Tippoo Sultan Bahadur* (London, 1794), pp. 478–9; A. Beatson, *A View of the Origin and Conduct of the War with Tippoo Sultaun* (London, 1800), p. 158; R. Wigington, *The Firearms of Tipu Sultan 1783–1799* (Hatfield, 1992), pp. 7–8, 12, 32–3; Barua, 'Military Developments in India 1750–1850', p. 603. The great warlord Haidar Ali (Tipu's father) had laid Mysore's military foundations. Appreciating the advantages

of the flintlock over the matchlock, he employed European mercenaries—predominantly French—and must have established facilities for the maintenance and manufacture of such firearms. As early as 1769, following a lightning advance with his cavalry, Haidar's stunning assault on Madras forced a British surrender.

8   J. Weller, *Wellington in India* (London, 1972), pp. 192–4; C. Hibbert, *Wellington: A Personal History* (London, 1997), pp. 42–3.

9   Barua, 'Military Developments in India 1750–1850', pp. 612–13; C. E. Carrington, *The British Overseas: Exploits of a Nation of Shopkeepers* (Cambridge, 1950), p. 438. It is also notable that by the 1790s, Indian polities like the Maratha Confederacy and Mysore were capable of manufacturing their own gunpowder. In the aftermath of the siege of Seringapatam, 520,000 lbs of locally made gunpowder was found in the fortress. During the 1857 Mutiny-Rebellion, the rebels were still able to produce it in many places, for saltpetre remained abundant in the region, refined and manufactured in Bengal and parts of northern India. In the twilight of a 'Gunpowder Age' that preceded the late nineteenth-century invention of smokeless explosives, 'saltpetre was an essential constituent of explosives and India had almost a monopoly of supplies'—supplies that made their way even to the firearms, armies, and battlefields of the American Civil War. See Watt, *The Commercial Products of India*, pp. 972–5.

10   Anon., 'Journal of an Excursion to the Native Provinces of Java', *JIAEA*, 9 (1854), p. 81; Raffles, *History of Java*, vol. 1, p. 296; Carey, *Babad Dipanagara*, pp. 274–6.

11   J. A. de Moor, 'Warmakers in the Archipelago: Dutch Expeditions in Nineteenth Century Indonesia', in de Moor and Wesseling (eds) *Imperialism and War*, pp. 51–2.

12   L. M. Crowell, 'Logistics in the Madras Army circa 1830', *War & Society*, 10:2 (1992), pp. 1–4, 22–4.

13   R. O. Winstedt (ed.) *Malaya* (London, 1925), p. 25. 'Sumatras' are violent squalls of monsoonal wind and rain from the west, typical of the Straits of Malacca and Singapore.

14   Begbie, *The Malayan Peninsula*, pp. 187–9, 220.

15   Saxton, 'The War in Malacca', p. 363.

16   *Hikayat Abdullah*, p. 259; Saxton, 'The War in Malacca', pp. 354–5.

17   Begbie, *The Malayan Peninsula*, pp. 179, 214, 253–7; Andaya and Andaya, *A History of Malaysia*, p. 144; Gullick, *Indigenous Political Systems of Western Malaya*, pp. 120–2. Begbie, the subaltern in the Madras Native Infantry who commanded the six-pounders and sappers during the conflict, acknowledged that '500 [Malays] in the jungle were equivalent to twenty times that number in the plain'.

18   Begbie, *The Malayan Peninsula*, pp. 248–50; cf. Chew, 'The Naning War 1831–1832', pp. 356, 382.

19   Maj. V. Hough, *A Narrative of the March and Operations of the Bengal Column of the Army of Indus in the Expedition to Afghanistan in the Years 1838–1839* (London, 1841); Col. Firebrace, 'On the Errors and Faults in Our Military System', *Colburn's United Service Magazine*, 1 (1843), pp. 55–6; Col. Blacker, 'Cabool', *Colburn's United Service Magazine*, 2 (1843), p. 173.

20   'Personal Narrative of Mootee Ram', *United Service Magazine and Naval and Military Journal*, 2 (1842), p. 409.

21 'An Officer in Her Majesty's Service', 'Journal of the Last Three Months' Operations in Afghanistan', *Colburn's United Service Magazine*, 1 (1843), p. 335; Lieutenant H. M. Greenwood, *Narrative of the Late Victorious Campaign in Afghanistan under General Pollock* (London, 1844).

22 de Moor and Wesseling (eds) *Imperialism and War*, p. 6.

23 Kennedy, *The Rise and Fall of the Great Powers*, p. 150.

24 M. Howard, 'Colonial Wars and European Wars', in de Moor and Wesseling (eds) *Imperialism and War*, p. 221; Kennedy, *The Rise and Fall of the Great Powers*, p. 150. Horatio Herbert Kitchener (1850–1916) rose to national prominence during his second tour of duty in the Sudan (1886–99), being appointed Governor of the British Red Sea territories with the brevet rank of Colonel in 1886, and later Commander-in-Chief (*Sirdar*) of the Egyptian Army in 1892, first with the rank of British Army Brigadier and then Major-General. It was shortly after this appointment that Captain Lugard, fresh from his own tour of duty as British Military Administrator of Uganda, warned Kitchener about small arms proliferation across East Africa, along the Great Lakes and upper reaches of the Nile, into the Sudan. As *Sirdar* of the Egyptian Army, Kitchener moved swiftly to crush the Sudanese Dervishes—separatist forces of the Mahdi—in the Battle of Omdurman (1898). In so doing, he demonstrated the efficacy of the new magazine rifle: the Lee-Enfields of the British opened fire at 2,000 yards and stopped the Dervishes at 800 yards, whereas the Martinis of the Egyptians opened at 1,000 yards and stopped them at 500. He then ordered the occupation of nearby Khartoum, which he transformed into an oasis of good governance in the Sudan. Kitchener's success was rewarded by his elevation to the peerage as Baron (later Earl) Kitchener of Khartoum, and promotion to the highest Army rank of Field Marshal (1910) before, finally, his stint in government as Secretary of State for War (1914–16).

25 Napier's despatch, 12 May 1868, reported in *London Gazette*, 30 June 1868.

26 Caulk, 'Firearms and Princely Power in Ethiopia in the Nineteenth Century', pp. 619–30.

27 R. H. Rainero, 'The Battle of Adowa on 1st March 1896: A Reappraisal', in de Moor and Wesseling (eds) *Imperialism and War*, pp. 189–200; Beachey, 'The Arms Trade in East Africa', p. 462. Experts continue to debate the estimated size of the Ethiopian forces, ranging from a minimum of 73,000 to a maximum of well over 100,000.

28 The Boers also possessed Maxim 'pom-poms' and Krupp artillery. This is a case where indigenous weaponry was arguably superior to British weaponry in the *modern* form, though it is probably instructive to note that the indigenous population was also of European origin. See T. Pakenham, *The Boer War* (London, 1979), pp. 41, 196.

29 'Armaments for the Boers', *Arms and Explosives*, February 1901, p. 19. Another report sets the figure at over 150,000 rifles, excluding 'countless others which had been steadily accumulating from prehistoric times'. See 'The Boer and his Rifle', *Arms and Explosives*, March 1902, p. 36.

30 'The Boer and his Rifle', *Arms and Explosives*, March 1902, p. 37.

31 Keppel, *Gun-running and the Indian North-West Frontier*, pp. 28–9.

32 Moreman, 'The British and Indian Armies and North West Frontier Warfare, 1849–1914', pp. 40, 46–7, 58.

33 de Moor, 'Warmakers in the Archipelago', in de Moor and Wesseling (eds) *Imperialism and War*, pp. 63–5.

34 de Moor and Wesseling (eds) *Imperialism and War*, p. 6.

35 'Birmingham...', *The Ironmonger*, February 1879, pp. 209–10; Guy, 'A Note on Firearms in the Zulu Kingdom', p. 559.

36 See A. Duminy and C. Ballard (eds) *The Anglo-Zulu War: New Perspectives* (Pieter-maritzburg, 1981); Parker (ed.) *Cambridge Illustrated History of Warfare*, p. 245.

37 M. Howard, *War in European History* (Oxford, 1976), p. 121.

38 C. Trebilcock, 'War and the Failure of Industrial Mobilization: 1899 and 1914', in J. M. Winter (ed.) *War and Economic Development: Essays in Memory of David Joslin* (Cambridge, 1975), pp. 139–64; Pakenham, *The Boer War*, p. 574; Parker (ed.) *Cambridge Illustrated History of Warfare*, p. 252. There were many committees of inquiry, and even a Royal Commission on the War in South Africa (1903–4), which drew most of the right lessons. But this was yet another example of the bureaucratic constraint and institutional complacency which prevents learning from experience: the lessons were shelved and forgotten between 1904 and 1914. These included lessons on the industrial mobilization of military resources; in 1915, as in 1899, the British Army experienced acute problems in the supply of ammunition—a very nearly disastrous shortage of shells and fuses—on account of poor judgement and coordination between state and private arms production. The central tactical lesson of the Boer War also appeared to have eluded the military high command: it was that the smokeless, long-range, high velocity, small-bore magazine bullet from rifle or machine-gun—combined with the trench—had decisively swung the balance against attack in favour of defence.

39 J. Lonsdale, 'The Conquest State of Kenya', in de Moor and Wesseling (eds) *Imperialism and War*, pp. 87–120; Mungeam, *British Rule in Kenya*, pp. 72–4; K. Njama and D. L. Barnett, *Mau Mau from Within: Autobiography and Analysis of Kenya's Peasant Revolt* (New York, 1966), pp. 174–6; W. Itote, '*Mau Mau' General* (Nairobi, 1967), pp. 105–6; G. Gikoyo, *We Fought for Freedom* (Nairobi, 1979), p. 55; M. S. Clough, *Mau Mau Memoirs: History, Memory, and Politics* (Boulder, 1998), pp. 134, 148–53. The violence of the rebellion and the brutality of its suppression would even result in deliberate, systematic destruction of Foreign Office documents in 1961 at the direction of Iain Macleod, then Secretary of State for the Colonies, to prevent their eventual transfer into the public domain, where they might be exploited for propaganda purposes by the post-colonial regime. See I. Cobain, O. Bowcott, and R. Norton Taylor, 'Britain destroyed records of colonial crimes', *The Guardian*, 18 April 2012.

40 C. Fourniau, 'Colonial Wars before 1914: The Case of France in Indo-China', and Y-G. Paillard, 'The French Expedition to Madagascar in 1895: Program and Results', in de Moor and Wesseling (eds) *Imperialism and War*, pp. 72–86, 168–88.

41 See R. A. Hunt and R. H. Schultz, Jr. (eds) *Lessons from an Unconventional War* (New York, 1982). For instance, the Viet Cong with the bicycle and the AK-47 outfought the gunships of the U.S. Air Cavalry.

42 N. M. Kamrany and L. B. Poullada, 'The Potential of Afghanistan's Society and Institutions to Resist Soviet Penetration and Domination', Modelling Research Group, University of Southern California, January 1985, pp. 10–20, 58; and *The Military Balance 1985–1986* (London, 1985), p. 119; *Afghanistan Forum*, September 1985, pp. 16–17; and 'Caravans on Moonless Nights: How the CIA Supports and Supplies the Anti-Soviet Guerrillas', *Time Magazine*, 11 June 1985. See also T. Karon, 'Why the War in Afghanistan is a Long Way From Over' and 'What We Learned in Shah-i-Kot', *Time Magazine*, 6 and 14 March 2002; and M. Elliott, 'The Valley of Death', *Time Magazine*, 9 March 2002.

Bibliography

Archival Sources

I. The Birmingham Gun Barrel Proof House, Banbury Street, Birmingham

Proof House Records:
Proceedings of the Guardians (PG);
Minute Books, 1855–1949
Miscellaneous Correspondence
Select Literature

Periodicals:
Arms and Explosives, 1892–1920
The Sporting Goods Review

II. Birmingham Reference Library (Local History Section), Birmingham

'Barbara Smith Archive':
Annotated bibliography of sources on Birmingham's industrial history, 1858–1900

III. Cambridge University Library, West Road, Cambridge

Government Publications:
Hansard's Parliamentary Debates
House of Commons Parliamentary Papers
Reports from Committees of the House of Commons
World Military Expenditures and Arms Transfers, United States Arms Control and
   Disarmament Agency (ACDA)

Periodicals:
Journal of the Royal United Service Institution, 1857–1914
United Service Magazine and Naval and Military Journal, 1829–41, continued as
   Colburn's United Service Magazine, 1842–1920

IV. The British Library (Oriental and India Office Collections), St. Pancras, London

India Office Records:
E/4, Despatches to Bombay, Madras and Bengal
P/D/74, Bombay Political and Secret Consultations/Proceedings
L/AG/18/1/30–34 & L/AG/18/2/1–35, Cash Journals and Ledgers of the East
   India Company & Miscellaneous Trade Statistics and Accounts, *c.*1730–1860
L/MAR/C, Minutes of the East India Company Committee of Shipping
L/MIL, Records of the India Office Military Department: Military Collections,
   *c.*1850–1920s (L/MIL/7/4, 48, 158, 267, 403); Committee Proceedings,

Compilations, and Miscellaneous Matters (L/MIL/1, L/MIL/5); Regional Sources, including the Northwest Frontier (L/MIL/17/13), the Persian Gulf (L/MIL/17/15), Africa (L/MIL/17/17), and Southeast Asia (L/MIL/17/19)

L/P&S, Records of the India Office Political and Secret Department: Home Correspondence, 1807–1911 (L/P&S/3); Secret, Political, Secret and Political Correspondence with India, 1756–1874, 1792–1874, 1875–1911 (L/P&S/5–7); Correspondence on Areas Outside India, 1781–1911 (L/P&S/9); Political and Secret Subject Files, 1902–20 (L/P&S/10/101–426); Political and Secret Memoranda, 1840–1920 (L/P&S/18/A–D)

R/15/1-6, Records of the British Residency and Agencies in the Persian Gulf, 1763–1914

Private Papers:
Curzon MSS, Papers of George Nathaniel Curzon, 1st Marquess Curzon of Kedleston
Elgin MSS, Papers of Victor Bruce, 9th Earl of Elgin
Kilbracken MSS, Papers of Arthur Godley, 1st Baron Kilbracken

Periodicals:
Proceedings (later *Journal*) *of the United Service Institution of India*, 1871–1914

V. The British Library (Newspaper Library), Colindale Avenue, London

Birmingham Daily Post
Daily Telegraph
London Gazette
Pall Mall Gazette
Straits Settlements Government Gazette
The Ironmonger
The Times
The Times of India

VI. The Public Record Office, Kew, London

Admiralty Records:
ADM/16/1089, Correspondence on Arms Traffic in the Persian Gulf

Board of Ordnance Records:
WO 44 & 46, Letter Books (containing Official Correspondence and Documents)
WO 47, Minute Books (pertaining to the Board, or a Committee, or Surveyor General)
WO 51 & 52, Bill Books (detailing Officials' payments, suppliers' invoices, etc)
WO 55, General Archive (containing warrants, deliveries to, and receipts from, gunmakers)

Colonial Office Records:
CO 879, Correspondence on South and East Africa

Board of Customs Records:
CUST 9/3 & 9/107, Reports on British Firearms Exports

Foreign Office Records:
FO 2, Correspondence on West Africa, Uganda and Central Africa from 1893
FO 12 & 71, Correspondence relating to Borneo and Sulu, 1842–1905
FO 41, Correspondence on East India, 1776–97
FO 54, Correspondence on Muscat, 1834–1905
FO 60 & 248–251, Correspondence and Consular/Embassy Archives on Persia, 1807–1905
FO 84, Correspondence on Slave Trade and African Departments, 1816–92
FO 107, Correspondence relating to East Africa and Zanzibar, 1893–1905
FO 371 & 428, Confidential Print/Correspondence on Arms Traffic from 1906

Records of the Royal African Company:
T 70/75–150, Minute Books
T 70/930–32, Invoice Books

War Office Records:
Director of Army Contracts (DAC) Reports, 1889–1910

Private Papers:
Egremont MSS, Papers of Charles Wyndham, 2nd Earl of Egremont

VII. National Archives of Singapore, Canning Rise, Singapore

European MSS Collection:
Records of the East India Company and India Office
Straits Settlements Records (SSR)

VIII. The National Library, Victoria Street, Singapore

Straits Settlements Records
Singapore Free Press (SFP)

IX. National University of Singapore Central and History Department Libraries

Published Government Records:
Colonial Office Records (CO 144 & 273)
Straits Settlements Factory Records

Periodicals:
Journal of the Indian Archipelago and Eastern Asia (JIAEA) (ed. J. R. Logan), 1847–63

Published Documentary Collections:
Cowan, C. D. (ed.), 'Early Penang and the Rise of Singapore 1805–1832: Documents from the Manuscript Records of the East India Company', *Journal of the Malayan Branch, Royal Asiatic Society*, 23:2 (1950)

The Royal Orders of Burma, 1598–1885: Part 6, 1807–1810, edited and translated by Than Tun (Kyoto, 1987)

X. Library of the S. Rajaratnam School of International Studies, Nanyang Technological University, Singapore

Military-Strategic Reports and Reviews:
Jane's Intelligence Review
Jane's Terrorism and Security Monitor
MarEx Newsletter
Reports of the Environmental Investigation Agency
Small Arms Survey
Stockholm International Peace Research Institute (SIPRI) Yearbook
Terrorism Monitor
The Military Balance

Newspapers:
The Straits Times
Today

Secondary Sources

I. General

Adas, M. *Machines as the Measure of Men: Science, Technology, and Ideologies of Western Dominance* (Ithaca, 1989)
—— *Prophets of Rebellion: Millenarian Protest Movements Against the European Colonial Order* (Cambridge, 1987)
Allen, G. C. *The Industrial Development of Birmingham and the Black Country* (1929; reprinted London, 1966)
Bailey, D. W. *British Military Longarms, 1715–1865* (London, 1986)
Bailey, D. W. and Nie, D. A. *English Gunmakers: The Birmingham and Provincial Gun Trade in the Eighteenth and Nineteenth Century* (London, 1978)
Bayly, C. A. (ed.) *Atlas of the British Empire* (London, 1989)
—— *Imperial Meridian: The British Empire and the World 1780–1830* (London, 1989)
—— *The Birth of the Modern World 1780–1914: Global Connections and Comparisons* (Oxford, 2004)
Berend, I. T. *The European Periphery and Industrialization 1780–1914* (Paris, 1982)
Bitzinger, R. A. (ed.) *The Modern Defence Industry: Political, Economic, and Technological Issues* (Santa Barbara, 2009)
Black, J. *European Warfare 1660–1815* (London, 1994)
Blackbourn, D. *Fontana History of Germany 1780–1918: The Long Nineteenth Century* (London, 1997)
Blacker, J. 'Understanding the Revolution in Military Affairs: A Guide to America's Twenty-first Century Defence', *Defence Working Paper No. 3*, Progressive Policy Institute (Washington, DC, 1997)
Blackmore, H. L. *A Dictionary of London Gunmakers* (Oxford, 1986)
—— *British Military Firearms, 1650–1850*, revised edition (London, 1994)

Bonn International Center for Conversion, *Conversion Survey 1997: Global Disarmament and Disposal of Surplus Weapons* (Oxford, 1997)

Bose, S. *A Hundred Horizons: The Indian Ocean in the Age of Global Empire* (Cambridge, Massachusetts, 2006)

Boutwell, J., Klare, M. T. and Reed, L. W. (eds) *Lethal Commerce: The Global Trade in Small Arms and Light Weapons* (Cambridge, Massachusetts, 1995)

Boxer, C. R. *The Dutch Seaborne Empire 1600–1800* (New York, 1965)

—— *The Portuguese Seaborne Empire 1415–1825* (New York, 1969)

Braudel, F. *The Mediterranean and the Mediterranean World in the Age of Philip II*, translated by S. Reynolds, 2 volumes (London, 1966)

Brewer, J. *The Sinews of Power: War, Money and the English State, 1688–1783* (Cambridge, Massachusetts, 1988)

Burne, A. H. 'Cannons at Crécy', *English Historical Review*, 77 (1962)

Cain, P. J. and Hopkins, A. G. *British Imperialism: Innovation and Expansion 1688–1914* (London, 1993)

Callwell, C. E. *Small Wars: Their Principles and Practice* (London, 1906)

Capie, D. *Small Arms Production and Transfers in Southeast Asia* (Canberra, 2002)

Carrington, C. E. *The British Overseas: Exploits of a Nation of Shopkeepers* (Cambridge, 1950)

Chamberlain, M. E. *'Pax Britannica'?: British Foreign Policy, 1789–1914* (London, 1988)

Chandra, S. (ed.) *The Indian Ocean: Explorations in History, Commerce and Politics* (Delhi, 1987)

Chaudhuri, K. N. *Asia before Europe: Economy and Civilization of the Indian Ocean from the Rise of Islam to 1750* (Cambridge, 1990)

—— *Trade and Civilization in the Indian Ocean: An Economic History from the Rise of Islam to 1750* (Cambridge, 1985)

Cipolla, C. *Guns, Sails and Empires: Technological Innovation in the Early Phases of European Expansion 1400–1700* (New York, 1965)

Clarence-Smith, W. G. (ed.) *The Economics of the Indian Ocean Slave Trade in the Nineteenth Century* (London, 1989)

Coleman, D. C. 'Gentlemen and Players', *Economic History Review*, 26:1 (1973)

Collier, B. *Arms and the Men: The Arms Trade and Governments* (London, 1980)

Collins, B. and Robbins, K. (eds) *British Culture and Economic Decline* (London, 1990)

Cooper, N. *The Business of Death: Britain's Arms Trade at Home and Abroad* (London, 1997)

Cordingly, D. (ed.) *Pirates: Terror on the High Seas—from the Caribbean to the South China Sea* (Atlanta, 1996)

Cornish, P. *The Arms Trade and Europe* (London, 1995)

Corvisier, A. and Childs, J. (eds) *A Dictionary of Military History and the Art of War* (Oxford, 1994)

Davenport-Hines, R. P. T. *Dudley Docker: The Life and Times of a Trade Warrior* (Cambridge, 1984)

de Moor, J. A. and Wesseling, H. L. (eds) *Imperialism and War* (Leiden, 1989)

de Vries, J. 'The Industrious Revolution and the Industrial Revolution', *Journal of Economic History*, 54 (1994)

Dijk, W. O. 'The VOC's Gunpowder Factory, c.1620–1660', *The International Institute for Asian Studies Newsletter Online*, 26 (November 2001)

Elbaum, B. and Lazonick, W. *The Decline of the British Economy* (Oxford, 1986)

Elleman, B., Forbes, A. and Rosenberg, D. (eds) *Piracy and Maritime Crime: Historical and Modern Case Studies* (Newport, Rhode Island, 2010)

Eltis, D. *The Military Revolution in Sixteenth-Century Europe* (London, 1995)

Fieldhouse, D. K. *Economics and Empire 1830–1914* (London, 1973)

Forbes, V. L. *The Maritime Boundaries of the Indian Ocean Region* (Singapore, 1995)

Friedberg, A. L. *The Weary Titan: Britain and the Experience of Relative Decline* (Princeton, 1988)

Fries, R. I. 'British Response to the American System: The Case of the Small-Arms Industry after 1850', *Technology and Culture*, 16 (1975)

Galbraith, J. S. 'The "Turbulent Frontier" as a Factor in British Expansion', *Comparative Studies in Society and History*, 2 (1960)

Gallagher, J. and Robinson, R. 'The Imperialism of Free Trade', *Economic History Review*, 6:1 (1953)

Gemery, H. A. and Hogendorn, J. S. 'Technological Change, Slavery and the Slave Trade', in C. Dewey and A. G. Hopkins (eds) *The Imperial Impact: Studies in the Economic History of India and Africa* (London, 1978)

Gillard, D. *The Struggle for Asia, 1828–1914* (London, 1977)

Gosse, P. *The History of Piracy* (New York, 1934)

Graham, G. S. *Great Britain and the Indian Ocean: A Study in Maritime Enterprise 1810–1850* (Oxford, 1967)

Gunaratna, R. (ed.) *Terrorism in the Asia-Pacific: Threat and Response* (Singapore, 2003)

Hall, R. *Empires of the Monsoon: A History of the Indian Ocean and its Invaders* (London, 1996)

Hanson, C. E. *The Northwest Gun* (Lincoln, Nebraska, 1956)

Harding, D. F. *Small Arms of the East India Company 1600–1856*, vol. 1: *Procurement and Design*; and vol. 2: *Catalogue of Patterns* (London, 1997)

Harlow, V. T. *The Founding of the Second British Empire 1763–1793*, vol. 1: *Discovery and Revolution* (London, 1952)

—— *The Founding of the Second British Empire 1763–1793*, vol. 2: *New Continents and Changing Values* (London, 1964)

Harris, C. (ed.) *The History of the Birmingham Gun-barrel Proof House*, 2nd edition (Birmingham, 1949)

Haycock, R. and Neilson, K. (eds) *Men, Machines, and War* (Waterloo, Ontario, 1988)

Headrick, D. R. *The Tentacles of Progress: Technology Transfer in the Age of Imperialism* (New York, 1988)

—— *The Tools of Empire: Technology and European Imperialism in the Nineteenth Century* (New York, 1981)

Hibbert, C. *Wellington: A Personal History* (London, 1997)

Hopkins, A. G. (ed.) *Globalization in World History* (London, 2002)

Howard, M. *War in European History* (Oxford, 1976)

Hunt, R. A. and Schultz, R. H. (eds) *Lessons from an Unconventional War* (New York, 1982)

Hyam, R. *Britain's Imperial Century* (London, 1993)

Israel, J. *The Dutch Republic: Its Rise, Greatness, and Fall 1477–1806* (Oxford, 1995)

Kaldor, M. *The Baroque Arsenal* (London, 1982)

Kaplan, R. *Monsoon: The Indian Ocean and the Future of American Power* (New York, 2010)

Kennedy, P. *The Rise and Fall of the Great Powers: Economic Change and Military Conflict from 1500 to 2000* (New York, 1987)

Kiernan, V. G. *European Empires from Conquest to Collapse, 1815–1960* (London, 1982)

Kleinen, J. and Osseweijer, M. (eds) *Pirates, Ports, and Coasts in Asia: Historical and Contemporary Perspectives* (Singapore, 2010)

Knox, M. and Murray W. (eds) *The Dynamics of Military Revolution 1300–2050* (Cambridge, 2001)

Kolodziej, E. A. *Making and Marketing Arms: The French Experience and Its Implications for the International System* (Princeton, 1987)

Krause, K. *Arms and the State: Patterns of Military Production and Trade* (Cambridge, 1992)

Kuitenbrouwer, M. *The Netherlands and the Rise of Modern Imperialism: Colonies and Foreign Policy, 1870–1902* (New York, 1991)

Laurance, E. J. *The International Arms Trade* (New York, 1992)

Leach, D. E. *Arms for Empire: A Military History of the British Colonies in North America, 1607–1763* (New York, 1973)

Levine, A. L. *Industrial Retardation in Britain 1880–1914* (London, 1967)

Lorge, P. A. *The Asian Military Revolution: From Gunpowder to the Bomb* (Cambridge, 2008)

Lumpe, L. (ed.) *Running Guns: The Global Black Market in Small Arms* (London, 2000)

Marder, A. J. 'The English Armament Industry and Navalism in the Nineties', *Pacific Historical Review* (1938)

Markham, G. *Guns of the Empire: Firearms of the British Soldier, 1837–1987* (London, 1990)

Markusen, A. R. and Costigan, S. S. (eds) *Arming the Future: A Defence Industry for the Twenty-first Century* (New York, 1999)

Marshall, P. J. 'Western Arms in Maritime Asia in the Early Phases of Expansion', *Modern Asian Studies*, 14:1 (1980)

Martin, E. B. and C. P. *Cargoes of the East: The Ports, Trade and Culture of the Arabian Seas and Western Indian Ocean* (London, 1978)

McIntyre, W. D. *The Imperial Frontier in the Tropics, 1865–75* (London, 1967)

McNeill, W. H. *The Pursuit of Power: Technology, Armed Force, and Society since AD 1000* (Chicago, 1982)

McPherson, K. *The Indian Ocean: A History of People and the Sea* (Delhi, 1993)

Morrill, J. *The Nature of the English Revolution* (London, 1993)

Needham, J. *Science and Civilization in China, 5: Chemistry and Chemical Technology: 7: Military Technology: The Gunpowder Epic* (Cambridge, 1986)

Ness, G. D. and Stahl, W. 'Western Imperialist Armies in Asia', *Comparative Studies in Society and History*, 19:1 (1977)

Noel-Baker, P. *The Private Manufacture of Armaments* (London, 1936; reprinted 1972)

Orenstein, R. (ed.) *Elephants: The Deciding Decade* (New York, 1997)

Pam, D. O. *The Royal Small Arms Factory, Enfield and Its Workers* (Enfield, 1998)

Parker, G. (ed.) *Cambridge Illustrated History of Warfare: The Triumph of the West* (Cambridge, 1995)

—— *The Military Revolution: Military Innovation and the Rise of the West, 1500–1800* (Cambridge, 1988)

Parkinson, C. N. *Trade in the Eastern Seas, 1793–1813* (Cambridge, 1937)

—— *War in the Eastern Seas, 1793–1815* (London, 1954)

Payne-Gallway, R. *The Handgun in Relation to the Crossbow* (London, 1903)

Perrin, N. *Giving Up the Gun: Japan's Reversion to the Sword, 1543–1879* (Boston, Massachusetts, 1979)

Pollard, H. B. C. *A History of Firearms* (London, 1926)

Porter, P. *Military Orientalism: Eastern War Through Western Eyes* (New York, 2009)

Prescott, J. R. V. *Political Frontiers and Political Boundaries* (London, 1987)

Prince, A. E. 'The Importance of the Campaign of 1327', *English Historical Review*, 50 (1935)

Puype, J. P. and van der Hoeven, M. (eds) *The Arsenal of the World: The Dutch Arms Trade in the Seventeenth Century* (Amsterdam, 1996)

Ray, R. K. 'Asian Capital in the Age of European Expansion: The Rise of the Bazaar, 1800–1914', *Modern Asian Studies*, 29:3 (1995)

Reader, W. J. *Imperial Chemical Industries: A History*, vol. 1 (London, 1970)

Roads, C. H. *The British Soldier's Firearm, 1850–1864* (London, 1964)

Rogers, C. J. 'The Efficacy of the English Longbow: A Reply to Kelly DeVries', *War in History*, 5 (1998)

—— 'Military Revolutions of the Hundred Years' War', in C. J. Rogers (ed.) *The Military Revolution Debate* (Boulder, 1995)

Rosenberg, N. (ed.) *The American System of Manufactures* (Edinburgh, 1969)

Salzman, L. F. *English Industries of the Middle Ages* (1823; reprinted London, 1964)

Sampson, A. *The Arms Bazaar: The Companies, The Dealers, The Bribes: From Vickers to Lockheed* (London, 1977)

Saul, S. B. 'The Market and the Development of the Mechanical Engineering Industries in Britain, 1860–1914', *Economic History Review*, 20:1 (1967)

Showalter, D. *Railroads and Rifles: Soldiers, Technology and the Unification of Germany* (Hamden, 1975)

Skennerton, I. *The Lee-Enfield Story* (London, 1993)

Smith, B. M. D. 'The Galtons of Birmingham: Quaker Gun Merchants and Bankers, 1702–1831', *Business History*, 9:2 (1967)

Smith, M. R. *Harpers Ferry Armory and the New Technology: The Challenge of Change* (Ithaca, 1977)

Starkey, D., van Eyck van Heslinga, E. S. and de Moor, J. A. (eds) *Pirates and Privateers: New Perspectives on the War on Trade in the Eighteenth and Nineteenth Centuries* (Exeter, 1997)

Stone, L. (ed.) *An Imperial State at War: Britain from 1689 to 1815* (London, 1994)

Strachan, H. *From Waterloo to Balaclava: Tactics, Technology, and the British Army 1815–1854* (Cambridge, 1985)

—— 'Military Modernization 1789–1918', in T. C. W. Blanning (ed.) *The Oxford Illustrated History of Modern Europe* (Oxford, 1996)

Subrahmanyam, S. *The Portuguese Empire in Asia, 1500–1700* (London, 1993)

Supple, B. 'Fear of Failing: Economic History and the Decline of Britain', *Economic History Review*, 47:3 (1994)

Swope, K. M. 'Crouching Tigers, Secret Weapons: Military Technology Employed during the Sino-Japanese-Korean War, 1592–1598', *The Journal of Military History*, 69 (January 2005)

Tahtinen, D. R. *Arms in the Indian Ocean: Interests and Challenges* (Washington, DC, 1977)

Tan, A. T. H. (ed.) *The Global Arms Trade: A Handbook* (London, 2010)

Thucydides. *The Peloponnesian War*, translated by R. Warner (Harmondsworth, 1972)

Tout, T. F. 'Firearms in England in the Fourteenth Century', *English Historical Review*, 26 (1911)

Trebilcock, C. 'British Armaments and European Industrialization, 1890–1914', *Economic History Review*, 26:2 (1973)

—— '"Spin-off" in British Economic History: Armaments and Industry, 1760–1914', *Economic History Review*, 22:3 (1969)

—— *The Industrialization of the Continental Powers, 1780–1914* (London, 1981)

—— *The Vickers Brothers: Armaments and Enterprise 1854–1914* (London, 1977)

—— 'War and the Failure of Industrial Mobilization: 1899 and 1914', in J. M. Winter (ed.), *War and Economic Development: Essays in Memory of David Joslin* (Cambridge, 1975)

Trocki, C. A. *Opium, Empire and the Global Political Economy: A Study of the Asian Opium Trade* (London, 1999)

Tully, M. 'The Arms Trade and Political Instability in South Asia', *Churchill Review* (1995)

Vali, F. A. *Politics of the Indian Ocean Region* (New York, 1976)

Wallerstein, I. *The Modern World System, 2: The Consolidation of the European World Economy, 1600–1750* (New York, 1980)

II. Africa

Alpers, E. A. *Ivory and Slaves in East Central Africa: Changing Patterns of International Trade to the Later Nineteenth Century* (London, 1975)

Atmore, A., Chirenje, J. M. and Mudenge, S. I. 'Firearms in South Central Africa', *Journal of African History*, 12:4 (1971)

Atmore, A. and Sanders, P. 'Sotho Arms and Ammunition in the Nineteenth Century', *Journal of African History*, 12:4 (1971)

Beachey, R. W. 'The Arms Trade in East Africa in the Late Nineteenth Century', *Journal of African History*, 3:3 (1962)

Berg, G. M. 'The Sacred Musket: Tactics, Technology, and Power in 18th-Century Madagascar', *Comparative Studies in Society and History*, 27:2 (1985)

Birmingham, D. 'The Forest and Savanna of Central Africa', in J. E. Flint (ed.) *Cambridge History of Africa, 5: From c.1790 to c.1870* (Cambridge, 1976)

—— *Central Africa to 1870: Zambezia, Zaire and the South Atlantic* (Cambridge, 1981)

Birmingham, D. and Martin, P. M. *History of Central Africa*, 2 volumes (London, 1983)

Campbell, G. *An Economic History of Imperial Madagascar, 1750–1895* (Cambridge, 2005)

Caulk, R. A. 'Firearms and Princely Power in Ethiopia in the Nineteenth Century', *Journal of African History*, 13:4 (1972)

Ceulemans, R. P. F. *La Question Arabe et le Congo, 1883–1892* (Brussels, 1959)
Clancy-Smith, J. A. *Rebel and Saint: Muslim Notables, Popular Protest, Colonial Encounters (Algeria and Tunisia, 1800–1904)* (Berkeley, 1994)
Clough, M. S. *Mau Mau Memoirs: History, Memory, and Politics* (Boulder, 1998)
Cooper, F. *Plantation Slavery on the East Coast of Africa* (New Haven, 1977)
Coupland, R. *East Africa and Its Invaders* (Oxford, 1938)
—— *The Exploitation of East Africa 1856–1890: The Slave Trade and the Scramble*, 2nd edition (London, 1968)
Duminy, A. and Ballard, C. (eds) *The Anglo-Zulu War: New Perspectives* (Pietermaritzburg, 1981)
Echenberg, M. J. 'Late Nineteenth-Century Military Technology in Upper Volta', *Journal of African History*, 12:2 (1971)
Fisher, H. J. and Rowland, V. 'Firearms in the Central Sudan', *Journal of African History*, 12:2 (1971)
Goodfellow, C. F. *Great Britain and the South African Confederation (1870–1881)* (Cape Town, 1966)
Gray, R. 'Portuguese Musketeers on the Zambezi', *Journal of African History*, 12:4 (1971)
Guy, J. J. 'A Note on Firearms in the Zulu Kingdom with Special Reference to the Anglo-Zulu War, 1879', *Journal of African History*, 12:4 (1971)
Harman, N. *Bwana Stokesi and His African Conquests* (London, 1986)
Inikori, J. E. 'The Import of Firearms into West Africa, 1750–1807', *Journal of African History*, 18:3 (1977)
Kea, R. A. 'Firearms and Warfare on the Gold and Slave Coasts from the Sixteenth to the Nineteenth Centuries', *Journal of African History*, 12:2 (1971)
Law, R. 'Horses, Firearms and Political Power', *Past and Present*, 72 (1976)
Legassick, M. 'Firearms, Horses and Samorian Army Organization 1870–98', *Journal of African History*, 7:1 (1966)
Louis, W. R. 'The Stokes Affair and the Origins of the Anti-Congo Campaign, 1895–1896', *Revue Belge de Philologie et d'Histoire*, 43:2 (1965)
Low, D. A. 'Uganda: The Establishment of the Protectorate, 1894–1919', in V. Harlow and E. M. Chilver (eds) *History of East Africa*, vol. 2 (Oxford, 1965)
Luck, A. *Charles Stokes in Africa* (Nairobi, 1972)
Mangat, J. S. *A History of the Asians in East Africa c.1886 to 1945* (Oxford, 1969)
Marks, S. and Atmore, A. 'Firearms in Southern Africa: A Survey', *Journal of African History*, 12:4 (1971)
Martin, B. G. *Muslim Brotherhoods in Nineteenth-Century Africa* (Cambridge, 2003)
McIntyre, W. D. 'British Policy in West Africa: The Ashanti Expedition of 1873–4', *Historical Journal*, 5:1 (1962)
Merritt, H. P. 'Bismarck and the German Interest in East Africa', *Historical Journal*, 21 (1978)
Miers, S. 'Notes on the Arms Trade and Government Policy in Southern Africa between 1870 and 1890', *Journal of African History*, 12:4 (1971)
Mungeam, G. H. *British Rule in Kenya 1895–1912: The Establishment of Administration in the East Africa Protectorate* (Oxford, 1966)
Nicolini, B. 'Religion and Trade in the Indian Ocean: Zanzibar in the 1800s', *Newsletter of the International Institute for the Study of Islam in the Modern World* (July 1999)
Oliver, R. and Mathew, G. *History of East Africa*, vol. 1 (Oxford, 1963)

Pakenham, T. *The Boer War* (London, 1979)

Reid, R. *Political Power in Pre-Colonial Buganda: Economy, Society and Warfare in the Nineteenth Century* (Oxford, 2002)

—— *War in Pre-Colonial Eastern Africa: The Patterns and Meanings of State-Level Conflict in the Nineteenth Century* (Oxford, 2007)

Renault, F. *Tippo Tip* (Paris, 1987)

Richards, W. A. 'The Import of Firearms into West Africa in the Eighteenth Century', *Journal of African History*, 21:1 (1980)

Roberts, A. (ed.) *Tanzania before 1900* (Nairobi, 1968)

Rockel, S. J. '"A Nation of Porters": The Nyamwezi and the Labour Market in Nineteenth-Century Tanzania', *Journal of African History*, 41:2 (2000)

Sheriff, A. *Slaves, Spices and Ivory in Zanzibar: Integration of an East African Commercial Empire into the World Economy, 1770–1873* (Athens, Ohio, 1987)

Smaldone, J. P. 'Firearms in the Central Sudan: A Revaluation', *Journal of African History*, 13:4 (1972)

White, G. 'Firearms in Africa: An Introduction', *Journal of African History*, 12:2 (1971)

III. The Middle East

Abrahamian, E. *Iran between Two Revolutions* (Princeton, 1982)

Busch, B. C. *Britain and the Persian Gulf, 1894–1914* (Berkeley, 1967)

Elliott, M. 'The Valley of Death', *Time Magazine*, 9 March (2002)

Hill, R. *Egypt in the Sudan 1820–1882* (London, 1959)

Hodgson, M. G. S. *The Venture of Islam: Conscience and History in a World Civilization*, 3 vols. (Chicago, 1974)

Inalcik, H. 'The Socio-Political Effects of the Diffusion of Fire-arms in the Middle East', in V. J. Parry and M. E. Yapp (eds) *War, Technology and Society in the Middle East* (London, 1975)

Kelly, J. B. *Britain and the Persian Gulf 1795–1880* (Oxford, 1968)

Kumar, R. *India and the Persian Gulf Region 1858–1907: A Study in British Imperial Policy* (London, 1965)

Landen, R. G. *Oman since 1856: Disruptive Modernization in a Traditional Arab Society* (Princeton, 1967)

Owen, R. *The Middle East in the World Economy 1800–1914* (London, 1981)

Ralston, D. *Importing the European Army: The Introduction of European Military Techniques and Institutions into the Extra-European World, 1600–1914* (Chicago, 1990)

Rashid, A. *Taliban: Militant Islam, Oil and Fundamentalism in Central Asia* (London, 2000)

Rathmell, A. 'Policing the Arabian Gulf', *International Police Review* (January/February 1999)

IV. South Asia

Alavi, S. *The Sepoys and the Company* (Delhi, 1995)

Barua, P. 'Military Developments in India 1750–1850', *Journal of Military History*, 58:4 (1994)

Bayly, C. A. *Indian Society and the Making of the British Empire* (Cambridge, 1988)

Bidwell, S. *Swords for Hire: European Mercenaries in Eighteenth-century India* (London, 1971)

Brobst, P. J. *The Future of the Great Game: Sir Olaf Caroe, India's Independence, and the Defence of Asia* (Akron, Ohio, 2005)

Chaudhuri, S. B. 'The Enfield Rifle in the Indian Mutiny', *Bengal Past & Present*, 95:1 (1976)

Cooper, R. G. S. *The Anglo-Maratha Campaigns and the Contest for India: The Struggle for Control of the South Asian Military Economy* (Cambridge, 2003)

Crowell, L. M. 'Logistics in the Madras Army circa 1830', *War & Society*, 10:2 (1992)

Das Gupta, A. and Pearson, M. N. (eds) *India and the Indian Ocean 1500–1800* (Delhi, 1987)

Farooqui, A. 'Opium Enterprise and Colonial Intervention in Malwa and Western India, 1800–24', *Indian Economic and Social History Review*, 32:4 (1995)

Furber, H. *Rival Empires of Trade in the Orient 1600–1800* (Minneapolis, 1976)

Gordon, S. *Marathas, Marauders, and State Transformation in Eighteenth-Century India* (Delhi, 1994)

Karon, T. '"Why the War in Afghanistan is a Long Way From Over" and "What We Learned in Shah-i-Kot"', in *Time Magazine*, 6 and 14 March (2002)

Lenman, B. P. 'The Weapons of War in 18th-Century India', *Journal of the Society for Army Historical Research*, 46 (1968)

Marsden, P. *The Taliban: War and Religion in Afghanistan* (London, 2002)

Moreman, T. R. 'The British and Indian Armies and North West Frontier Warfare, 1849–1914', *Journal of Imperial and Commonwealth History*, 20:1 (1992)

Nightingale, P. *Trade and Empire in Western India 1784–1806* (Cambridge, 1970)

Peers, D. M. 'Between Mars and Mammon: The East India Company and Efforts to Reform its Army, 1796–1832', *Historical Journal*, 33 (1990)

—— *Between Mars and Mammon: Colonial Armies and the Garrison State in India, 1819–1835* (London, 1995)

Singh, S. B. *European Agency Houses in Bengal, 1783–1833* (Calcutta, 1966)

Sykes, Sir P. *A History of Afghanistan*, vol. 2 (London, 1940)

Talbot, I. *India and Pakistan* (London, 2000)

Watt, G. *The Commercial Products of India* (London, 1908)

Weller, J. *Wellington in India* (London, 1972)

Wigington, R. *The Firearms of Tipu Sultan 1783–1799* (Hatfield, 1992)

Young, H. A. *The East India Company's Arsenals and Manufactories* (Oxford, 1937)

V. Southeast Asia

Andaya, B. W. and Andaya, L. Y. *A History of Malaysia* (London, 1982)

Bastin, J. *The Native Policies of Sir Stamford Raffles in Sumatra and Java* (Oxford, 1957)

Blussé, L. 'Chinese Century: The Eighteenth Century in the China Sea Region', *Archipel*, 58 (Paris, 1999)

Boomgard, P. (ed.) *A World of Water: Rain, Rivers and Seas in Southeast Asian Histories* (Singapore, 2007)

Borschberg, P. *The Singapore and Melaka Straits: Violence, Security and Diplomacy in the Seventeenth Century* (Singapore, 2010)

Cady, J. F. *A History of Modern Burma* (Ithaca, 1958)

—— *The Roots of French Imperialism in Eastern Asia* (Ithaca, 1954)

Carey, P. B. R. *Babad Dipanagara: An Account of the Outbreak of the Java War 1825–1830* (Kuala Lumpur, 1981)

Charney, M. W. *Southeast Asian Warfare, 1300–1900* (Leiden, 2004)

Chew, E. C. T. and Lee, E. (eds) *A History of Singapore* (Singapore, 1991)

Chew, E. M. 'The Naning War 1831–1832: Colonial Authority and Malay Resistance in the Early Period of British Expansion', *Modern Asian Studies*, 32:2 (1998)

Elliott, D. 'Pirates of the South China Sea', *Newsweek Magazine*, 5 July (1999)

Gullick, J. M. *Indigenous Political Systems of Western Malaya*, revised edition (London, 1988)

Hall, D. G. E. *A History of South-East Asia* (London, 1964)

Harfield, A. *British and Indian Armies in the East Indies 1685–1935* (Chippenham, 1984)

Htin Aung, M. *Lord Randolf Churchill and the Dancing Peacock: British Conquest of Burma 1885* (Delhi, 1990)

Lee, K. H. *The Sultanate of Aceh: Relations with the British, 1760–1824* (Kuala Lumpur, 1995)

Lieberman, V. B. *Burmese Administrative Cycles: Anarchy and Conquest, c.1580–1760* (Princeton, 1987)

—— 'Local Integration and Eurasian Analogies: Structuring Southeast Asian History', *Modern Asian Studies*, 27:3 (1993)

Maxwell, W. G. and Gibson, W. S. (eds) *Treaties and Engagements Affecting the Malay States and Borneo* (London, 1924)

Mills, L. A. *British Malaya 1824–1867* (Kuala Lumpur, 1966)

Osborne, M. E. *The French Presence in Cochin China and Cambodia; Rule and Response, 1859–1905* (Ithaca, 1969)

Ramakrishna, K. and Tan, S. S. (eds) *After Bali: The Threat of Terrorism in Southeast Asia* (Singapore, 2003)

Reid, A. and Castles, L. (eds) *Pre-Colonial State Systems in Southeast Asia: The Malay Peninsula, Sumatra, Bali-Lombok, South Celebes* (Kuala Lumpur, 1975)

Reid, A. 'Europe and Southeast Asia: The Military Balance', James Cook University of North Queensland, Occasional Paper 16 (Townsville, Queensland, 1982)

Rubin, A. P. *Piracy, Paramountcy and Protectorates* (Kuala Lumpur, 1974)

Rutter, O. *The Pirate Wind: Tales of the Sea-Robbers of Malaya* (Oxford, 1930; reprinted 1986)

SarDesai, D. R. *British Trade and Expansion in Southeast Asia 1830–1914* (Delhi, 1977)

Tagliacozzo, E. *Secret Trades, Porous Borders: Smuggling and States Along a Southeast Asian Frontier, 1865–1915* (New Haven, 2005)

Tarling, N. *Anglo-Dutch Rivalry in the Malay World, 1780–1824* (Cambridge, 1962)

—— *Piracy and Politics in the Malay World: A Study of British Imperialism in Nineteenth-Century South-East Asia* (Melbourne, 1963)

—— (ed.) *The Cambridge History of Southeast Asia*, vol. 1: *From Early Times to c.1800* (Cambridge, 1992)

Thant, M. *The Making of Modern Burma* (Cambridge, 2001)

Trocki, C. A. *Prince of Pirates: The Temenggongs and the Development of Johor and Singapore 1784–1885* (Singapore, 1979)

—— *Opium and Empire: Chinese Society in Colonial Singapore 1800–1910* (New York, 1990)

—— *Opium, Empire and the Global Political Economy: A Study of the Asian Opium Trade* (London, 1999)

Turnbull, C. M. *A History of Singapore 1819–1975* (Kuala Lumpur, 1977)

—— *The Straits Settlements 1826–67: Indian Presidency to Crown Colony* (London, 1972)

Warren, J. F. *Iranun and Balangingi: Globalization, Maritime Raiding and the Birth of Ethnicity* (Singapore, 2002)

—— *The Sulu Zone 1768–1898: The Dynamics of External Trade, Slavery, and Ethnicity in the Transformation of a Southeast Asian Maritime State* (Singapore, 1981)

Winstedt, R. O. (ed.) *Malaya* (London, 1925)

Wong, L. K. 'The Trade of Singapore, 1819–69', *Journal of the Malayan Branch, Royal Asiatic Society*, 33:4 (1960)

Wurtzburg, C. E. *Raffles of the Eastern Isles* (London, 1954)

Letters, memoirs, and diaries

Abdullah bin Abdul Kadir, *The Hikayat Abdullah*, translated and edited by A. H. Hill (Kuala Lumpur, 1970)

Anderson, J. *Acheen and the Ports on the North and East Coasts of Sumatra* (London, 1840; reprinted Kuala Lumpur, 1971)

—— *Mission to the East Coast of Sumatra in 1823* (London, 1826; reprinted Oxford, 1971)

'Artifex' and 'Opifex' (pseud.) *The Causes of Decay in a British Industry* (London, 1907)

Auber, P. *An Analysis of the Constitution of the East India Company and of the Laws Passed by Parliament for the Government of Their Affairs, at Home and Abroad* (London, 1826)

Austin, Brigadier-General H. H. 'Gun-running in the Gulf', *Blackwood's Magazine*, 208 (1920)

Barker, W. C. 'Narrative of a Journey to Shoa', in G. W. Forrest (ed.) *Travels and Journals Preserved in the Bombay Secretariat* (Bombay, 1902)

Beatson, A. *A View of the Origin and Conduct of the War with Tippoo Sultaun* (London, 1800)

Begbie, P. J. *The Malayan Peninsula* (Madras, 1834; reprinted Kuala Lumpur, 1967)

Belcher, Captain E. *Voyage of H.M.S. 'Samarang'*, vol. 1 (London, 1848)

Buckley, C. B. *An Anecdotal History of Old Times in Singapore* (Singapore, 1902; reprinted 1969)

Burton, R. F. *The Lake Regions of Central Africa*, 2 volumes (1860; reprinted New York, 1962)

Cameron, V. L. *Across Africa*, vol. 1 (London, 1877)

Chanler, W. A. *Through Jungle and Desert: Travels in Eastern Africa* (London, 1896)

Collin Davies, C. *The Problem of the North-West Frontier, 1890–1908* (Cambridge, 1932)

Compton, H. *A Particular Account of the European Military Adventurers of Hindustan from 1784–1803* (London, 1893)

Crawfurd, J. *Journal of an Embassy from the Governor-General of India to the Courts of Siam and Cochin-China*, vol. 2 (London, 1830)

Dalrymple, A. *Reprint From Dalrymple's Oriental Repertory, 1791–7, of Portions Relating to Burma* (Rangoon, 1926)

Decle, L. 'The Murder in Africa', *New Review* (December 1895)

Earl, G. W. *The Eastern Seas, or Voyages and Adventures in the Indian Archipelago in 1832, 1833, 1834* (London, 1837)

Gikoyo, G. *We Fought for Freedom* (Nairobi, 1979)

Goodman, J. D. 'The Birmingham Gun Trade', in S. Timmins (ed.) *The Resources, Products, and Industrial History of Birmingham and the Midland Hardware District* (London, 1866)

Goodwin, J. *The Newdigates of Arbury: Early Memorials of the Birmingham Gun Trade* (Birmingham, 1869)

Greener, W. *The Present Proof Company. The Bane of the Gun Trade: A Letter Addressed to the Masters, and Journeymen Gun Makers of the Kingdom* (Birmingham, 1845)

Greenwood, Lieutenant H. M. *Narrative of the Late Victorious Campaign in Afghanistan under General Pollock; with Recollections from Seven Years' Service in India* (London, 1844)

Hackwood, F. W. *Wednesbury Workshops* (Birmingham, 1889)

Hardinge, C. (Lord Hardinge of Penshurst) *Old Diplomacy* (London, 1947)

Hough, Major V. *A Narrative of the March and Operations of the Bengal Column of the Army of Indus in the Expedition to Afghanistan in the Years 1838–1839* (London, 1841)

Hutton, W. *An History of Birmingham*, 6th edition (Birmingham, 1835)

Itote, W. *'Mau Mau' General* (Nairobi, 1967)

Keppel, A. *Gun-running and the Indian North-West Frontier* (London, 1911)

Keppel, Captain H. *A Visit to the Indian Archipelago in H.M.S. 'Maeander'*, vol. 1 (London, 1853)

—— *Expedition to Borneo of H.M.S. 'Dido'*, vol. 2 (London, 1847)

Krapf, J. L. *Travels, Researches and Missionary Labours, during an Eighteen Years' Residence in Eastern Africa* (London, 1860)

Landers, E. *The Birmingham Gun Trade* (Birmingham, 1869)

Livingstone, D. *Narrative of an Expedition to the Zambesi and Its Tributaries and of the Discovery of the Lakes Shirwa and Nyassa, 1858–1864* (London, 1865)

—— *The Last Journals of David Livingstone in Central Africa, from 1865 to His Death*, edited by H. Waller, 2 volumes (London, 1874)

Lugard, F. D. *The Rise of Our East African Empire* (Edinburgh, 1893)

—— *Diaries*, edited by M. F. Perham and M. Bull, 4 volumes (London, 1959)

Milburn, W. *Oriental Commerce*, 2 volumes (London, 1813)

Moor, J. H. *Notices of the Indian Archipelago and Adjacent Countries* (Singapore, 1837, reprinted London, 1968)

Moor, Lieutenant E. *A Narrative of the Operations of Captain Little's Detachment... during the Late Confederacy in India, against the Nawab Tippoo Sultan Bahadur* (London, 1794)

Morfitt, J. 'Sketches of Birmingham Industries', in S. J. Pratt, *Harvest-Home*, vol. 1 (London, 1805)

Mundy, G. R. *Narrative of Events in Borneo and Celebes*, vol. 2 (London, 1848)

Newbold, T. J. *Political and Statistical Account of the British Settlements in the Straits of Malacca*, 2 volumes (London, 1839; reprinted Kuala Lumpur, 1971)

Njama, K. and Barnett, D. L. *Mau Mau from Within: Autobiography and Analysis of Kenya's Peasant Revolt* (New York, 1966)

Parsons, B. (attributed) *Observations on the Manufacture of Firearms for Military Purposes, etc.* (London, 1829)

Pennell, T. L. *Among the Wild Tribes of the Afghan Frontier: A Record of Sixteen Years' Close Intercourse with the Natives of the Indian Marches* (London, 1909)

Pires, T. *The Suma Oriental of Tomé Pires: An Account of the East, from the Red Sea to Japan, Written in Malacca and India in 1512–1515*, translated and edited by A. Cortesão (London, 1944)

Pickering, D. (ed.) *The Statutes at Large: from Magna Charta to... 1761*, vol. 5 (Cambridge, 1763)

Postlethwaite, J. R. P. *I Look Back* (London, 1947)

Prosser, R. B. *Birmingham Inventions and Inventors, Being a Contribution to the Industrial History of Birmingham* (Birmingham, 1881)

Raffles, Lady S. *Memoir of the Life and Public Services of Sir Thomas Stamford Raffles* (London, 1830; reprinted Singapore, 1991)

Raffles, Sir T. S. *The History of Java*, 2 volumes (London, 1817; reprinted Kuala Lumpur, 1965)

Shaw, S. *History of Staffordshire*, vol. 2 (London, 1801)

St. John, H. *The Indian Archipelago*, vol. 2 (London, 1835)

St. John, S. *Life in the Forests of the Far East*, vol. 2 (London, 1862)

—— *The Life of Sir James Brooke, Rajah of Sarawak* (Edinburgh, 1879)

Sketchley, J. *The Directory of Birmingham, Wolverhampton and Walsall*, 3rd edition (Birmingham, 1767)

Speke, J. H. *What Led to the Discovery of the Source of the Nile* (1863; reprinted New York, 1967)

Spurrier, W. J. 'Degenerates Regenerated: The Guns and Rifles of Birmingham', *The Birmingham Magazine of Arts and Industries*, 4:3 (1902)

Stanley, R. and Neame, A. (eds) *The Exploration Diaries of H. M. Stanley* (London, 1961)

Sullivan, G. L. *Dhow Chasing in Zanzibar Waters* (London, 1873)

Wallace, A. R. 'On the Trade between the Eastern Archipelago and New Guinea and Its Islands', *Proceedings of the Royal Geographical Society of London*, 6:2 (1862)

West, W. *The History, Topography and Directory of Warwickshire* (Birmingham, 1830)

Wilson, Lieutenant-Colonel Sir A. T. *South-west Persia: A Political Officer's Diary 1907–1914* (London, 1941)

—— *The Persian Gulf: An Historical Sketch from the Earliest Times to the Beginning of the Twentieth Century* (London, 1928; reprinted 1954)

Unpublished dissertations and papers

Al-Wuhaibi, A. M. F. 'Oman under Sultans Taimur and Sa'id 1913–1970' (Ph.D. thesis, University of Cambridge, 1995)

Bastable, M. J. 'Arms and the State: The History of Sir William G. Armstrong & Co., 1854–1914' (Ph.D. thesis, University of Toronto, 1990)

Chander, S. 'From a Pre-Colonial Order to a Princely State: Hyderabad in Transition' (Ph.D. thesis, University of Cambridge, 1987)

Cooper, R. G. S. 'Cross-cultural Conflict Analysis: The "Reality" of British Victory in the Second Anglo-Maratha War 1803–1805' (Ph.D. thesis, University of Cambridge, 1992)

Reber, A. L. 'The Sulu World in the Eighteenth and Early Nineteenth Centuries: A Historiographical Problem in British Writings on Malay Piracy' (M.A. thesis, Cornell University, 1966)

Richards, W. A. 'The Birmingham Gun Manufactory of Farmer and Galton and the Slave Trade' (M.A. thesis, University of Birmingham, 1972)

Young, D. W. 'History of the Birmingham Gun Trade' (M.Com. thesis, University of Birmingham, 1936)

Index